Polymer Crystallization

Polymer Crystallization

Methods, Characterization, and Applications

Edited by

Jyotishkumar Parameswaranpillai, Jenny Jacob, Senthilkumar Krishnasamy, Aswathy Jayakumar, and Nishar Hameed

WILEY-VCH

Editors

Dr. Jyotishkumar Parameswaranpillai
Alliance University
Department of Science, Faculty of
Science & Technology
Chandapura-Anekal Main Road
Bengaluru, Karnataka
India

Dr. Jenny Jacob
Mar Athanasios College for Advanced
Studies
(MACFAST) School of Biosciences
Pathanamthitta
689101 Tiruvalla, Kerala
India

Dr. Senthilkumar Krishnasamy
PSG Institute of Technology and
Applied Research
Department of Mechanical Engineering
Coimbatore
Tamil Nadu
India

Dr. Aswathy Jayakumar
Kyung Hee University
Department of Food and Nutrition
BioNanocomposite Research Center
Kyungheedae-ro Dongdaemun-gu,
Seoul
Republic of Korea

Dr. Nishar Hameed
Swinburne Research Institutes
Mechanical Engineering & Product
Design
Melbourne Hawthorn campus
John Street
3122 Hawthorn
Australia

Cover Image: © Sebastian Janicki/
Shutterstock

All books published by **WILEY-VCH** are carefully produced. Nevertheless, authors, editors, and publisher do not warrant the information contained in these books, including this book, to be free of errors. Readers are advised to keep in mind that statements, data, illustrations, procedural details or other items may inadvertently be inaccurate.

Library of Congress Card No.: applied for

British Library Cataloguing-in-Publication Data
A catalogue record for this book is available from the British Library.

Bibliographic information published by the Deutsche Nationalbibliothek
The Deutsche Nationalbibliothek lists this publication in the Deutsche Nationalbibliografie; detailed bibliographic data are available on the Internet at <http://dnb.d-nb.de>.

© 2023 WILEY-VCH GmbH, Boschstraße 12, 69469 Weinheim, Germany

All rights reserved (including those of translation into other languages). No part of this book may be reproduced in any form – by photoprinting, microfilm, or any other means – nor transmitted or translated into a machine language without written permission from the publishers. Registered names, trademarks, etc. used in this book, even when not specifically marked as such, are not to be considered unprotected by law.

Print ISBN: 978-3-527-35081-0
ePDF ISBN: 978-3-527-83922-3
ePub ISBN: 978-3-527-83923-0
oBook ISBN: 978-3-527-83924-7

Typesetting Straive, Chennai, India
Printing and Binding CPI Group (UK) Ltd, Croydon, CR0 4YY

Contents

Preface *xi*
Editor Biography *xiii*

1 Introduction to Polymer Crystallization *1*
   N.M. Nurazzi, M.N.F. Norrrahim, S.S. Shazleen, M.M. Harussani,
   F.A. Sabaruddin, and M.R.M. Asyraf
1.1 Introduction *1*
1.2 Degree of Crystallinity *3*
1.3 Thermodynamics on the Crystallization of Polymers Characteristics *4*
1.4 Polymer Crystallization Mechanism *5*
1.4.1 Strain-Induced Crystallization of Polymer *5*
1.4.2 Crystallization of Polymer from Solution *7*
1.5 Applications of Crystalline Polymer *8*
   References *10*

2 Characterization of Polymer Crystallization by Using Thermal Analysis *13*
   Kai Yang, Xiuling Zhang, Mohanapriya Venkataraman, Jakub Wiener, and
   Jiri Militky
2.1 Introduction *13*
2.2 Basic Principle *14*
2.2.1 General Idea *14*
2.2.2 Application of DSC Method *15*
2.3 Characterization of Polymer Crystallization According to Isothermal Crystallization Process *16*
2.3.1 Performance of Isothermal Crystallization Process *16*
2.3.2 Analysis of Isothermal Crystallization Process *16*
2.3.2.1 Crystal Geometry *17*
2.3.2.2 Characterization of Crystallization Rate *18*
2.3.2.3 Characterization of Crystallization Activation Energy *18*
2.3.3 Isothermal Crystallization of Some Polymer Composites *19*
2.4 Characterization of Polymer Non-isothermal Crystallization Process *20*
2.4.1 Basics of Nonlinear Crystallization Modeling *20*

2.4.2	Performance of Non-isothermal Crystallization Process	20
2.4.3	Analysis of Crystal Geometry During Non-isothermal Crystallization Process	21
2.4.3.1	Jeziorny-Modified Avrami Equation	21
2.4.3.2	Ozawa Model	21
2.4.3.3	Mo model	25
2.4.4	Determination of Crystallization Activation Energy (E)	26
2.4.5	Analysis of Relative Crystallinity	27
2.5	Conclusion	27
	Acknowledgment	28
	Abbreviations	28
	References	28

3 Crystallization Behavior of Polypropylene and Its Blends and Composites 33
Daniela Mileva, Davide Tranchida, Enrico Carmeli, Dietrich Gloger, and Markus Gahleitner

3.1	Introduction – Polypropylene Crystallinity in Perspective	33
3.2	Chain Structure and Molecular Weight Effects for iPP Crystallinity and Polymorphism	37
3.3	Nucleation of iPP	42
3.4	Crystallization in Multiphase Copolymers, Blends, and Composites	47
3.5	Processing Effects and Resulting Properties	54
3.6	Investigation Methods for PP Crystallization and Morphology	60
	Acknowledgments	64
	References	65

4 Crystallization of PE and PE-Based Blends, and Composites 87
Amirhosein Sarafpour, Gholamreza Pircheraghi, Farzad Gholami, Rouhollah Shami-Zadeh, and Farzad Jani

4.1	An Introduction to Polyethylene, Its Crystallization, and Kinetics	87
4.1.1	Basics of Structure and Morphology	87
4.1.2	Theory of Crystallization and Its Kinetics	92
4.2	Experimental Study on Crystallization Kinetics of Polyethylene	93
4.2.1	Isothermal Crystallization	93
4.2.2	Non-isothermal Crystallization	96
4.3	Nucleation Theory	99
4.4	Crystal Growth	100
4.5	PE Blends and Co-crystallization	103
4.6	PE Nanocomposites	109
4.7	Summary	112
	References	112

5	**Crystallization of PLA and Its Blends and Composites** 121
	Jesús M. Quiroz-Castillo, Ana D. Cabrera-González, Luis A. Val-Félix, and Tomás J. Madera-Santana
5.1	Introduction 121
5.2	Crystallization of Macromolecules 123
5.2.1	Improvement of PLA Crystallization Kinetics 126
5.3	Polylactic Acid Nucleation 130
5.3.1	Inorganic Nucleating Agents 130
5.3.2	Organic Nucleating Agents 133
5.4	Polylactic Acid Blends 136
5.4.1	Polylactic Acid Binary Blends with Biopolymers–Starch and PHAs 136
5.4.2	Polylactic Acid Binary Blends with Biodegradable Polymers – PCL, PBAT, and PBS 138
5.5	Polylactic Acid Composites 139
5.5.1	Polylactic Acid – Natural Fiber Composites 139
5.5.2	Polylactic Acid – Nanocomposites 140
5.6	Conclusions 143
	References 144

6	**Crystallization in PLLA-Based Blends, and Composites** 161
	Pratick Samanta and Bhanu Nandan
6.1	Introduction 161
6.2	Chemical and Crystal Structure of PLLA 162
6.3	PLLA Properties: Glass Transition and Melting Temperature 162
6.3.1	Glass Transition Temperature 162
6.3.2	Melting Temperature 163
6.4	PLLA Crystallization 163
6.4.1	PLLA Crystallization Study Through Spherulite Growth 163
6.4.2	Lauritzen and Hoffman Theory in PLLA Crystallization 164
6.4.3	Crystallization Kinetics Through Calorimetry Study 166
6.5	Crystallization of PLLA in Blends 168
6.6	Crystallization of PLLA in Nanocomposites 172
6.7	Crystallization of PLLA in Block Copolymer 175
6.8	Crystallization of PLLA After Adding Nucleating Agents 178
6.9	PLLA Plasticization 182
6.10	Conclusion and Future Outlook 182
	References 183

7	**Crystallization in PCL-Based Blends and Composites** 195
	Madhushree Hegde, Akshatha Chandrashekar, Mouna Nataraja, Jineesh A. Gopi, Niranjana Prabhu, and Jyotishkumar Parameswaranpillai
7.1	Introduction 195

7.2	Crystallinity of PCL and the Factors Affecting Crystallinity	195
7.3	Crystalline Behavior of PCL-Based Multiphase Polymer Systems	199
7.3.1	Crystallization Behavior of Blends of PCL	199
7.3.2	Crystallization Behavior of Block Copolymers of PCL	202
7.3.3	Effect of Fillers on the Crystalline Behavior of PCL	203
7.4	Conclusion	207
	References	208

8 Crystallization and Shape Memory Effect 215
Shiji Mathew

8.1	Introduction	215
8.2	Shape Memory Cycle	216
8.3	Mechanism of Shape Memory Effect	217
8.4	Types of Shape Memory Polymers	218
8.5	Biomedical Applications of Shape Memory Polymers	218
8.5.1	Tissue Engineering	218
8.5.2	Bone Engineering	220
8.5.3	Medical Stents	221
8.5.4	Drug Delivery Application	222
8.5.5	SMPs as Self-Healing Materials	222
8.5.6	Vascular Embolization	226
8.6	Conclusion	227
	References	227

9 3D Printing of Crystalline Polymers 233
Hiriyalu S. Ashrith, Tamalapura P. Jeevan, and Hanume Gowda V. Divya

9.1	Introduction	233
9.2	3D Printing Materials and Processes	234
9.2.1	Nylon and Polyamides	234
9.2.2	Polyethylene	238
9.2.3	Polyethylene Terephthalate	240
9.2.4	Polypropylene	241
9.2.5	Polylactic Acid	243
9.3	Characterization of 3D-Printed Crystalline Polymers	244
9.3.1	Mechanical Properties/Mechanical Characteristics	244
9.3.2	Thermal Properties/Thermal Characteristics	246
9.3.3	Tribological Properties/Tribological Characteristics	247
9.4	Conclusion	248
	References	250

10 Crystallization from Anisotropic Polymer Melts 255
Daniel P. da Silva, James J. Holt, Supatra Pratumshat, Paula Pascoal-Faria, Artur Mateus, and Geoffrey R. Mitchell

10.1	Introduction	255
10.2	Evaluating Anisotropy	256

10.3	Crystallization During Deformation of Networks	258
10.4	Sheared Polymer Melts	260
10.5	Crystallization During Injection Molding	264
10.6	Sheared Polymer Melts with Nucleating Agents	266
10.7	Sheared Polymer Melts with Nanoparticles	271
10.8	3D Printing Using Extrusion	272
10.8.1	In-Situ Studies of Polymer Crystallization During 3D Printing	273
10.9	Morphology Mapping	275
10.10	Discussion	276
	Acknowledgments	277
	References	277

11 Molecular Simulations of Polymer Crystallization 283
Yijing Nie and Jianlong Wen

11.1	Introduction	283
11.2	Establishment of Polymer Simulation Systems	283
11.2.1	MC Simulations	284
11.2.2	MD Simulations	284
11.2.2.1	United Atom Chain Model	285
11.2.2.2	Coarse-Grained Polymer Model	285
11.3	Polymer Crystallization at Quiescent State	285
11.3.1	Crystal Nucleation	285
11.3.2	Intramolecular Nucleation Model	287
11.4	Nanofiller-Induced Polymer Crystallization	288
11.4.1	Nanofiller-Induced Homopolymer Crystallization	288
11.4.2	Nanofiller-Induced Copolymer Crystallization	291
11.4.2.1	Nanofiller-Induced Block Copolymer Crystallization	291
11.4.2.2	Random Copolymer Nanocomposite Crystallization	293
11.4.3	Crystallization of Polymers Grafted on Nanofillers	293
11.5	Effect of Grafting Density	293
11.6	Effect of Chain Length	293
11.7	Effect of Interfacial Interactions	295
11.8	Stereocomplex Crystallization of Polymer Blends	295
11.8.1	Simulation Details	296
11.8.2	Effects of Different Methods	297
11.8.2.1	Effect of Chain Length	297
11.8.2.2	Effect of Stretching	298
11.8.2.3	Effect of Nanofillers	298
11.8.2.4	Effect of Chain Topology	299
11.8.2.5	Effect of Chain Structure	300
11.9	Flow-Induced Polymer Crystallization	301
11.9.1	Flow-Induced Polymer Nucleation	301
11.9.2	Stretch-Induced Crystalline Structure Changes	306
11.10	Summary	308
	References	309

12	**Application, Recycling, Environmental and Safety Issues, and Future Prospects of Crystalline Polymer Composites** *323*	
	Busra Cetiner, Havva Baskan-Bayrak, and Burcu S. Okan	
12.1	Introduction *323*	
12.2	Crystalline Polymers and Composites *324*	
12.2.1	Crystalline Polymers *324*	
12.2.2	Crystalline Polymer Composites *326*	
12.2.2.1	Crystalline Polymer Composites with Organic Reinforcements *328*	
12.2.2.2	Crystalline Polymer Composites with Inorganic Reinforcements *329*	
12.2.2.3	Crystalline Polymer Composites with Natural Reinforcements *330*	
12.3	Applications of Crystalline Polymer Composites *331*	
12.3.1	Automotive Applications of Crystalline Polymer Composites *331*	
12.3.2	Biomedical Applications of Crystalline Polymer Composites *334*	
12.3.3	Defense and Aerospace Applications of Crystalline Polymer Composites *335*	
12.3.4	Other Applications of Crystalline Polymer Composites *339*	
12.4	Recycling, Environmental, and Safety Issues of Crystalline Polymer Composites *340*	
12.4.1	Recycling of Glass Fiber-Reinforced Crystalline Polymer Composites *340*	
12.4.2	Recycling of Carbon Fiber-Reinforced Crystalline Polymer Composites *341*	
12.4.3	Recycling of Carbon Nanotubes-Reinforced Crystalline Polymer Composites *342*	
12.4.4	Recycling of Natural Fiber-Reinforced Crystalline Polymer Composites *343*	
12.4.5	Environmental Impact and Safety Issues of Crystalline Polymer Composites *343*	
12.5	Future Prospects of Crystalline Polymer Composites *344*	
12.6	Conclusions *345*	
	References *345*	

Index *359*

Preface

Polymer crystallization is one of the main factors controlling the properties of crystalline or semi-crystalline polymers. The parameters such as molecular weight, arrangement of polymer chains, the interaction between the polymer chains, chain folding, branching affect the polymer crystallization (orientation). Traditional semi-crystalline polymers are polyolefins (e.g. polypropylene, polyethylene), polyamides (e.g. nylon), and polyesters (e.g. *polyethylene terephthalate*). The crystallinity of the polymers can be controlled by changing the thermal parameters (heating and cooling rates), blending, modifying the polymer chain length, etc. Different methodologies can be used to evaluate the crystallinity, growth, size, and other features of crystals in polymers, such as polarized optical microscopy, X-ray diffraction (XRD), Fourier transform infrared spectroscopy (FTIR), Raman spectroscopy, differential scanning calorimetry (DSC), and nuclear magnetic resonance, to name a few. There are many theories and mechanisms for polymer crystallization that were suggested for a better understanding of the crystallization kinetics/process and studying its impact on the properties of the polymer. In terms of application, crystalline polymers are used in automobiles, aircraft, toys, biomedical devices, household applications, construction, building, etc., due to their high strength and load-bearing capacity.

This book comprises 12 chapters. Chapter 1, "Introduction to Polymer Crystallization," gives a basic introduction to polymer crystallization. Chapter 2, "Characterization of Polymer Crystallization by Using Thermal Analysis," discusses the isothermal and non-isothermal crystallization mechanisms of polymers and polymer composites in detail. Chapter 3, "Crystallization Behavior of Polypropylene and Its Blends and Composites," gives a complete picture of polypropylene crystallinity. Chapter 4, "Crystallization of PE and PE-Based Blends, and Composites," discusses the structure, morphology, crystallization kinetics, and theory of crystallization in PE blends and composites. Chapter 5, "Crystallization of PLA and its Blends and Composites," gives a detailed overview of the crystallization kinetics of PLA, PLA-based blends, and composites. Chapter 6, "Crystallization in PLLA-Based Blends, and Composites," gives a detailed overview of the structure, properties, and crystallization behavior of PLLA (one type of optical isomer of PLA) based systems. Chapter 7, "Crystallization in PCL-Based Blends and Composites," gives a detailed outline of the crystalline behavior of PCL, PCL-based blends, and composites. Chapter 8, "Crystallization and Shape Memory Effect," examines the interrelationship between crystallinity and shape memory effect in polymers and their potential biomedical applications. Chapter 9, "3D Printing of Crystalline Polymers," examines the mechanical, thermal, and tribological characteristics of 3D-printed crystalline polymers. Chapter 10, "Crystallization from Anisotropic Polymer Melts," discusses in detail the impact of anisotropy in the melt phase on the morphology of the semi-crystalline polymers. Chapter 11, "Molecular Simulations of Polymer Crystallization," discusses the usefulness of molecular simulations as a tool for a complete understanding of the mechanisms of polymer crystallization. Chapter 12, "Application, Recycling, Environmental and Safety Issues, and Future Prospects of Crystalline Polymer Composites," discusses the applications (automotive, biomedical, defense, aerospace, etc.), recycling, environmental issues, and prospects of crystalline polymer composites. The priceless information on all the areas of polymer crystallization will make this book a one-stop reference for academicians, scientists, professors, researchers, students, and those who are interested in understanding the fundamentals and advancements in the crystallization of polymers.

Thanks to the authors for their contribution.

14 February 2023

Jyotishkumar Parameswaranpillai (India)
Jenny Jacob (India)
Senthilkumar Krishnasamy (India)
Aswathy Jayakumar (Republic of Korea)
Nishar Hameed (Australia)

Editor Biography

Dr. Jyotishkumar Parameswaranpillai is currently an associate professor at Alliance University, Bangalore. He received his PhD in Chemistry (Polymer Science and Technology) from Mahatma Gandhi University, Kottayam, India, in 2012. He has research experience in various international laboratories such as the Leibniz Institute of Polymer Research Dresden (IPF), Germany; Catholic University of Leuven, Belgium; University of Potsdam, Germany; and King Mongkut's University of Technology North Bangkok (KMUTNB), Thailand. He has more than 260 international publications. He is a frequent invited and keynote speaker and a reviewer for more than 100 international journals, book proposals, and international conferences. He received numerous awards and recognitions including the prestigious INSPIRE Faculty Award 2011, Kerala State Award for the Best Young Scientist 2016, and Best Researcher Award 2019 from King Mongkut's University of Technology North Bangkok. He is named among the world's Top 2% of the most-cited scientists in the Single Year Citation Impact (2020, 2021) by Stanford University. His research interests include polymer coatings, shape memory polymers, antimicrobial polymer films, green composites, nanostructured materials, water purification, polymer blends, and high-performance composites. https://scholar.google.co.in/citations?user=MWeOvlQAAAAJ&hl=en

Dr. Jenny Jacob is an Associate Professor at Mar Athanasios College for Advanced Studies Tiruvalla (MACFAST), India. She has published many high-impact journal papers and book chapters. She is an experienced academician with research interests in phytomedicine, alternative medicine, enzyme inhibition, nanomaterial drug delivery systems, antimicrobial resistance.

Dr. Aswathy Jayakumar is currently working as Post Doctoral Researcher at Kyung Hee University, Republic of Korea. She has completed her PhD from Mahatma Gandhi University, Kottayam, India. She has published around 60 international publications. She completed a postdoctoral fellowship at King Mongkut's University of Technology North Bangkok (KMUTNB).

Dr. Senthilkumar Krishnasamy is currently working as an associate professor at PSG Institute of Technology and Applied Research. He had been working in the Department of Mechanical Engineering, Kalasalingam Academy of Research and Education (KARE), India, from 2010 to 2018 (October). He graduated as a bachelor in mechanical engineering from Anna University, Chennai, in 2005. He then chose to continue his master's studies and graduated with a master's degree in CAD/CAM from Anna University, Tirunelveli, in 2009. He obtained his PhD from the Department of Mechanical Engineering – Kalasalingam University (2016). He has completed his postdoctoral fellowship at King Mongkut's University of Technology (KMUTNB) North Bangkok and Universiti Putra Malaysia, Serdang, Selangor, Malaysia, under the research topics of "Sisal composites and fabrication of eco-friendly hybrid green composites on tribological properties in a medium-scale application" and "Experimental investigations on mechanical, morphological, thermal and structural properties of kenaf fibre/mat epoxy composites," respectively. His area of research interests includes modification and treatment of natural fibers, nanocomposites, hybrid reinforced polymer composites. He has published research papers in international journals, book chapters, and conferences in the field of natural fiber composites. He is also editing books from different publishers. https://scholar.google.com/citations?user=DG7lhXMAAAAJ&hl=en.

Dr. Nishar Hameed is an associate professor at Swinburne University of Technology, Australia. He has published more than 100 high-impact journal papers, 6 book chapters, 3 edited books, and 2 patents.

1

Introduction to Polymer Crystallization

N.M. Nurazzi[1,2], M.N.F. Norrrahim[3], S.S. Shazleen[4], M.M. Harussani[5], F.A. Sabaruddin[6], and M.R.M. Asyraf[7,8]

[1] Universiti Sains Malaysia, School of Industrial Technology, Bioresource Technology Division, 11800 Penang, Malaysia
[2] Universiti Sains Malaysia, School of Industrial Technology, Green Biopolymer, Coatings & Packaging Cluster, 11800 Penang, Malaysia
[3] Universiti Pertahanan Nasional Malaysia (UPNM), Research Centre for Chemical Defence, Kem Perdana Sungai Besi, 57000 Kuala Lumpur, Malaysia
[4] Universiti Putra Malaysia, Institute of Tropical Forestry and Forest Products (INTROP), 43400 Serdang, Selangor, Malaysia
[5] Tokyo Institute of Technology, School of Environment and Society, Department of Transdisciplinary Science and Engineering, Meguro, Tokyo 152-8552, Japan
[6] Universiti Putra Malaysia, Faculty of Biotechnology and Biomolecular Sciences, 43400 Serdang, Selangor, Malaysia
[7] Universiti Teknologi Malaysia, Engineering Design Research Group (EDRG), Faculty of Engineering, School of Mechanical Engineering, 81310, Johor Bahru, Johor, Malaysia
[8] Universiti Teknologi Malaysia, Centre for Advanced Composite Materials (CACM), 81310, Johor Bahru, Johor, Malaysia

1.1 Introduction

Long-chain molecule polymeric materials have benefited from the use of crystallization as a fundamental thermodynamic phase transition in condensed matter physics of pure substances. Keller made the electron microscope findings on polyethylene (PE) single crystals grown in diluted solutions in 1957, following the synthesis of high-density PE with the development of Ziegler–Natta catalysts, thus developed the chain-folding model [1]. Since then, the discovery of diverse polymer crystal morphologies has been aided by the chain-folding concept. Nowadays, semi-crystalline polymers, such as polyolefins, polyesters, and polyamides, account for more than two thirds of all synthetic polymer products produced worldwide due to their numerous uses in our everyday lives. The degree of crystallinity, which normally ranges between 10% and 80%, describes the proportion of organized polymer molecules [2]. Only small-molecule materials, which are often brittle materials, can attain the greater value of crystallinity.

Hu asserts that the chemical structures of repeating units of polymer can be categorized using two distinct contributions to the perseverance of melting points: intramolecular interactions of collinear connection energy of bonds on the chain for

Polymer Crystallization: Methods, Characterization, and Applications, First Edition.
Edited by Jyotishkumar Parameswaranpillai, Jenny Jacob, Senthilkumar Krishnasamy, Aswathy Jayakumar, and Nishar Hameed.
© 2023 WILEY-VCH GmbH. Published 2023 by WILEY-VCH GmbH.

thermodynamic adaptability and intermolecular interactions of local bond–bond interactions for the parallel-packing of two neighboring bonds in the conventional lattice models for parallel-packing order [3]. As a result, the melting temperatures of polymers with repeating units that favor greater stiffness or more dense/stronger packing are typically higher. Techniques used to evaluate the crystallinity of polymers include density measurement, X-ray diffraction (XRD), infrared spectroscopy, differential scanning calorimetry (DSC), and nuclear magnetic resonance (NMR) [4, 5].

Referring to Zhang et al., the mechanical and optical performance of crystalline polymers like PE and polyethylene terephthalate (PET) corresponds with molding parameters that are strongly influenced by their crystallinity [6]. Crystalline polymers undergo stress at freezing and retain stress from crystallization, according to Kato et al. [7]. Due to the lack of appropriate methods for quantitatively evaluating these transitions, the micro-mechanical forces during polymer crystallization remain a highly discussed topic. Up until now, the forms of proof have been theoretical, indirect experimental, or empirical discussions [7]. There are several experimental methodologies and approaches to estimate the amplitude of micro-mechanical forces during polymer crystallization to limit and avoid material failure owing to these forces. This includes non-destructive test [8], destructive test [9], and computer simulation [10]. Between these, non-destructive techniques have been employed to examine the physical relaxation of components during heating and determine their initial stress state, such as holographic interferometry and synchrotron XRD research. Despite the benefits of these techniques being non-destructive, neither a qualitative computation nor a stress visualization can be completed instantly.

Approximately 30–60% of the substance was comprised of polymer crystals, which ranged in size from a few nanometers to several, randomly oriented in space. Because crystalline polymers could withstand loads and act in diverse directions like reinforced rubber, as well as because macromolecules were often much longer than the crystal dimensions. The fundamental understanding that crystals might function as cross-linkers similar to those in cross-linked rubbers [11]. The tensile, microhardness, and compression behavior patterns of semi-crystalline polymers (Figure 1.1b) have been significantly influenced by micro-mechanical forces throughout polymer crystallization through tie chain portions, which appear to be molecular connections between individual crystallites from the perspective of the molecular topology of the amorphous phase (Figure 1.1a). Additionally, tie chain polymer crystallization improves fracture toughness and slow crack propagation resistance [12, 13].

Most molecular-level descriptions of the semi-crystalline phase are based on topological properties, including the theories of tie chain segments, loop segments, tails, and the alternating of crystalline and amorphous domains [14]. Olsson et al. claim that interface Monte Carlo moves are utilized to relocate sites and change chain connections on the atoms and chains in the amorphous domain to produce new loops, tails, and bridges. The resulting samples' crystalline components are still faultless, that is, devoid of twins or dislocations. According to reports, these faults weaken the critical shear stress and weaken slide processes. As a result, the models under consideration are idealizations of a true semi-crystalline PE material, and

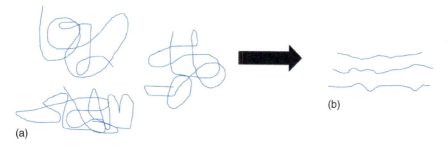

Figure 1.1 The arrangement of polymer molecular chains (a) in amorphous and (b) in semi-crystalline polymers state.

the anticipated resistance to crystal yielding is anticipated to be larger than what has actually been empirically observed [15].

1.2 Degree of Crystallinity

The degree of crystallinity determines how ordered a solid is structurally; the more crystalline a polymer is, the more regularly its chains are aligned, and the arrangement of atoms or molecules is repeatable and consistent. The degree of crystallization of polymer materials has a big impact on their characteristics. In terms of performance, a molded part is stiffer, stronger, but also more brittle the more crystallization there is. Hardness, density, transparency, and diffusion are all significantly influenced by the degree of crystallinity. Chemical composition and thermal history, such as cooling conditions during manufacturing fabrication process and post-thermal treatment, have an impact on the degree of crystallization. However, the characteristics are also influenced by the size of the structural units or the molecular orientation in addition to the degree of crystallinity [16, 17]. In general, a higher degree of crystallinity is typically the result of variables that make polymers more regular and organized because fewer short branches allow molecules to pack more tightly together. Syndiotactic and isotactic polymers have a higher degree of stereoregularity than atactic polymers, but the polymers are also more organized and have regular copolymer structures [18]. Based on the study by Yao et al., it was discovered that a rise in crystallinity directly correlated with an improvement in mechanical characteristics by examining the effects of various crystallization parameters, such as crystal shape ratio and crystallinity [19]. The PET crystal structure ratios did not, however, substantially enhance the mechanical characteristics. Furthermore, at a higher isothermal temperature, considerably higher than the T_g, the crystallinity of PET foam will be strongly increased. Slow crystallization can be used to explain the increase in crystalline content at higher temperatures, which promotes regular chain folding and subsequently reduces topological disorder at the surface of the crystallites. According to Jonas et al., the relationship between the service temperature and crystallinity is strong within the experimental range of 10–150 °C. When the operating temperature is close to or higher than T_g, migration causes isothermal-induced crystal perfection, and

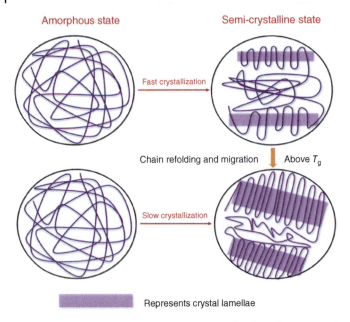

Figure 1.2 The schematic diagram of the mechanism of crystallization enhancement formation from the amorphous state to crystalline in isothermal treatment above T_g. Source: Adapted from Yao et al. [19].

rejection of the structural faults at the crystal's surface causes a rise in the crystalline phase content [20]. The mechanism of crystallization enhanced development from the amorphous state to the crystalline state in isothermal treatment above T_g is schematically depicted in Figure 1.2. The delayed crystallization promotes better crystal lamella development and chain refolding, as seen in Figure 1.2.

1.3 Thermodynamics on the Crystallization of Polymers Characteristics

The partial alignment of the molecular chains in polymer materials can cause crystallization. Amorphous and crystalline domains coexist in these thin lamellar formations, which are created on the scale of nanometers when molecular chains change from a high-entropy random coil state to a reduced-entropy partially folded (semi-crystalline) state [1]. The majority of the solid-state attributes created by polymer materials were impacted by crystallization. Although there is a significant thermo-mechanical dependence in polymer crystallization, one of the major difficulties is controlling the semi-crystalline state precisely. In order to manage the crystalline behavior of the polymers for practical applications, processing factors such as (i) the crystallization temperature [21], (ii) cooling rate [22], and (iii) the application of high shear strain [23] are varied.

Generally, as the crystallization temperature rises, so does the thickness of the crystalline lamellae. Slower cooling rates lead to greater crystallinity, and applying shear stress or shear strain speeds up the nucleation and crystallization of polymer structures [24, 25]. Additionally, the presence of shear stress or shear strain will lead to shish-kebab morphologies of the crystalline polymer and provide rise to varied crystalline morphologies [26]. Crystalline polymers typically exhibit spherulitic structure [27] and fiber formation [28, 29] in the absence of shear. It is true that the crystallization of polymers results in non-equilibrium states. While there is a sizable disparity between melting temperatures (T_m) and crystallization temperatures, there is no thermodynamic phase cohabitation for semi-crystalline polymers (T_c).

From condensed matter physics perspective, the close packing of molecules necessary to create the crystalline lattice order is typically caused by intermolecular interactions. Polyolefins, such as isotactic polypropylene (PP), organize their internal rotations to generate helices in their crystalline states, notably zigzag 2/1 helices for PE and also twisting 3/1 helices, by minimizing their local intramolecular interaction potentials [30]. Consequently, the rigid-rod helices highlight the anisotropic characteristics of intermolecular interactions: the local intermolecular interactions between two rods differ significantly depending on whether they are packed parallel or crossing each other. This leads us to the macromolecular component of the thermodynamic forces that drive polymer crystallization, which is illustrated by the interactions between local chains of macromolecules parallelly packed together [31]. For instance, the stereo-optical sequence regularity of polymers with strong intermolecular interactions like polyvinyl chloride (PVC) and polyacrylonitrile (PAN) may be compromised during crystallization [32]. According to a different theory of protein folding, the lengthy hydrogen-bonding interactions further along the chain are what cause extreme β-folding for the crystalline sequence, whereas the short-range hydrogen-bonding interactions along the chain correspond to intermolecular interactions in polyamide crystals.

Conclusively, the most essential factor in the parallel packing of polymers during their crystalline phase is chain connectedness. Therefore, even though the melting enthalpy and intermolecular interactions of polyolefins may be influenced by intramolecular interactions, considerations from anisotropic intermolecular interactions favor parallel packing as the thermodynamic driving forces for polymer crystallization in accordance with the nature of condensed matter physics [3].

1.4 Polymer Crystallization Mechanism

1.4.1 Strain-Induced Crystallization of Polymer

The development of a highly oriented crystalline phase has a favorable effect on the material's mechanical behavior in many of these applications. The development of extended crystals in the direction of extension during fiber spinning significantly boosts the fiber's strength. The melt is exposed to bi-axial extension during the film-blowing process, and the films have crystals orientated on the plane, giving them the appropriate mechanical characteristics. The invention of a special blow molding procedure that guarantees that the polymer is bi-axially oriented has made it possible to use polyester bottles to accommodate carbonated beverages. Injection molding, for example, the production of a highly oriented outer layer might result in readily cleaved articles. Orientation can also have a negative effect on the mechanical behavior of articles [33].

According to Nitta, a melt-crystallized polymer displays an alternating two-phase structure made up of layers of amorphous material and crystalline lamellae that resemble plates (Figure 1.3a). Folded chain crystallites made up of partially stretched conformations emerge when a polymer molecule's contour length is noticeably greater than the typical lamellar thickness of the order of 10 nm and the chain axes within the lamellae are generally normal to the face of the lamellae [34]. The oriented skin is the layer that is closest to the wall. It is preceded by a partially oriented fine-grained layer with isotropic structural morphology in stress-free areas close to the die's center [35]. Solid-state drawing is typically done during the production of polymer films utilizing a high-speed drawing technique under flowing melt conditions. The manner in which polymer molecules crystallize in the drawing solid and flowing melt determines the structure and characteristics of these polymeric products (Figure 1.3b).

At high temperatures, polymers above their melting point are modeled as viscoelastic liquids. The solid phase can be either amorphous or semi-crystalline depending on the molecular composition and cooling rate. While polymers with regular structures can crystallize because the chains are too regular to allow for regular packing, those with irregular structures cannot. The rate of crystallization is typically zero at the T_m and T_g states and achieves its highest at a temperature in between these two when polymers typically crystallize. The glass transition temperature is the point below which the polymer molecules cease to be mobile and turn "frozen" or also called vitrified. As a result of their high rates of crystallization at temperatures below the melting point and the inability to cool polymers like PE quickly enough to temperatures below the glass transition temperature without significant crystallization occurring, polymers like PE have always been

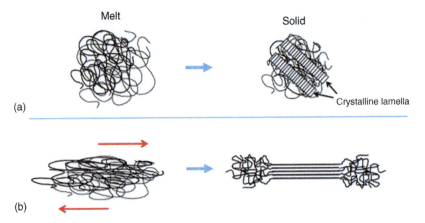

Figure 1.3 Schematic illustrations of crystallization (a) from an equilibrium melt and (b) in the drawing solid and flowing melt.

semi-crystalline in solid form. The melt must be chilled gradually for significant crystallization in polymers like PET, which crystallize slowly. These polymers retain an amorphous state if they are cooled below their glass transition temperature. Crystallization is triggered by the deformation when amorphous PET is subsequently distorted at temperatures barely above the glass transition temperature. The majority of PET products are produced by deforming at these temperatures because the amount and orientation of the crystalline phase can be regulated, allowing for precise control of the final solid's mechanical properties.

1.4.2 Crystallization of Polymer from Solution

Depending on the degree of dilution, polymer crystallization can occur from a solution or by evaporating solvent. In diluted solutions, the molecular chains have no interaction with one another and exist as isolated polymer coils in the solution. Solvent evaporation causes the concentration of the solvent to rise, which encourages molecular chain interaction and the potential for crystallization, such as when a melt crystallizes [36]. The highest level of polymer crystallinity might be achieved through crystallization from the solution. For instance, when crystallizing from a diluted solution, extremely linear PE can produce single crystals resembling platelets with a thickness of 10–20 nm. Using a solvent that dissolves individual monomers but not the final polymer, precipitation is a distinct procedure. After a certain level of polymerization, the semi-crystalline, polymerized product precipitates out of the solution.

According to Huang et al., the kinetics of crystallization from solvent evaporation as well as thermodynamics determine the crystal structure and morphology of polymers. To better comprehend the crystallization process and resulting final structure of polymers, several kinetic parameters were applied to a model system [37]. The migration of polymer chains to the crystal growth front and the rate of crystal development, which may be altered in solution crystallization by modifying the rate of solvent evaporation, are two opposing processes that influence the formation of crystal structure and morphology. By altering the kinetic process, Huang and his colleagues have investigated the crystalline form and structure of poly(L-lactide) (PLLA) in a PLLA–chloroform mixture. The findings led to the identification of the three stages of the PLLA crystallization process: solvent adsorption, surface gel formation, and crystallization. The tiny chloroform molecules that were continually adsorbed into the PLLA samples ignited the solvent adsorption. As shown in Figure 1.4a, the formation of surface gels and even local PLLA–chloroform solutions was caused by the adsorption of chloroform onto the PLLA film's surface, diffusion of the PLLA segments, and hydrodynamic flow. Because the amount of solvent adsorbed into each layer of PLLA decreased along the direction perpendicular to the substrate, which is coupled to the amount of solvent adsorbed as well as the migration of PLLA segments, a concentration gradient of PLLA chains was also produced. PLLA then crystallized as a result of its concentration fluctuation, which was connected to a shifting concentration gradient and nonlinear solvent evaporation kinetics. As a result of solvent evaporation at that point, PLLA crystal lamellae began to form around the nuclei from the PLLA–chloroform system (Figure 1.4b).

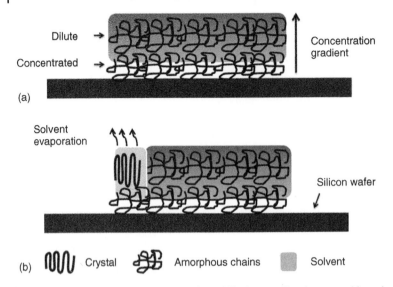

Figure 1.4 Diagrammatic representation of PLLA crystallization caused by solvent evaporation and PLLA concentration gradient. In (a), solvent is adsorbed onto the surface of the film to create a PLLA solution with concentration gradient. In (b), solvent evaporation drives PLLA crystal nucleation and growth. Source: Reproduced from ref. [37].

1.5 Applications of Crystalline Polymer

All polymers have some degree of crystallinity. As has been previously discussed, crystallinity of polymer has a significant influence on its properties, with more crystalline polymers having chains that are more regularly aligned. By increasing the degree of crystallinity, the density and hardness of the material are increased. This is due to the fact that more intermolecular bonds are formed when it is in the crystalline phase. Therefore, the polymer gets stronger and its deformation can result in the higher strength owing to oriented chains [38]. Besides, highly crystalline polymers are stiff, less susceptible to solvent penetration, have high melting points, are barriers to moisture and gases, and are resistant to oil and grease [39]. For instance, PP, PE, nylon, syndiotactic polystyrene, and Kevlar. Even though crystallinity makes a polymer strong, it also lessens its resistance to impact.

Conversely, amorphous polymers are softer, have lower melting points, and are more permeable to solvents. Some highly amorphous polymers include polycarbonate, poly(methyl methacrylate), polyisoprene, and polybutadiene. While semi-crystalline polymers, on the other hand, have both crystalline and amorphous areas. Most plastics benefit from semi-crystallinity because it combines the flexibility of amorphous polymers with the strength of crystalline polymers, making this form of polymer both strong and flexible. Semi-crystalline polymers have a limited heat tolerance before softening and bending. Yet, semi-crystalline plastics

Table 1.1 Difference in general properties of highly crystalline, semi-crystalline, and amorphous polymers [40, 42].

Properties	Type of polymer		
	Highly crystalline	Semi-crystalline	Amorphous
Hardness	Hard	Hard	Soft
Melting point	High and sharp melting point	High and sharp melting point	No distinct melting point and softens over a broad range of temperature
Mechanical	High strength	High strength	Low strength
	High fatigue and wear resistance	Good fatigue and wear resistance	Poor fatigue and wear resistance
Clarity	Opaque to visible light	Translucent	Tend to be translucent or transparent
Resistance to chemical	High	High	Low
Gas permeability	Low	Low	High
Arrangement of molecules	Regular and uniformly packed molecules	Regular and uniformly packed molecules	Random
Ideal application	Ideal for long exposure and high strength applications such as in structural applications	Ideal for applications that need high strength and low friction and have an environment that experiences any repeated cyclic loading and chemical contact	Ideal for applications that require high dimensional accuracy and stability with a transparent, overall good appearance, low to zero mechanical abuse, and chemical contact

have a propensity to quickly shift from a solid state to a low-viscosity liquid once the melting point is achieved [40].

In the industrial sector, crystallization kinetics are a crucial factor to take into account while designing a polymer for a certain application because it will affect the final polymer product [41]. For instance, flexibility at low to ambient temperatures is required for many applications of polymers and polymer coatings. Amorphous polymers are the best option in this situation since they have some resistance to elasticity and impact. In contrast, a polymer with more crystallinity may be favored when hardness and rigidity are needed. The difference between highly crystalline, semi-crystalline, and amorphous polymers as well as their ideal applications are highlighted in Table 1.1.

References

1. Keller, A. (1957). A note on single crystals in polymers: evidence for a folded chain configuration. *Philosophical Magazine* 2 (21): 1171–1175.
2. Ehrenstein, G.W. (2012). *Polymeric Materials: Structure, Properties, Applications*. Carl Hanser Verlag GmbH Co KG.
3. Hu, W.B. (2022). Polymer features in crystallization. *Chinese Journal of Polymer Science* 40: 545–555.
4. Razali, N.A.M., Sohaimi, R.M., Othman, R.N.I.R. et al. (2022). Comparative study on extraction of cellulose fiber from rice straw waste from chemo-mechanical and pulping method. *Polymers* 14 (3): 387.
5. Maus, A., Hertlein, C., and Saalwächter, K. (2006). A robust proton NMR method to investigate hard/soft ratios, crystallinity, and component mobility in polymers. *Macromolecular Chemistry and Physics* 207 (13): 1150–1158.
6. Zhang, Y., Ben Jar, P.-Y., Xue, S., and Li, L. (2019). Quantification of strain-induced damage in semi-crystalline polymers: a review. *Journal of Materials Science* 54 (1): 62–82.
7. Kato, S., Furukawa, S., Aoki, D. et al. (2021). Crystallization-induced mechanofluorescence for visualization of polymer crystallization. *Nature Communications* 12 (1): 1–7.
8. Adams, A. (2019). Non-destructive analysis of polymers and polymer-based materials by compact NMR. *Magnetic Resonance Imaging* 56: 119–125.
9. Striemann, P., Hülsbusch, D., Niedermeier, M., and Walther, F. (2020). Optimization and quality evaluation of the interlayer bonding performance of additively manufactured polymer structures. *Polymers* 12 (5): 1166.
10. Ramos, J., Vega, J.F., and Martínez-Salazar, J. (2018). Predicting experimental results for polyethylene by computer simulation. *European Polymer Journal* 99: 298–331.
11. Zhang, M.C., Guo, B.H., and Xu, J. (2016). A review on polymer crystallization theories. *Crystals* 7 (1): 4.
12. Gholami, F., Pircheraghi, G., Rashedi, R., and Sepahi, A. (2019). Correlation between isothermal crystallization properties and slow crack growth resistance of polyethylene pipe materials. *Polymer Testing* 80: 106128.
13. Bretz, P.E., Hertzberg, R.W., and Manson, J.A. (1982). The effect of molecular weight on fatigue crack propagation in nylon 66 and polyacetal. *Journal of Applied Polymer Science* 27 (5): 1707–1717.
14. Adhikari, S. and Muthukumar, M. (2019). Theory of statistics of ties, loops, and tails in semicrystalline polymers. *Journal of Chemical Physics* 151 (11): 114905.
15. Olsson, P.A.T., in't Veld, P.J., Andreasson, E. et al. (2018). All-atomic and coarse-grained molecular dynamics investigation of deformation in semi-crystalline lamellar polyethylene. *Polymer* 153: 305–316.
16. NETZSCH (2022). Crystallinity/degree of crystallinity. https://analyzing-testing.netzsch.com/en/training-know-how/glossary/crystallinity-degree-of-crystallinity (accessed 06 March 2023).

17 Nurazzi, N.M., Khalina, A., Sapuan, S.M. et al. (2017). Curing behaviour of unsaturated polyester resin and interfacial shear stress of sugar palm fibre. *Journal of Mechanical Engineering and Sciences* 11 (2): 2650.

18 Lodge, T.P. and Hiemenz, P.C. (2020). *Polymer Chemistry*. CRC Press.

19 Yao, S., Hu, D., Xi, Z. et al. (2020). Effect of crystallization on tensile mechanical properties of PET foam: experiment and model prediction. *Polymer Testing* 90: 106649.

20 Jonas, A.M., Russell, T.P., and Yoon, D.Y. (1995). Synchrotron X-ray scattering studies of crystallization of poly(ether-ether-ketone) from the glass and structural changes during subsequent heating–cooling processes. *Macromolecules* 28 (25): 8491–8503.

21 Schuster, M. (2022). Determination of the linear viscoelastic material behaviour of interlayers with semi-crystalline structures. *Challenging Glass Conference Proceedings* 8.

22 Benedetti, L., Brulé, B., Decreamer, N. et al. (2019). Shrinkage behaviour of semi-crystalline polymers in laser sintering: PEKK and PA12. *Materials & Design* 181: 107906.

23 Moghim, M.H., Nahvi Bayani, A., and Eqra, R. (2020). Strain-rate-dependent mechanical properties of polypropylene separator for lithium-ion batteries. *Polymer International* 69 (6): 545–551.

24 Ma, Z., Balzano, L., and Peters, G.W.M. (2016). Dissolution and re-emergence of flow-induced shish in polyethylene with a broad molecular weight distribution. *Macromolecules* 49 (7): 2724–2730.

25 Wang, Z., Ma, Z., and Li, L. (2016). Flow-induced crystallization of polymers: molecular and thermodynamic considerations. *Macromolecules* 49 (5): 1505–1517.

26 Kimata, S., Sakurai, T., Nozue, Y. et al. (2007). Molecular basis of the shish-kebab morphology in polymer crystallization. *Science* 316 (5827): 1014–1017.

27 Balzano, L., Ma, Z., Cavallo, D. et al. (2016). Molecular aspects of the formation of shish-kebab in isotactic polypropylene. *Macromolecules* 49 (10): 3799–3809.

28 Breese, D.R. and Beaucage, G. (2004). A review of modeling approaches for oriented semi-crystalline polymers. *Current Opinion in Solid State and Materials Science* 8 (6): 439–448.

29 Kim, I.H., Yun, T., Kim, J.-E. et al. (2018). Mussel-inspired defect engineering of graphene liquid crystalline fibers for synergistic enhancement of mechanical strength and electrical conductivity. *Advanced Materials* 30 (40): 1803267.

30 Natta, G. and Corradini, P. (1967). Structure and properties of isotactic polypropylene. In: *Stereoregular Polymers and Stereospecific Polymerizations* (ed. G. Natta and F. Danusso), 743–746. Elsevier.

31 Hu, W. and Frenkel, D. (2005). Polymer crystallization driven by anisotropic interactions. In: *Interphases and Mesophases in Polymer Crystallization III* (ed. G. Allegra), 1–35. Springer.

32 Tadokoro, H. (1984). Structure and properties of crystalline polymers. *Polymer* 25 (2): 147–164.

33 Rao, I.J. and Rajagopal, K.R. (2001). A study of strain-induced crystallization of polymers. *International Journal of Solids and Structures* 38 (6–7): 1149–1167.

34 Nitta, K.H. (2016). On the orientation-induced crystallization of polymers. *Polymers* 8 (6): 229.

35 Kumaraswamy, G. (2005). Crystallization of polymers from stressed melts. *Journal of Macromolecular Science, Part C Polymer Reviews* 45 (4): 375–397.

36 Lehmann, J. (1966). Die Beobachtung der Kristallisation hochpolymerer Substanzen aus der Lösung durch Kernspinresonanz. *Kolloid Zeitschrift & Zeitschrift fur Polymere* 212 (2): 167–168.

37 Huang, S., Li, H., Wen, H. et al. (2014). Solvent micro-evaporation and concentration gradient synergistically induced crystallization of poly(L-lactide) and ring banded supra-structures with radial periodic variation of thickness. *CrystEngComm* 16 (1): 94–101.

38 Immergut, E.H. and Mark, H.F. (1965). *Principles of Plasticization*. ACS Publications.

39 Mark, J.E. (2007). *Physical Properties of Polymers Handbook*, vol. 1076. Springer.

40 Sheiko, S.S., Vainilovitch, I.S., and Magonov, S.N. (1991). FTIR spectroscopy of polymer films under uniaxial stretching: 2. Amorphous and semicrystalline poly (ethyleneterephthalate). *Polymer Bulletin* 25: 499–506.

41 Séguéla, R. (2007). On the natural draw ratio of semi-crystalline polymers: review of the mechanical, physical and molecular aspects. *Macromolecular Materials and Engineering* 292 (3): 235–244.

42 Lin, Y., Bilotti, E., Bastiaansen, C.W. et al. (2020). Transparent semi-crystalline polymeric materials and their nanocomposites: a review. *Polymer Engineering and Science* 60 (10): 2351–2376.

2

Characterization of Polymer Crystallization by Using Thermal Analysis

Kai Yang, Xiuling Zhang, Mohanapriya Venkataraman, Jakub Wiener, and Jiri Militky

Technical University of Liberec, Faculty of Textile Engineering, Department of Material Engineering, Studentska 1402/2, Liberec 461 17, Czech Republic

2.1 Introduction

The kinetics of crystallization is one important factor in polymer processing, which affects final thermal property, mechanical property, etc. The crystallization process starts with nucleation followed by crystals growth. Heterogeneous nucleation is more common in polymers and can be accelerated by adding nucleation agents. Crystallization processes can be measured via non-isothermal crystallization, cold crystallization, and isothermal crystallization. The crystallization of polymers usually consists of two stages: nucleation and growth of crystals [1–4]. The stage of crystal growth can be analyzed in categories of (secondary) nucleation on the surface of already existing crystals. Crystallization can be regarded as processes of nucleation that take place on a crystal's bulk and on its surface. Correspondingly, both stages affect the final crystallization behavior. The material could be brittle when the slow primary crystallization of polymer occurs and a few nuclei become large spherulites. Besides, it is reported that the heterogenous nucleation of polymer composites is enhanced, which can result in different mechanical and thermal behaviors. Since the crystallization process is dependent on various parameters (e.g. cooling rate, cooling temperature, etc.), it is necessary to know the kinetics of polymer crystallization to control final properties or reveal crystallization mechanism.

To characterize the polymer crystallization, various methods have been applied, including computation method, X-ray diffraction (XRD) [5, 6], Fourier-transform infrared spectroscopy (FTIR) [7], RAMAN spectroscopy, polarized optical microscopy (POM) [8], differential scanning calorimetry (DSC) [9], and so on. The computation of polymer crystallization is a theoretical method and can accurately provide theoretical information [10]. However, the difference between theoretical assumption and actual crystallization is different. Besides, computation studies need high performance computing systems. The XRD can provide data related to the crystalline structure, crystal size, and crystallinity. The FTIR and

---

Polymer Crystallization: Methods, Characterization, and Applications, First Edition.
Edited by Jyotishkumar Parameswaranpillai, Jenny Jacob, Senthilkumar Krishnasamy, Aswathy Jayakumar, and Nishar Hameed.
© 2023 WILEY-VCH GmbH. Published 2023 by WILEY-VCH GmbH.

RAMAN spectrum can detect the configuration of macromolecules during phase transition. Although such methods can provide much information about polymer crystallization, it is hard to have the model to characterize and predict it. The POM can count the number of crystals and measure the crystal size. Based on the change in crystal numbers and their size with time, the prediction of polymer crystallization is achieved. However, the performance of POM to characterize the polymer crystallization requires preferred experimental conditions.

The DSC method is now usually used to characterize polymer crystallization, where the sample is placed in the enclosed environment. To characterize the polymer crystallization, only the plots of heat flow, temperature, and time are required. The collection of data using the DSC method has been proposed by ICTAC Kinetics Committee [11]. In this chapter, we are aiming to present the characterization of polymer crystallization by using DSC method. Both isothermal crystallization mechanism and non-isothermal crystallization mechanism are discussed.

2.2 Basic Principle

2.2.1 General Idea

Simplified thermal and dynamic description of crystallization kinetics following equation (2.1) is based on constitutive system of equations defining the relations between rate of crystallization ($d\alpha/dt$), rate of temperature changes (dT/dt), and state of investigated system (α, T). The $k(T)$ represents temperature term model, and $f(\alpha)$ represents kinetic term model. The overall transformation can be affected by several processes, and Eq. (2.2) can be used. The d represents number of steps during the process. Here, we only focus on one-step thermal dynamical description of crystallization kinetics.

$$\frac{d\alpha}{dt} = k(T)f(\alpha) \tag{2.1}$$

$$\frac{d\alpha}{dt} = \sum_{1}^{d} k(T)_i f(\alpha_i)_i \tag{2.2}$$

- Temperature term model $k(T)$

  Generally, $k(T)$ is expressed by Arrhenius model following equation (2.3), where A is pre-factor, R is gas constant, and E is activation energy.

$$k(T) = A \exp\left(-\frac{E}{RT}\right) \tag{2.3}$$

- Kinetic term model $f(\alpha)$

  The $f(\alpha)$ is connected to kinetic models, including accelerating model (Eq. (2.4)), deaccelerating model (Eq. (2.5)), and sigmodal model (Eq. (2.6)). Especially,

n represents the reaction order. In detail, the accelerating model is typically suitable for thermal degradation of polymer under oxygen atmosphere. The deaccelerating model is more popular for processes where the conversion rates decrease with reaction time. The thermal degradation of polymer under nitrogen atmosphere and polymer crystallization can be described. The sigmodal model characterizes the process where initial stage and final stage have accelerating behavior and deaccelerating behavior, respectively. Besides, the empirical model equation (2.7) is proposed, where a, b, and c are constants, and the common models with a value, b value, and c value are given in Ref. [11].

$$f(\alpha) = n\alpha^{n(n-1)} \tag{2.4}$$

$$f(\alpha) = (1-\alpha)^n \tag{2.5}$$

$$f(\alpha) = n(1-\alpha)[-\ln(1-\alpha)]^{(n-1)/n} \tag{2.6}$$

$$f(\alpha) = \alpha^a (1-\alpha)^b [-\ln(1-\alpha)]^c \tag{2.7}$$

2.2.2 Application of DSC Method

DSC results are function of heating or cooling, characterized by heating/cooling rate φ (K min^{-1}). By selecting isothermal crystallization method or non-isothermal crystallization method, the kinetics of polymer crystallization can be described.

The characterization of polymer crystallization via DSC method is based on the parametrized models composed from time (t) dependent (kinetic) and temperature (T) dependent (thermodynamic) terms that are predicting conversion degree (α). Time-dependent term has as response the rate constant of crystallization $k(T)$ dependent on the temperature. The common kinetic models are shown in reference [11]. Based on kinetic models, both isothermal crystallization and non-isothermal crystallization processes are successfully described. The isothermal crystallization of polymer is based on the observation of crystallization with time, which neglects the effect of temperature and simplifies the model to characterize the polymer crystallization. The crystallization rate constant $k(T)$ is one of the most important results from isothermal crystallization. Besides, it is highly repeatable for results of isothermal crystallization since the polymer is totally melted before start of isothermal crystallization. For the non-isothermal crystallization of polymer, it closely matches the one from industrial production point. Effective information can be provided from non-isothermal crystallization method, especially for effect of temperature field on crystallization. So, the isothermal crystallization mechanism and non-isothermal crystallization mechanism are discussed in Sections 2.3 and 2.4.

2.3 Characterization of Polymer Crystallization According to Isothermal Crystallization Process

2.3.1 Performance of Isothermal Crystallization Process

The isothermal crystallization model created from the DSC measurement follows the four steps (Figure 2.1):

(i) In step i, the samples are heated from initial temperature (T_0) to higher final temperature (T_f) with heating rate φ. Usually, the linear heating rate is used, and φ is 10 K min^{-1}. The T_f value should be at least about 30 °C higher than melting temperature of sample.
(ii) Then, the samples are kept at T_f for five minutes or longer to totally erase the thermal history.
(iii) Then, the cooling of sample from T_f to a specific crystallization temperature (T_c) with high cooling rate is immediately performed.
(iv) After step (iii), the sample is kept at T_c for five minutes or longer. The step (iv) is the isothermal crystallization of sample, and the heating flow with time is recorded.

Then, the four steps are repeated with different T_c values, which are shown in Figure 2.1. Especially, the step (v) is the following step (i) after the previous step (iv) for the new start of cycles. For step (iv), the heat flow change with the time t is recorded. As a result, the relative crystallinity degree (X_t) value can be obtained by following equation (2.8).

$$X_t = \frac{\int_0^{t_i} H \, dt}{\int_0^{\infty} H \, dt} \qquad (2.8)$$

2.3.2 Analysis of Isothermal Crystallization Process

Typically, the empirical Avrami equations (2.9a) and (2.9b) are used to describe the isothermal crystallization analysis. More convenient form of Avrami model has form.

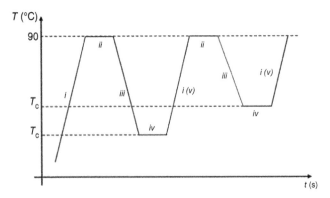

Figure 2.1 Isothermal procedure of the DSC measurement for the prepared samples.

2.3 Characterization of Polymer Crystallization According to Isothermal Crystallization Process

$$X_t = 1 - \exp(-Kt^n) \qquad (2.9a)$$

$$X_t = 1 - \exp[-(kt)^n] \qquad (2.9b)$$

The X_t is relative crystalline degree, and here is as a function of time t (min), k $((1/\min)^{1/n})$ is rate constant connected with nucleation growth, K (1/min) is standard rate constant of kinetic processes, and n is characterizing the crystal geometry [12]. Then, Eq. (2.10) the linearized form of Eq. (2.9a) is frequently used, and both K and n values can be obtained.

$$\log[-[\ln(1 - X_t)]] = \log(K) + n\,\log(t) \qquad (2.10)$$

Although it is acceptable to characterize isothermal crystallization, the error in measurements should be noticed. These models are not assuming real situations where the experimentally based data X_t are subjected to errors of measurement. In majority of cases, for parameter K, n estimation uses linear regression by least squares because the $\log[-[\ln(1-X_t)]]$ is linearly related to the $\log(t)$. The intercept of this straight line is $\log(K)$ and the slope is equal to n. Estimated values can be found here, far from values obtained by more correct methods. In sequel, we will follow this rough approach because it is quick and direct way to obtain parameter estimates and is used in articles describing their physical interpretation.

In addition, crystallization rate and crystallization activation energy is related to selected X_t. The extensions of Avrami equation for isothermal crystallization should be discussed, or other models can be used [13]. Main problem is estimation of model parameters K and n from experimental data subjected to errors by using statistically correct approach based on nature of error terms. It is shown that Avrami model fits well the data generated by different models, but estimated parameter values are not correct. Arnaldo T. Lorenzo et al. [13] systematically revealed basic principle for applying the Avrami equation to describe the isothermal crystallization from four fields, including the initial time, the selection of baseline, the cooling effect, and the crystallization range. Especially, it was found that the cooling rate during the isothermal crystallization should be at least 60 °C min^{-1} to avoid the effect of the previous non-isothermal crystallization. Besides, the crystallization range was proposed to range from 3% to 20%. By using Avrami equation, three parameters are proposed to characterize the isothermal crystallization process, including crystal geometry, crystallization rate, and crystallization activation energy.

2.3.2.1 Crystal Geometry

The n exponent represents the crystal geometry during isothermal crystallization process. The Avrami equation can well predict the primary nucleation and crystal growth [13]. By fitting the primary linear part of plots of $\log[-[\ln(1-X_t)]]$ against $\log(t)$, the n exponent is obtained. Besides, it is proposed that estimated n values may have different physical meanings for various crystallizations, which are affected by nucleation type and control of crystal growth. Table 2.1 lists the physical meaning of n values for different crystallizations.

Table 2.1 Growth of different crystals characterized by n exponent.

n exponent	Crystal geometry	Type of nucleation	Control of crystal growth
0.5	One dimension	Athermal nucleation	Diffusion-control
1	One dimension	Athermal nucleation	Nucleation-control
1.5	One dimension	Thermal nucleation	Diffusion-control
2	One dimension	Thermal nucleation	Nucleation-control
1	Two dimensions	Athermal nucleation	Diffusion-control
2	Two dimensions	Athermal nucleation	Nucleation-control
2	Two dimensions	Thermal nucleation	Diffusion-control
3	Two dimensions	Thermal nucleation	Nucleation-control
1.5	Three dimensions	Athermal nucleation	Diffusion-control
2.5	Three dimensions	Thermal nucleation	Diffusion-control
3	Three dimensions	Athermal nucleation	Nucleation-control
4	Three dimensions	Thermal nucleation	Nucleation-control

Source: Adapted from Cheng and Wunderlich [14].

2.3.2.2 Characterization of Crystallization Rate

The crystallization half-time ($t_{1/2}$) is one major characteristic of the crystallization rate, which can be obtained by following equation (2.11).

$$t_{1/2} = (\ln 2/K)^{1/n} \qquad (2.11)$$

After obtaining $t_{1/2}$ values from Eq. (2.11), it is also necessary to compare them with $t_{1/2}$ values from isoconversional curves. The isoconversional principle states that the reaction rate at a constant extent of crystallinity is only a function of temperature [15]. Furthermore, list of measured $t_{1/2}$ values and calculated $t_{1/2}$ values should be discussed.

2.3.2.3 Characterization of Crystallization Activation Energy

The crystallization process of all the samples is considered to be thermally activated. Therefore, the crystallization activation energy is evaluated from temperature dependence of rate constant k on temperature T by using the Arrhenius equation [15, 16], expressed by Eq. (2.3). For rough estimation of the logarithmic transformation, Eq. (2.12) is used. The Linearization equation (2.12) is very rough because of wide range of rate constants in the multiplicative model of measurement errors (constant variation coefficient is valid). Moreover, using linear least squares for estimation of $\ln A$ and E leads to high correlation between estimates of these parameters (multicollinearity) and false interpretation (parameters are in fact independent but their estimates are strongly dependent).

$$\ln(K) = \ln A - \frac{E}{RT} \qquad (2.12)$$

In most cases, the higher E values correspond to delayed crystallization, and smaller E values suggest fast crystallization. While for some research work, change of E values is not consistent with change of $t_{1/2}$. To better understand the reason, it should be noticed that polymer crystallization includes nucleation and diffusion of polymer chains. The E values characterize overall crystallization, while the heterogenous nucleation makes the crystallization process complicated.

2.3.3 Isothermal Crystallization of Some Polymer Composites

In Section 2.3.2, basic principle of Avrami equation to characterize crystallization process is described. Here, the mini-review is to reveal the relevant research works. For example, Juan Li et al. studied the crystallization mechanism of polyamide 6 (PA 6)/multiwalled carbon nanotubes (MCNTs) composites [17]. The introduction of MCNTs in the PA 6 matrix significantly enhanced the isothermal crystallization rate. Besides, the PA 6/MCNTs composites consisted of 2D and 3D crystals while the pure PA 6 only had 3D crystals, which was observed that the n value of PA 6/MCNTs composites was around 2 while the n value of PA 6/MCNTs composites was beyond 3. Besides, the E_a values of PA 6/MCNTs composites were higher than those of pure PA 6. It was proposed that the movement of PA 6 molecules during isothermal crystallization was strongly confined due to presence of MCNTs. Nigel Coburn et al. also studied crystallization mechanism of polypropylene (PP)/MCNTs via isothermal method [18]. The introduction of MCNTs in PP matrix enhanced the isothermal crystallization rate while slightly affecting the crystal geometry. Strawhecker and Manias systematically studied the crystallization behavior of polyethylene oxide (PEO)/clay nanocomposites [19]. The PEO crystallization is hindered by clays while the overall crystallization rate is higher in the composites when compared with the pure polymer. More crystallites of PEO are created because of the clay, and heterogenous nucleation is enhanced. The similar phenome is also found in PEO/nanoplate composites [19]. The contradiction is attributed to the interaction between polyethylene glycol (PEG) and clay, causing the formation of amorphous PEG in the vicinity of clay layers. To explain the difference, two factors are taken into consideration for crystallization of polymer composites [20, 21]. The semi-crystalline morphology, which is used to describe any variations in the quantity, size, quality, and distribution of the polymer crystals, is the first and most crucial feature. Second, a lot of research explaining the function of the interfacial polymer, also known as the so-called rigid amorphous fraction (RAF) [20], within the overall polymer nanocomposites (PNC) performance has been published during the past 10 years. RAF is divided into two groups: that which is present at the filler surface and that which is present around the crystals [22]. Recently, Klementina Pusnik Cresnar et al. systematically studied effect of low amounts of silver (Ag), zinc oxide (ZnO), and titanium dioxide (TiO$_2$) nanoparticles on crystallization behavior of polylactic acid (PLA) and proposed that such fillers in PLA matrix resulted in the different heterogeneous nucleation and the diffusion of PLA chains [23]. The similar results were also confirmed in our published work related to isothermal crystallization of PEG/copper (Cu) particle composites [9].

2.4 Characterization of Polymer Non-isothermal Crystallization Process

2.4.1 Basics of Nonlinear Crystallization Modeling

For non-isothermal crystallization data treatment, the differential rate equation (analogy with mass transfer) is often factorized to the temperature and kinetic terms, which is same as Eq. (2.1). The response variable (conversional degree or relative crystallinity degree) α in range $(0 < \alpha < 1)$ is function of crystallization time t and temperature T. The difference is that the heating/cooling rate follows Eq. (2.13), and correspondingly, it is found that $T(t) = T_0 + \varphi t$.

$$\frac{dT}{dt} = \varphi \tag{2.13}$$

To simply find the relationship between α, t, and T, the integral $g(\alpha)$ for $f(\alpha)$ is used and is expressed in Eqs. (2.14), (2.15), and (2.16), respectively. The integral $g(\alpha)$ is the foundation for various integral methods to characterize polymer crystallization, which are shown in Section 2.4.4.

$$g(\alpha) = \int_0^\alpha \frac{d\alpha}{f(\alpha)} \tag{2.14}$$

$$g(\alpha) = A \int_0^t \exp\left(\frac{-E}{RT}\right) dt \tag{2.15}$$

$$g(\alpha) = \frac{A}{\varphi} \int_0^T \exp\left(\frac{-E}{RT}\right) dT \tag{2.16}$$

2.4.2 Performance of Non-isothermal Crystallization Process

The non-isothermal crystallization process for characterization of polymer crystallization is performed by using various heating/cooling rates (φ). At least three φ values should be applied. It is found that the peak temperature T_{pc} corresponding to the maximum crystallization rate shifted to a lower temperature with increasing φ. Correspondingly, the retardation effect of φ on the polymer crystallization process is proposed.

2.4.3 Analysis of Crystal Geometry During Non-isothermal Crystallization Process

To determine the crystal geometry of polymer during the non-isothermal crystallization process, various methods have been proposed, including Jeziorny-modified Arvami equation, Ozawa model, and combined Avrami and Ozawa model (Mo model).

2.4.3.1 Jeziorny-Modified Avrami Equation

By assuming that crystallization temperature is one constant, Avrami equation can describe the primary stage of non-isothermal crystallization by following equation (2.17) [24]. By comparing with Eq. (2.9a), Z is inadequate to describe the crystallization since the retardation effect of φ on crystallization is found. After having linear equation (2.18), n value is obtained and represents crystal growth geometry. Besides, the modified rate parameter Z_c is proposed and calculated according to Eq. (2.19) [24].

$$1 - X_t = \exp\left(Z_t^n\right) \tag{2.17}$$

$$\log[\ln(1 - X_t)] = \log(Z_t) + n\log(t) \tag{2.18}$$

$$\log(Z_c) = \frac{\log(Z_t)}{\varphi} \tag{2.19}$$

Similar to one of isothermal crystallization processes, three linear parts are obviously found. Better to characterize the overall crystallization process, it should be careful to select X_t range to characterize primary crystallization mechanism. Figure 2.2 presents different fitting results for same PEG by selecting different X_t ranges. The X_t ranging from 15% to 80% is better to characterize PEG crystallization.

Although the Jeziorny-modified Avrami equation has been used to characterize the crystallization mechanism of various polymers or polymer composites, it is still argued that the n value and Z_c value do not have clear physical meaning for dynamic crystallization. So, it is hard to estimate the crystal geometry by following physical meaning of n value from isothermal crystallization. It is also reported that the Jeziorny-modified Avrami equation retrieves the incorrect n values [25].

2.4.3.2 Ozawa Model

The Ozawa model is developed based on Arvami theory, where the effect of φ on the crystallization is considered, and φ is assumed as a constant during the whole non-isothermal crystallization process [26]. Equation (2.20) is proposed to describe the crystallization during the non-isothermal process. Then, the relationship between X_t and T is expressed as Eq. (2.21). Here, X_t is the function of relative crystalline degree with temperature. The k_c represents the cooling function and is related to overall crystallization rate. The m is the estimated Ozawa exponent, which represents the dimension of crystal growth.

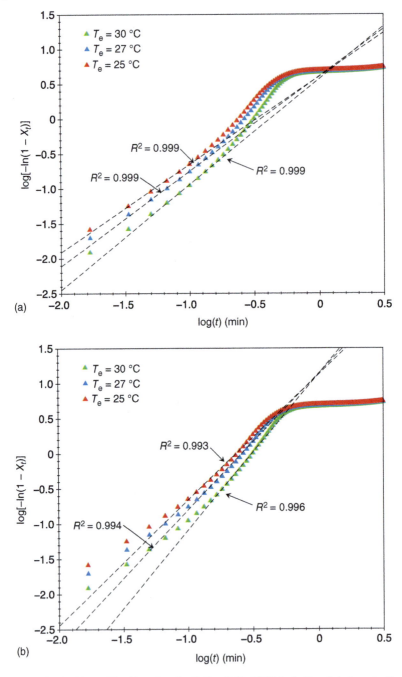

Figure 2.2 Plots of log(t) against log[ln(1 − X_t)] of PEG (a, b: the plots by selecting X_t values ranging from 3% to 20% and 15% to 80%, respectively). Source: Reproduced with permission from Yang et al. [9]/Elsevier.

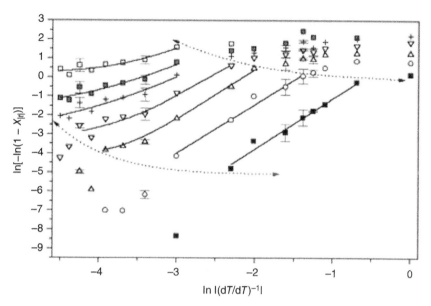

Figure 2.3 Successful fitting of Ozawa model for PET. Source: Reproduced with permission from Sajkiewicz et al. [27]/Elsevier.

$$X_t = 1 - \exp(-k_c/\varphi^m) \quad (2.20)$$

$$\ln[1 - \ln(1 - X_t)] = \ln(k_c) - m \ln(\varphi) \quad (2.21)$$

For example, P. Sajkiewicz et al. used the Ozawa model to describe the crystallization of polyethyleneterephtalate (PET) [27]. Besides, it reveals that the application of Ozawa model requires suitable φ values. The crystallization process of PET tended to be different from the one predicted by Ozawa model (Figure 2.3). Wengui Weng et al. also failed to describe the crystallization mechanism of PA/foliated graphite via Ozawa model [28] (Figure 2.4). It was also found in our previous work and other research work [9, 29]. It is attributed to the fact that the non-isothermal crystallization process is dynamic and the crystallization rate cannot be a constant. The crystallization rate should be a function of crystallization time and cooling rate. The nonlinearity is observed when the process approaches out of the primary crystallization. It is also argued that the difference between measured crystallization process and predicted crystallization process may be caused by spatial constraints of spherulitic growth.

Based on the theoretical drawbacks of Ozawa model, Chuah et al. [30] proposed a modified Ozawa model by following equations (2.22)–(2.26). Both λ and T_1 are empirical constants. It is assumed that the extreme point of the pertinent $\partial X_t/\partial T$ curve occurs at the point $T = T_\varphi$. As a result, the linear plot of the T_φ value against the $\left(\frac{m \ln \varphi}{\lambda}\right)$ value was found, and the modified Ozawa exponent (m) was obtained.

$$\ln(k_c) = \lambda(T - T_1) \quad (2.22)$$

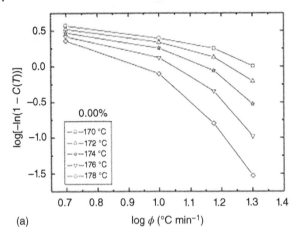
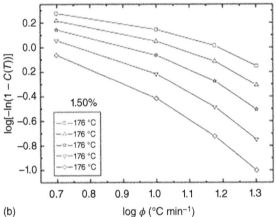

Figure 2.4 Failure of Ozawa model to characterize crystallization of PA (a) and PA/foliated graphite nanocomposites (b). Source: Reproduced with permission from Weng et al. [28]/Elsevier. (Notice: $C(T)$ is X_t.)

$$(\partial^2 X_t/\partial T^2)_{T_\varphi} = 0 \tag{2.23}$$

$$k_c(T_\varphi) = \varphi^m \tag{2.24}$$

$$\ln[-\ln(1 - X_t)] = \lambda(T - T_\varphi) \tag{2.25}$$

$$T_\varphi = (m \ln \varphi)/(\lambda) + T_1 \tag{2.26}$$

By applying modified Ozawa model, it is proposed that the range of X_t should range from 1% to 40% and only be valid for the crystallization regime I. For example, Wengui Weng et al. succeeded in describing the crystallization mechanism of PA/foliated graphite via modified Ozawa model [28] (Figure 2.5). It was also revealed that the obtained m values of PA/foliated graphite composites during non-isothermal crystallization process were comparable with n values obtained by Arvami equation from isothermal crystallization.

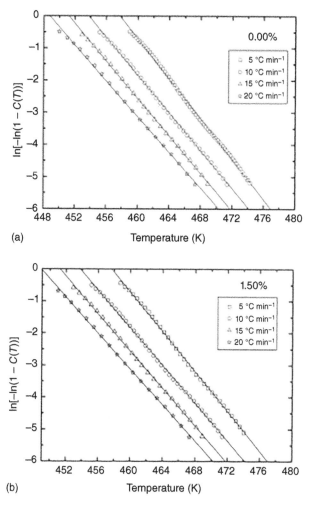

Figure 2.5 Success of modified Ozawa model to characterize crystallization of PA (a) and PA/foliated graphite nanocomposites (b). Source: Reproduced with permission from Weng et al. [28]/Elsevier. (Notice: C(T) is X_t.)

2.4.3.3 Mo model

From the abovementioned description, both Avrami equation and Ozawa model have drawbacks to describe the non-isothermal crystallization. Aimed at finding a suitable method to characterize the polymer crystallization during non-isothermal process, Mo and coworkers have proposed a thermal-kinetic model to reveal the non-isothermal crystallization mechanism of polymer by combining Avrami equation and Ozawa model [31]. By selecting certain X_t values, Eqs. (2.19) and (2.22) are combined into Eqs. (2.27)–(2.31) [32]. $F(T)$ is related to crystallization process, and a is the ratio between Arvami exponent and Ozawa exponent.

$$\log Z_t + n \log t = \log k_c - m \log \varphi \tag{2.27}$$

$$\log \varphi = \left(\frac{1}{m}\right) \log \left(\frac{k_c}{Z_t}\right) - \left(\frac{n}{m}\right) \log t \tag{2.28}$$

$$\log \varphi = \log F(T) - a \log t \tag{2.29}$$

$$F(T) = \left[\frac{k_c}{Z_t}\right]^{1/m} \tag{2.30}$$

$$a = \frac{n}{m} \tag{2.31}$$

Although a linear relationship between $\log\varphi$ and $\log t$ can be found, it is argued that $F(T)$ and a cannot have a straightforward interpretation for their actual physical meaning. By combining with analysis of Jeziorny-modified Arvami equation, the m can be obtained, and corresponding crystal geometry can be estimated. The crystallization mechanisms of various polymers or polymer composites during the non-isothermal process have been studied, including PEG, PEG/Cu composites, PA 6, PA 6/MCNTs, polyvinylidene fluoride (PVDF), PP/MCNTs, and so on [9, 18, 33]. Besides, it is suggested that the plots of $\log\varphi$ against $\log t$ can give hints of change in the crystallization mechanism [33] by observing different fitting results. By investigating the crystallization mechanism of PVDF via Mo method, it is revealed that the transition point or so-called turning point corresponds to the transition of regimes I and II.

2.4.4 Determination of Crystallization Activation Energy (E)

Generally, kinetic model-free methods are applied to characterize the crystallization process. By referring to Eqs. (2.14)–(2.16), the integral $g(\alpha)$ is very useful,. After having simple arrangements, Eq. (2.32) is obtained by following isoconversional principle. Both B and C are the parameters that are dependent on the integral type, and i corresponds to different temperature programs. For example, the equation with $B = 0$ and $C = 1.052$ is the Kissinger–Akahira–Sunrose (KAS) method, which is expressed in Eq. (2.33). The equation with $B = 0$ and $C = 1$ is the Ozawa model, which is expressed in Eq. (2.34).

Apart from integral isoconversional method, the differential isoconversional methods are applicable for characterization of polymer crystallization during non-isothermal processes [34]. The classical Ozawa, Flynn, and Wall (OFW) method is usually used, which is expressed in Eq. (2.35). The parameter with subscript α is the value with selected conversional degree.

$$\ln\left(\frac{\varphi_i}{T^B_{\alpha,i}}\right) = \text{Const} - C\left(\frac{E}{RT}\right) \tag{2.32}$$

$$\ln \varphi = \text{const} - 1.052 \frac{E}{RT} \tag{2.33}$$

$$\ln \frac{\varphi}{T^2} = \text{const} - \frac{E}{RT} \tag{2.34}$$

$$\ln\left[\varphi_i \left(\frac{d\alpha}{dt}\right)_i\right] = \ln(f(\alpha)A_\alpha) - \frac{E_\alpha}{RT_i} \tag{2.35}$$

It is noticed that the model-free kinetic methods are based on integration performed with respect to the temperature. Correspondingly, the model-free kinetic methods are valid only when the estimated E is independent of the degree of crystallization. However, the crystallization of polymers or polymer composites could be more complicated, and energy barrier for crystallization may be changed with conversion [35]. To roughly estimate the polymer crystallization, Kissinger method is proposed and is expressed as Eq. (2.36) [11, 35, 36]. The advantage of the Kissinger method is to obtain the activation energy regardless of the actual complicated kinetic process, where T_p is peak temperature.

$$\frac{d\left[\ln \varphi/T_p^2\right]}{d(1/T_p)} = -\frac{E}{R} \tag{2.36}$$

Although a great progress in crystallization activation energy has been achieved, it is argued for the physical meaning of the crystallization activation energy. On one hand, it is interpreted that the measured crystallization activation energy represents the energy barrier for crystallization. On other hand, it is proposed that the measured crystallization activation energy is just the apparent value to describe the kinetic reaction rate. In addition, the measured crystallization activation energy values are different when different models are applied. So, how to accurately interpret the crystallization activation energy should be focused.

2.4.5 Analysis of Relative Crystallinity

Apart from the characterization of crystallization mechanism of polymer composites, the relative crystallinity can be also evaluated. The relative crystallinity is usually used to characterize the confined polymer in nano- or meso-pores or other composites. By following equations (2.37) and (2.38), the relative crystalline degree (χ) can be obtained, where ΔH is the measured enthalpy value, ΔH^T is the theoretical enthalpy value, and p is the content of polymer in the polymer composite [37, 38]. The χ values smaller than 100% suggested the confined crystallization of paraffin wax (PW) in the composites. Besides, the heating/cooling rate influences the relative crystallinity. For example, little effect of heating/cooling rate on the relative crystallinity of PEG is proposed [9] while smaller relative crystallinity of PW is found when higher heating/cooling rate is applied [39]. The main reason is that the reaction time for macromolecular chains is different. Highly self-crystallization behavior supports stable relative crystallinity. So, the dependence of relative crystallinity on heating/cooling rate should be specified in research work as well.

$$\Delta H^T = \Delta H \times p \tag{2.37}$$

$$\chi = \Delta H/\Delta H^T \times 100\% \tag{2.38}$$

2.5 Conclusion

In this chapter, we present the basic principle and common methods to characterize polymer crystallization using thermal method. By using the DSC data, the polymer crystallization can be well characterized based on either isothermal crystallization or non-isothermal crystallization. The crystal geometry, crystallization rate, and crystallization activation energy can be obtained. Although such methods (e.g. Avrami equation, Ozawa model, and Mo method) have successfully described the crystallization mechanism of various polymers or polymer composites, the crystallization of polymer composites (e.g. polymer composites with nanoparticles as fillers) becomes more complicated, which results in the deviation of measured one from the predicted one. For example, interface between molecular chains and nanoparticles has been proven to affect the crystallization and alter the crystallization activation energy. So, it is necessary to combine DSC with other technologies to discover more details.

Acknowledgment

This work is supported by the project "Advanced structures for thermal insulation in extreme conditions" (Reg. No. 21-32510M) granted by the Czech Science Foundation (GACR).

Abbreviations

CS	chitosan
EG	expanded graphite
MCNT	multiwalled carbon nanotube
PA	polyamide
PEG	polyethylene glycol
PEO	polyethylene oxide
PET	polyethyleneterephtalate (common abbreviation is polyester)
PP	polypropylene
PVDF	polyvinylidene fluoride
PW	paraffin wax

References

1 Sangroniz, L., Cavallo, D., and Müller, A.J. (2020). Self-nucleation effects on polymer crystallization. *Macromolecules* **53** (12): 4581–4604.
2 Habel, C., Maiz, J., Olmedo-Martínez, J.L. et al. (2020). Competition between nucleation and confinement in the crystallization of poly(ethylene glycol)/large aspect ratio hectorite nanocomposites. *Polymer* **202**: 122734.
3 Liu, G., Müller, A.J., and Wang, D. (2021). Confined crystallization of polymers within nanopores. *Accounts of Chemical Research* **54** (15): 3028–3038.
4 Altorbaq, A.S., Jimenez, A.M., Pribyl, J. et al. (2021). Polymer spherulitic growth kinetics mediated by nanoparticle assemblies. *Macromolecules* **54** (2): 1063–1072.

5 Liu, Y., Cui, L., Guan, F. et al. (2007). Crystalline morphology and polymorphic phase transitions in electrospun nylon-6 nanofibers. *Macromolecules* **40** (17): 6283–6290.

6 Yang, K., Venkataraman, M., Karpiskova, J. et al. (2021). Structural analysis of embedding polyethylene glycol in silica aerogel. *Microporous and Mesoporous Materials* **310**: 110636.

7 Lee, K.-H., Kim, K.-W., Pesapane, A. et al. (2008). Polarized FT-IR study of macroscopically oriented electrospun nylon-6 nanofibers. *Macromolecules* **41** (4): 1494–1498.

8 Schick, C., Androsch, R., and Schmelzer, J.W.P. (2017). Homogeneous crystal nucleation in polymers. *Journal of Physics: Condensed Matter* **29** (45): 453002.

9 Yang, K., Venkataraman, M., Wiener, J. et al. (2022). Crystallization mechanism of micro flake Cu particle-filled poly(ethylene glycol) composites. *Thermochimica Acta* **710**: 179172.

10 Yamamoto, T. (2009). Computer modeling of polymer crystallization – toward computer-assisted materials' design. *Polymer* **50** (9): 1975–1985.

11 Vyazovkin, S., Burnham, A.K., Criado, J.M. et al. (2011). ICTAC Kinetics Committee recommendations for performing kinetic computations on thermal analysis data. *Thermochimica Acta* **520** (1–2): 1–19.

12 Avrami, M. (1940). Kinetics of phase change. II. Transformation–time relations for random distribution of nuclei. *Journal of Chemical Physics* **8** (2): 212–224.

13 Lorenzo, A.T., Arnal, M.L., Albuerne, J., and Müller, A.J. (2007). DSC isothermal polymer crystallization kinetics measurements and the use of the Avrami equation to fit the data: guidelines to avoid common problems. *Polymer Testing* **26** (2): 222–231.

14 Cheng, S.Z.D. and Wunderlich, B. (1988). Modification of the Avrami treatment of crystallization to account for nucleus and interface. *Macromolecules* **21** (11): 3327–3328.

15 Jackson, K.A. (2004). *Kinetic Processes: Crystal Growth, Diffusion, and Phase Transitions in Materials*, 2e. Wiley.

16 Vyazovkin, S. (2020). Activation energies and temperature dependencies of the rates of crystallization and melting of polymers. *Polymers-Basel* **12** (5): 1070.

17 Li, J., Fang, Z., Zhu, Y. et al. (2007). Isothermal crystallization kinetics and melting behavior of multiwalled carbon nanotubes/polyamide-6 composites. *Journal of Applied Polymer Science* **105** (6): 3531–3542.

18 Coburn, N., Douglas, P., Kaya, D. et al. (2018). Isothermal and non-isothermal crystallization kinetics of composites of poly(propylene) and MWCNTs. *Advanced Industrial and Engineering Polymer Research* **1** (1): 99–110.

19 Strawhecker, K.E. and Manias, E. (2003). Crystallization behavior of poly(ethylene oxide) in the presence of Na^+ montmorillonite fillers. *Chemistry of Materials* **15** (4): 844–849.

20 Wurm, A., Ismail, M., Kretzschmar, B. et al. (2010). Retarded crystallization in polyamide/layered silicates nanocomposites caused by an immobilized interphase. *Macromolecules* **43** (3): 1480–1487.

21 Terzopoulou, Z., Klonos, P.A., Kyritsis, A. et al. (2019). Interfacial interactions, crystallization and molecular mobility in nanocomposites of poly(lactic acid) filled with new hybrid inclusions based on graphene oxide and silica nanoparticles. *Polymer* **166**: 1–12.

22 Sargsyan, A., Tonoyan, A., Davtyan, S., and Schick, C. (2007). The amount of immobilized polymer in PMMA SiO_2 nanocomposites determined from calorimetric data. *European Polymer Journal* **43** (8): 3113–3127.

23 Črešnar, K.P., Zemljič, L.F., Papadopoulos, L. et al. (2021). Effects of Ag, ZnO and TiO_2 nanoparticles at low contents on the crystallization, semicrystalline morphology, interfacial phenomena and segmental dynamics of PLA. *Materials Today Communications* **27**: 102192.

24 Jeziorny, A. (1978). Parameters characterizing the kinetics of the non-isothermal crystallization of poly(ethylene terephthalate) determined by d.s.c. *Polymer* **19** (10): 1142–1144.

25 Zhang, Z., Xiao, C., and Dong, Z. (2007). Comparison of the Ozawa and modified Avrami models of polymer crystallization under nonisothermal conditions using a computer simulation method. *Thermochimica Acta* **466** (1–2): 22–28.

26 Koga, N. (2013). Ozawa's kinetic method for analyzing thermoanalytical curves. *Journal of Thermal Analysis and Calorimetry* **113** (3): 1527–1541.

27 Sajkiewicz, P., Carpaneto, L., and Wasiak, A. (2001). Application of the Ozawa model to non-isothermal crystallization of poly(ethylene terephthalate). *Polymer* **42** (12): 5365–5370.

28 Weng, W., Chen, G., and Wu, D. (2003). Crystallization kinetics and melting behaviors of nylon 6/foliated graphite nanocomposites. *Polymer* **44** (26): 8119–8132.

29 Qiu, S. and Qiu, Z. (2016). Crystallization kinetics and morphology of poly(ethylene suberate). *Journal of Applied Polymer Science* **133** (12): https://doi.org/10.1002/app.43086.

30 Chuah, K.P., Gan, S.N., and Chee, K.K. (1999). Determination of Avrami exponent by differential scanning calorimetry for non-isothermal crystallization of polymers. *Polymer* **40** (1): 253–259.

31 Liu, T., Mo, Z., Wang, S., and Zhang, H. (1997). Nonisothermal melt and cold crystallization kinetics of poly(aryl ether ether ketone ketone). *Polymer Engineering and Science* **37** (3): 568–575.

32 Vyazovkin, S. (2018). Nonisothermal crystallization of polymers: getting more out of kinetic analysis of differential scanning calorimetry data. *Polymer Crystallization* **1** (2): https://doi.org/10.1002/pcr2.10003.

33 Song, J., Mo, Z., Lu, C. et al. (2009). The turning point on plots of $\log \varphi$ and $\log t$ of Mo's equation. *Polymer International* **58** (7): 807–810.

34 Criado, J.M., Sánchez-Jiménez, P.E., and Pérez-Maqueda, L.A. (2008). Critical study of the isoconversional methods of kinetic analysis. *Journal of Thermal Analysis and Calorimetry* **92** (1): 199–203.

35 Šimon, P., Thomas, P., Dubaj, T. et al. (2014). The mathematical incorrectness of the integral isoconversional methods in case of variable activation energy and the consequences. *Journal of Thermal Analysis and Calorimetry* **115** (1): 853–859.

36 Wellen, R.M.R. and Canedo, E.L. (2014). On the Kissinger equation and the estimate of activation energies for non-isothermal cold crystallization of PET. *Polymer Testing* **40**: 33–38.

37 Yang, K., Wiener, J., Venkataraman, M. et al. (2021). Thermal analysis of PEG/metal particle-coated viscose fabric. *Polymer Testing* **100**: 107231.

38 Yang, K., Zhang, X., Venkataraman, M. et al. (2023). Structural analysis of phase change materials (PCMs)/expanded graphite (EG) composites and their thermal behavior under hot and humid conditions. *ChemPlusChem*. https://doi.org/10.1002/cplu.202300081.

39 Yang, K., Zhang, X., Wiener, J. et al. (2022). Nanofibrous membranes in multi-layer fabrics to avoid PCM leakages. *ChemNanoMat* 8 (10): e202200352.

3

Crystallization Behavior of Polypropylene and Its Blends and Composites

Daniela Mileva, Davide Tranchida, Enrico Carmeli, Dietrich Gloger, and Markus Gahleitner

Borealis Polyolefine GmbH, Innovation Headquarters, St. Peter-Strasse 25, 4021 Linz, Austria

3.1 Introduction – Polypropylene Crystallinity in Perspective

In 1954, Giulio Natta discovered the suitability of the combinations of transition metal complexes with an aluminum alkyl, developed only shortly before by Karl Ziegler for the low-pressure polymerization of ethylene, for the stereospecific polymerization of propylene. This was the origin of isotactic polypropylene (iPP), starting with products having an isotacticity of only around 40%, which was increased to 80% within one year through modification of the catalyst [1, 2]. While this allowed Montecatini to establish the first production facility for iPP in 1957, ultimately leading to one of the most widely produced thermoplastic polymers with a global production volume of more than 70 million tons per year in 2021, it also was a decisive point for polymer crystallization.

Looking back, one can clearly see that with the effect of chain branching for polyethylene (PE) and the stereostructure for polypropylene (PP), two crucial control factors for crystallinity, mechanical and optical properties of polymers had been discovered [3]. Moreover, the basic structure of most crystallizing polymers was also discovered in the same period, with Andrew Keller presenting his concept for chain folding in PE crystals in 1957 [4]. In contrast to the planar zig-zag structure of PE, however, iPP adapts a helical conformation in crystallization, similar to other poly-α-olefins like poly-1-butene. As Figure 3.1 shows, this forms the lowest one of several levels of the structural organization of an injection-molded iPP part [5]. The unit cell shown here relates to the α-modification, which is the dominant one for iPP homopolymers and impact copolymers (ICPs) under normal conditions. More than for other polymers, polymorphism – different materials with specific crystal structures formed by the same molecule [6, 7] – plays a major role for iPP.

While being less critical than in case of poly-1-butene [8], where re-crystallization of a metastable to a stable form causes massive shrinkage, also here industrial

---

Polymer Crystallization: Methods, Characterization, and Applications, First Edition.
Edited by Jyotishkumar Parameswaranpillai, Jenny Jacob, Senthilkumar Krishnasamy, Aswathy Jayakumar, and Nishar Hameed.
© 2023 WILEY-VCH GmbH. Published 2023 by WILEY-VCH GmbH.

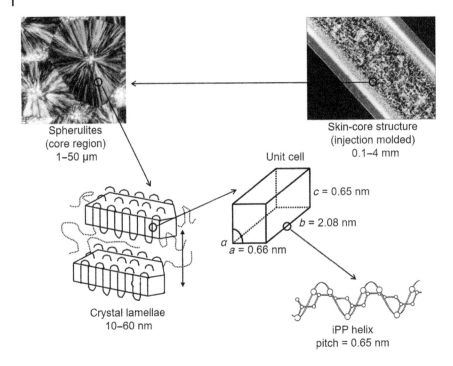

Figure 3.1 Characteristic hierarchy of morphological scales in iPP. The skin-core morphology of an injection-molded specimen is used to illustrate the morphology on the visual scale. Source: Designed based on Phillips and Wolkowicz [5] with own artwork and micrographs.

applicability depends on the formed crystal modification and structure. The polymorphism of iPP allows solidification in so far five documented crystal modifications from α to ε [9–12] and the mesomorphic (also called "smectic") state [10, 13]. Details about the generation of these modifications and their properties will be discussed below.

The chain configuration of PP, however, also allows two more basic steric configurations – syndiotactic polypropylene (sPP) and atactic polypropylene (aPP), as well as various combinations in terms of intra- and inter-chain variation, like controlled succession of aPP and iPP blocks in stereoblock polymers. aPP, originally a by-product of the older slurry-type polymerization processes for iPP, is a largely amorphous material, mostly used as a modifier for hot-melt adhesives and asphalt-based compositions [14]. While this by-product is mostly of very low molecular weight and wax-like in performance, targeted production with specific catalysts provides access to high molecular weight aPP types, which are elastomeric in nature [15]. Both have been tested in blends with iPP, resulting in a reduced crystallinity and enhanced flexibility, however with positive impact strength effects only above the glass transition temperature (T_g) of ~0 °C [16, 17]. A detailed look at the crystallization of these blends shows less crystallinity reduction than expected and a favored formation of the γ-modification at lower aPP contents in the blends, indicating miscibility and interaction on a molecular level. Similar in behavior to such blends, but superior in properties, are "low isotacticity" [18] PP grades with a meso pentad fraction between 40 and 60 mol% as commercialized by Idemitsu under the trade name "L-MODU." Like aPP, they can also be blended with iPP, however providing more stability due to long isotactic sequences capable of co-crystallization.

The syndiotactic version, sPP, enjoyed a certain period of high attention when its efficient production with single-site catalysts (SSCs) made it accessible in significant quantities, leading to its commercial production by Total [19]. Its pure application in films and flexible insulations was discussed, but the low crystallization speed and high degree of solubility in organic solvents limited its applicability [20]. Despite these technical limitations, sPP is a very interesting material showing a rich polymorphism, with four modifications designated by roman numerals I–IV (unlike iPP, where Greek letters are used) [21]. Of these polymorphs, only I, II, and IV contain chains in (two different) helical forms like iPP (see Figure 3.2), while in form III a trans-planar chain arrangement is found [23].

Form I is the thermodynamically stable phase of sPP obtained on cooling from non-oriented melts, characterized by an antichiral packing of left- and right-handed helical chains along both the a and b axes of an orthorhombic unit cell [22]. Its order is limited in practice, as is the crystallinity of such primarily crystallized samples. In contrast, form II is characterized by helices of the same chirality packed in an orthorhombic unit cell, being primarily formed from low stereoregularity materials under deformation, but also from an isotropic melt at high pressures or with nucleation [24]. Forms III and IV are metastable phases of sPP, the former being obtained on cold drawing of rapidly quenched samples, and the latter by exposing oriented samples of form III to solvents like benzene or toluene. Very recent work shows a "nodular" structure for sPP crystallized from quiescent quenched sPP melt [25], structurally similar to the mesophase of iPP [26]. Generally, transformations between the different polymorphs are more easily achieved than for iPP and do not require melting, but their effect on mechanical and optical performance is less well understood.

Blends of sPP and iPP have been studied frequently and with different targets [27–30]. Despite their structural similarity, the two polymers are immiscible and exhibit a two-phase structure over a wide concentration range, meaning that each of the polymers can be the dispersed phase and a co-continuous concentration range exists (see Figure 3.3). The crystalline interaction between the two polymers is very limited, with both melting and crystallization peaks (T_m and T_c, respectively) in differential scanning calorimetry (DSC) being clearly separated ($T_m(sPP) \sim 130\,°C$, $T_m(iPP) \sim 165\,°C$, $T_c(sPP) \sim 90\,°C$, $T_c(iPP) \sim 118\,°C$) and showing little concentration-induced variation [27, 29]. The sPP addition reduces the modulus of the blends following a nearly linear mixing rule, similar to the effect of impact modifiers, but increases toughness effectively only at higher concentrations [28].

While sPP is rather ductile at temperature levels above T_g, i.e. above 0 °C, it becomes very brittle below this due to its limited crystallinity. This motivated a study of sPP blends with ethylene–propylene rubber (EPR) and an ethylene–octene plastomer (EOC) [31]. The concentration effects of both elastomers are very similar to analogous iPP-based systems both in terms of morphology and mechanics, but

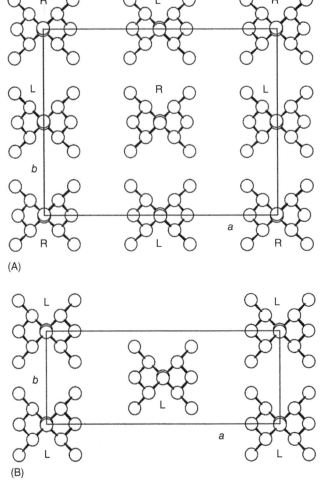

Figure 3.2 Two polymorphic forms of sPP with chains in s(2/1)2 helical conformation: (A) form I and (B) form II. Source: Reproduced with permission from De Rosa et al. [22], © 1988, American Chemical Society.

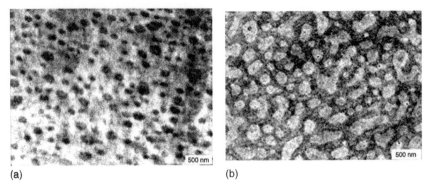

Figure 3.3 Morphology of iPP/sPP blends with a weight ratio of (a) 75/25 and (b) 40/60. Source: Bourbigot et al. [30]/The Allen Institute for Artificial Intelligence.

the modulus level is very low, as the starting material only offers 340 MPa. Not too much can consequently be expected in terms of practical applicability of such blends.

3.2 Chain Structure and Molecular Weight Effects for iPP Crystallinity and Polymorphism

Before coming to its molecular origin, the polymorphism of iPP needs to be discussed in more detail. The alpha (α) monoclinic form of iPP, found by Natta and Corradini [32], exhibits lamellar-shaped crystals with unique branching of radial and nearly tangential orientation of lamellae, mostly but not exclusively occurring in spherulitic growth [33]. In the crystal unit cell, the propylene macromolecule adopts a chain conformation of left-handed and right-handed ternary helices. The packing of helices is such that every helix interacts only with helices from the opposite hand. In Figure 3.4, the crystal structure and morphology of the monoclinic α-phase of iPP are presented together with the most relevant other polymorphs.

The beta (β) phase of iPP produces well-individualized negatively birefringent spherulites. It is metastable in comparison to the α-phase (T_m = 428 K resp. 155 °C

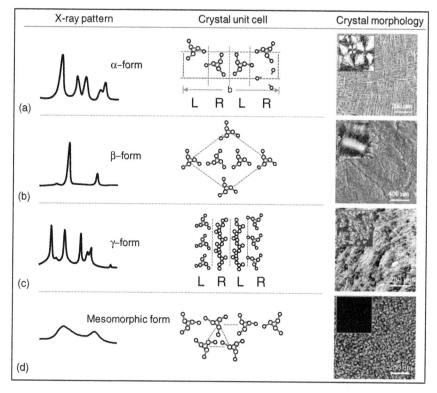

Figure 3.4 Wide-angle X-ray pattern (left), crystal unit cell (center), and crystal morphology (right) of monoclinic α-form (a), trigonal β-form (b), orthorhombic γ-form (c) and mesomorphic form (d) of iPP.

for β-iPP, related to $T_m = 438$ K resp. 165 °C for α-iPP, always for the homopolymer case) [34, 35]. The crystal structure was solved in 1994, independently by Meile et al. [36] and Padden and coworkers [37]. The same structural solution for this particular crystallographic form of iPP was reached, which is a trigonal cell with parameters $a = b = 0.11$ nm, $c = 0.65$ nm, containing three isochiral helices. The β-polymorph is formed either by selective nucleation (see Chapter 4) or crystallization at high pressure or shear; it is obviously favored by high isotacticity and the absence of chain defects [38, 39].

The gamma (γ) phase of iPP has remained rather elusive for many years. It was not usually observed as a different phase but co-crystallized with and within the α-phase spherulites. For the first time, Addink and Beintema observed an additional reflection in X-ray powder pattern of low molecular weight iPP [40]. According to electron microscopic studies, the γ crystals grow on the lateral "ac" faces of the α-iPP (α–γ lamellar branching). A structural similarity with the α-phase was expected, and a triclinic unit cell, corresponding to a slightly deformed α-phase unit cell, was long assumed for the γ-phase [41].

The real breakthrough in resolving the γ-phase structure dates from a paper published by Meile and Brückner in 1989 [42]. The authors re-evaluated the geometry of the proposed triclinic cell based on improved X-ray powder pattern data, noting that the triclinic cell is a sub-cell of a much larger face-centered orthorhombic unit cell, with parameters $a = 0.85$ nm, $b = 0.99$ nm, $c = 0.42$ nm. Based on this analysis, they proposed a structure in which the chain axes are not parallel to each other. The morphology of the γ-crystals does not differ significantly from the monoclinic crystals. In case of higher concentrations of γ-crystals, the length of the lamellae decreases in comparison to exclusively α-iPP [43]. Formation of the γ-phase is clearly favored by reduced stereoregularity and chain defects like regiodefects or comonomers [38, 43, 44].

Recently, it has been found that increasing the length of the branches in a propylene chain leads to development of new crystal modifications [45, 46]. In propylene–hexene copolymers, with hexene content larger than 10 mol%, a new polymorph melting at about 50 °C was observed [47]. The new trigonal or delta (δ) form of iPP allowed better incorporation of the hexene co-units in the crystal lattice. As a driving force for the crystallization of the new polymorph, the increased density was suggested due to the inclusion of hexene or pentene units in the crystal. A similar polymorphic modification was identified in oriented fibers of propylene copolymers with large concentrations of 1-butene [10].

Still, one more crystalline form of iPP has been presented by Lotz [48] for the rather specific case of highly stereo-defective PP based on a SSC. The epsilon (ε) form is nucleated by a crystal–crystal growth transition on parent α-phase crystals and has an orthorhombic unit cell with eight helices arranged in a near-double-tetragonal packing. Neither for this nor for the δ polymorph has any practical relevance been shown so far.

The mesomorphic form is a polymorph of iPP, which develops at very large supercoolings resp. under "quenching" situations. For the first time, Natta has shown that the PP chains can pack in a different way even if they maintain the same helix

3.2 Chain Structure and Molecular Weight Effects for iPP Crystallinity and Polymorphism

symmetry [32]. He showed that rapid cooling of iPP resulted in a paracrystalline smectic modification, in which right- and left-handed helices are placed at random. The density of this modification is lower than that of the monoclinic modification. It was also recognized that the packing of the chains perpendicularly to their axes is more disordered than along the axes. For this reason, Natta, Peraldo, and Corradini gave the name "smectic" to this form, indicating a degree of order higher than that of an ideal nematic liquid crystalline phase, in which the only degree of order is the molecular parallelism [49]. Further elucidation of the mesophase morphology revealed formation of particle-like crystals [50]. The nodular domains observed in very thin quenched films were not organized into high-ordered spherulitic superstructure [25, 51, 52]. Similar to the γ-form, reduced chain regularity enhances the possibility of formation of the mesomorphic form [53, 54].

It has been discussed before that the highest degree of crystallinity, related to the highest crystal growth rate, can be expected for chains of perfect stereo- and regio-regularity [55], with polymers of syndiotactic and low-isotacticity structure being more limited. Even for iPP homopolymers, differences in the chain structure and, to a lesser extent, the molecular weight distribution (MWD) affect crystal growth and polymorphic expression, limiting the value of "universal" growth data for different crystal modifications [56].

Details of this regularity effect, including its logical extension into random copolymers of propylene with ethylene or higher α-olefins like butene or hexene, have been studied by numerous groups [53–55, 57–69], but only rarely include a relevant characterization of the studied polymers, the method of choice clearly being ^{13}C nuclear magnetic resonance (NMR) spectroscopy [55, 58, 63–65, 67, 68]. So, while all authors agree on the principle of melting point-, crystallinity-, and stiffness-reduction by increasing chain irregularities, the results are somewhat difficult to compare. Figure 3.5a gives an overview of the melting point from a standard DSC test as function of the total defect content as defined by Jeon et al. [65], including data from SSC [55, 63, 65] and Ziegler–Natta catalyst (ZNC) [62]. The relation is similar to the one shown by Fujiyama and Inata [61], as a function of the ethylene content only (see Figure 3.5b). The difference in both T_m and melting

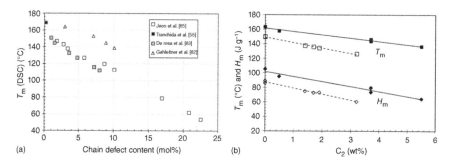

Figure 3.5 Effect of (a) total chain defect content on melting point, squares indicate SSC-PP [55, 63, 65] and triangles ZNC-PP [62], and (b) of C_2 content on melting point (squares) and enthalpy (diamonds) of C_2C_3 random copolymers based on metallocene (MC, open symbols) and ZN (full symbols) catalysts. Source: Data from Fujiyama and Inata [61].

enthalpy (H_m) between the two catalyst types is obviously related to the basic chain structure differences for the homopolymer. In view of the data of De Rosa et al. [63] and Jeon et al. [65], the effect of the regio-defects resulting from mis-inserted monomer units is clearly more relevant. Even if both series show a slightly different slope, the "near-perfect" iPP [55] is a suitable end point for both. Earlier papers in this area based on studies of ZNC-iPP fractions with different isotacticity and molecular weight [57–59] are difficult to relate to later studies, as fractions are necessarily different in their composition broadness, thus neglecting interactions occurring in practice.

Both diagrams in Figure 3.5 also show the effect of ethylene (C_2) as comonomer, which extends to crystallinity, mechanics, optics, and processing properties [41, 43, 53, 54, 62, 68, 70–73]. A recent review from our group [73] gives details on the effect of catalyst and comonomer type of iPP random copolymers in general, demonstrating that the type of comonomer insertion and the onset of phase separation at higher C_2 content (coinciding with the appearance of a second glass transition for the disperse copolymer phase) are critically relevant for mechanics. The improved clarity and reduced haze of random copolymers, however, are both dependent on the increasing formation of γ-phase under normal cooling rate and nucleation [41, 43, 70, 71] and the increasing formation of the mesomorphic phase at high cooling rates [53, 54, 72]. Figure 3.6 sums up this evolution for a series of one homo- and three copolymers analyzed in several papers [53, 54, 61, 68]. For slower cooling processes, the reduced growth rate causes lower crystallinity (coinciding with thinner crystal lamellae) and – at constant nucleus density – smaller spherulites, while the higher γ-phase content further enhances transparency and impact strength [73]. For fast-cooling situations, the transition from α/γ-to mesomorphic phase formation is shifted to lower rates, in the present series from 215 to 55 °C s^{-1} [53].

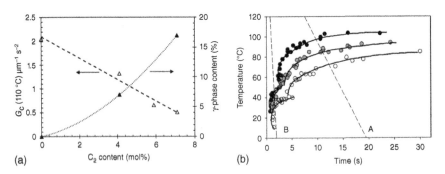

Figure 3.6 Effect of increasing C_2 content in ZNC-based PP random copolymers on (a) spherulitic growth rate G_c at 110 °C (open symbols) and γ-phase content (full symbols) on DSC specimen by wide angle X-ray scattering (WAXS). Source: Data from Fujiyama and Inata [61] and Agarwal et al. [68]. (b) On the continuous cooling transformation (CCT) curves of the homopolymer (white) and the random copolymers with 3.4 (grey) and 7.3 mol% C_2 (black), lines (A) and (B) indicating constant cooling rates of 10 and 100 °C s^{-1} resp. Source: Reproduced with permission from Cavallo et al. [53], © 2010, American Chemical Society.

3.2 Chain Structure and Molecular Weight Effects for iPP Crystallinity and Polymorphism

At the same time, these chain defects clearly reduce the possibility of β-phase formation, even in the presence of suitable nucleating agents [34, 39]. When considering the effect again as a function of cooling rate, this corresponds to a reduction of the rate at which β-phase formation stops and α-phase takes over. For iPP homopolymers, this has been shown to be around 100 °C s^{-1} [74], but it changes to 10 °C s^{-1} for a C_2C_3 random copolymer [71]. Varga [34] has called this the "prevention of β-phase formation in random copolymers," which is obviously a too strong term. It is interesting to observe that random copolymers with 1-butene or 1-hexene show some similarity regarding the transition to mesomorphic phase at higher cooling rates but no reduction of β-phase formation.

While there are numerous papers dealing with the effects of molecular weight and MWD on crystallization [75–83], especially on crystallization under flow or deformation [76, 78, 81, 83], it is not easy to sum up the respective effects as these are frequently combined with variations in chain structure (the underlying problem being the absence of isotacticity data in most cases). The next problem in determining molecular weight effects lies in the obviously more dominant effect of polydispersity, combined with the lack of sufficiently wide variations of melt flow rate (MFR) or weight average molecular weight (M_w). These are mostly found in visbreaking series [75, 80], in which the polydispersity (M_w/M_n) decreases with increasing distance to the parent polymer. When comparing this to a reactor-based MFR series, an inverse effect on modulus has been reported and shown in Figure 3.7a, related to differences in nucleation density in the original paper [77]. A more recent study links the nucleation density to M_w [83], but also here a combined difference in polydispersity exists. For an even longer visbreaking series a decrease of both T_m and H_m with increasing MFR has been observed [80].

Polydispersity, or more precisely the presence of well-dispersed molecules much longer than the average, is clearly more relevant for crystallinity and modulus [76, 78, 81, 83]. The enhanced formation of highly oriented skin layers with

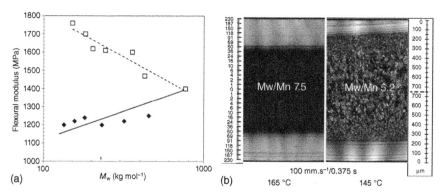

Figure 3.7 Effect of (a) weight average molecular weight (M_w) on flexural modulus for iPP homopolymers based on a reactor series (full symbols) and visbreaking (open symbols). Source: Data from Gahleitner et al. [77]. (b) Polydispersity on oriented structure formation. Source: Pantani et al. [81]/John Wiley & Sons.

dominant shish-kebab structures is obviously decisive here, as can be seen in the morphologies of model experiments by Pantani, Peters, and coworker in Figure 3.7b [81]: an increase in polydispersity not only gives a higher skin layer thickness and a finer core structure, the whole process is also shifted to higher temperatures. This effect is not only confirmed by other experiments and modeling [76, 78, 83], it has also been applied successfully in designing high stiffness homo- and copolymers with bi- or even trimodal MWD design [84].

Another peculiarity of iPP crystallization requiring discussion is the effect of long-chain branching (LCB) [85, 86]. While most commercial catalyst systems produce iPP of linear architecture, some metallocene catalysts can incorporate LCB via macro-monomer insertion reactions [87] or when using bifunctional comonomers [88]. The dominant route to LCB PP, is a post-polymerization modification, using a grafting reaction between linear iPP precursor chains. This reaction can use different coupling systems, being initiated by peroxides or other radical generators [89, 90], but also radiation-based processes have been applied successfully [91, 92]. LCB-PP, also called high melt strength (HMS) PP because of its increased melt strength and a certain degree of strain hardening in extensional flow similar to the behavior of low-density polyethylene (LDPE) has clear advantages in some conversion processes like foaming [93], extrusion coating, and thermoforming [86].

While rheological effects are clearly dominating these changes in processing behavior, LCB-PP also exhibits modified crystallization. The available literature on LCB-PP [94–97] shows enhanced crystallization temperature and higher nuclei density, but slower crystal growth rate (through overall higher crystallization rate) compared to linear iPP. The same is true for other LCB polymers compared to their linear version of same monomer chemistry. The consistency of the phenomenon across different LCB polymers suggests their crystallization behavior has a common cause, the branched-chain architecture. Most authors thus agree that branch points support the nucleation stage, increasing T_c, while chain diffusion to crystal growth fronts is slow, hindered by entanglements, causing slow crystal growth. Of practical interest are the possibility to use small amounts of LCB-PP for nucleation [94] and the positive reaction to flow fields [96], both showing similarity to the effects of very long molecules. The growth retardation goes, as in case of the random copolymers discussed above, together with a reduced β- and enhanced γ-phase formation [97].

3.3 Nucleation of iPP

The amount of literature on nucleation of iPP is enormous [98], simply resulting from the high diversity and technical relevance of this subject. No other polymer has received similar attention in this respect. Nucleation of iPP homo- and copolymers as well as its blends is used on industrial scale to improve the mechanical and optical properties [34, 35, 44, 70, 77, 99–102], but also the processing speed and stability [71, 103–106]. The present chapter will focus on some aspects of this complex matter, including the different types of nucleating agents and their selectivity toward the expression of different polymorphs of iPP.

3.3 Nucleation of iPP

Nucleating agents can be categorized in different ways: (i) by physical/chemical nature, (ii) by selectivity toward polymorphs, and (iii) by dispersion mechanism [98]. Regarding (i), one main difference is between particulate types including inorganic, low molecular weight organic and polymeric nucleating agents, and soluble nucleating agents being all of organic nature and low molecular weight. For (ii), there is a high number of α-nucleating agents, equally suitable to induce the γ-phase in polymers of lower chain regularity, and a range of β-all nucleating agents varying in selectivity and sensitivity. Finally, on (iii) one can distinguish between conventional dispersion (particulate mixing), dispersion by dissolution and re-crystallization from the polymer melt, and dispersion by polymerization. Most of these aspects will be discussed now.

The most relevant inorganic nucleating agent is talc, a silicate mineral also widely used as reinforcing filler for iPP homo- and copolymers [107–111]. Its effect on both crystallization and stiffness shown in Figure 3.8 results from a combination of good lattice matching toward the α-form of iPP [112] and its lamellar structure. The perpendicular growth of iPP lamellae on talc platelet surfaces has been evidenced by high-resolution microscopy [98, 111], maximizing stiffness (modulus) in a network-like fashion to which the high degree of orientation [109, 110] contributes. On the positive side, this is accompanied by lower shrinkage and higher heat deflection temperature, on the negative side by lower ductility and toughness above concentrations of ~5 wt%.

Figure 3.8 shows several further facts: (i) Calcite, the trigonal form of $CaCO_3$ with rather isotropic structure, has a very limited capacity for reinforcement (even below kaolin [107]) and shows practically no nucleation effect, making it a suitable reinforcement for β-nucleated compositions [113]. (ii) Short glass fibers (GF) – used here with a low amount of commercial polar modified PP as coupling agent – allow reaching the highest stiffness levels, but are not nucleating under quiescent conditions, despite their strong effect on flow-induced crystallization (FIC) [114–116] (this makes them also suitable for a combination with β-nucleation

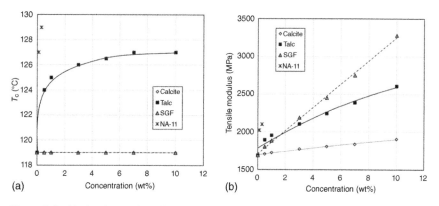

Figure 3.8 Nucleation and reinforcement of iPP by talc in comparison to calcite, glass fibers (short-cut glass fibers [SGF]), and an organophosphate nucleating agent (NA-11), effects for (a) crystallization temperature and (b) flexural modulus. Source: Own data based on PP homopolymer.

[117, 118], but limits their effect at very low concentrations). (iii) In relation to concentration, both nucleation and modulus increase by talc are still inferior to a highly efficient organic nucleating agent like the organophosphate NA-11, sodium 2,2′-methylene-bis-(4,6-di-*t*-butylphenylene)phosphate [100, 104, 119–121], which will be discussed further below. The latter fact, in combination with the poor optical performance of talc-nucleated iPP compositions, makes this mineral a kind of "economy solution" for stiffness increase, with the advantage of being non-extractable.

Of the big family of organic particulate nucleating agents for the α/γ-phase, only two representatives shall be discussed here: Benzoate derivatives [103, 122, 123] and organophosphates [100, 104, 119–125]. Next to those, there are numerous other candidates with often rather specific property combinations, examples include rosin-derivatives based on renewable sources [126] and derivatives of hexahydrophthalic acid showing potential for processing acceleration [127]. Finally, there are also substances with unintentional nucleating activity like pigments, the most famous example being phthalocyanine blue [128], known for causing shrinkage problems in otherwise non-nucleated iPP.

Benzoate salts, and also some substituted benzoate salts, have been in use for iPP nucleation for a very long time. For Na-benzoate itself, the main advantages are low cost and wide acceptance, even for medical packaging, but it is clearly limited in performance when compared to other particulate and soluble types [122, 123]. Different attempts at quantifying nucleation efficiency have been made, and Nagasawa et al. [122] even tried correlating nucleation density directly to Young's modulus (see Figure 3.9a), although they also found significant differences in particle shape and size distribution between the different nucleating agent types.

On top of size and shape, the actual dispersion quality plays a role in efficiency, as does the quality of lattice matching, which has been studied for the

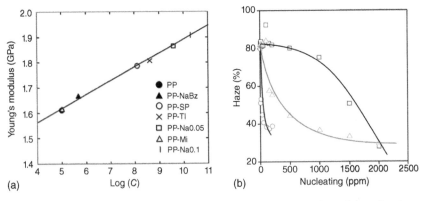

Figure 3.9 Mechanical and optical effects of nucleation, (a) Young's modulus of various polymer–nucleator mixtures in relation to nucleation density C. Source: Reproduced from Nagasawa et al. [122] © Elsevier. (b) Changes in haze of 2 mm thick iPP plates in the presence of three nucleating agents: (○) polymer PVCH, (Δ) DMDBS Millad 3988, (□) organophosphate NA-21. Source: Reproduced with permission from Menyhárd et al. [100] © Elsevier.

organophosphate NA-11 by Yoshimoto et al. [119]. Particles of this substance have a clearly acicular (needle-like) structure, further enhancing stiffness expression, and among the different possible salt ions, sodium has been found to be the most effective [124]; the efficiency is also higher in terms of thermodynamic interaction and empirical scales [123]. A second commercial example from this family is NA-21, a binary mixture of hydroxy aluminum bis(2,4,8,10-tetra-*trans*-butyl-6-hydroxy-12*H*-dibenzo[*d,g*][1,3,2]dioxaphosphocin-6-oxide) and lithium myristate [100, 125]. While giving less stiffness increase than NA-11, this type offers better optical performance, i.e. lower haze as required for thin-wall packaging applications. While, however, there is frequently a good correlation between nucleation efficiency and modulus, especially when considering concentration series, transparency, and haze show a more complex behavior. Menyhárd et al. give an explanation for this complexity (see Figure 3.9b) in their study comparing different types of nucleating agents [100], considering different levels of crystal structure organization.

The third type of particulate nucleating agents are polymeric, with poly(tetrafluoro ethylene) (PTFE) and poly(vinyl cyclohexane) (PVCH) being two well-known examples [100, 129–131]. Lattice matching has been demonstrated for PVCH [129], which gives very stable nucleation also at high cooling rates [131] and is very efficient at low concentrations (see Figure 3.9b).

Soluble nucleating agents are different in terms of activity and dispersion mechanism, frequently termed "clarifiers" for their specific activity toward haze reduction, or "organic gelators" due to the rheological phenomena associated with their application. There are two bigger chemical classes, sorbitol derivatives having a "butterfly-like" molecular structure [100–102, 105, 106, 132–136] and trisamides with a "windmill-like" structure [99, 137–139], nearly all presented structures being active for the α/γ-phase with one notable exception [138]. Unlike particulate nucleating agents, they have a rather low melting point and dissolve in the iPP melt when added during compounding or processing. Upon cooling, they form fibrillar crystals or at higher concentrations a network, which was observed by microscopy and rheology for di(benzylidene sorbitol) in ethylene glycol first [132]. Balzano et al. [133] studied the effects of temperature, concentration, and flow for 1,3:2,4-bis(3,4-dimethylbenzylidene)sorbitol (DMDBS, trade name Millad 3988) in detail, finding a transition between isotropic and flow-oriented morphologies to depend on the temperature at which shear is exerted on the composition. This temperature sensitivity is also observed in practice, and the enhancement of oriented skin layers by DMDBS has also been confirmed by other authors [71, 105]. Further development has led to another type with less sensitivity to processing temperature, 1,2,3-tridesoxy-4,6:5,7-bis-O-[(4-propylphenyl)methylene]nonitol (trade name Millad NX8000), which can consequently be applied in a wider processing window [134], but has equally been shown to develop a fibril network [135]. Like for other α-nucleating agents [70], the promotion of γ-phase formation in random copolymers [71] and for homopolymers under elevated pressure [136] has been demonstrated, which is likely the reason for impact strength improvement [101].

The trisamide family started with the systematic screening of a wide range of compounds [99] and also resulted in some commercial products like Irgaclear XT 386 (1,3,5-tris(2,2-dimethylpropionylamino)benzene). A specific advantage of this class is its high activity at very low concentrations [139].

Unlike α-nucleation, which enhances the formation of the dominant crystal form, β-nucleation creates an otherwise metastable form with enhanced mechanical performance. The sensitivity of its formation to chain irregularities [34, 39] and high cooling rate [71] has been mentioned above, but a more comprehensive review on structural and mechanical effects in β-nucleated iPP compositions is given by Grein [35].

Two traditional types of β-nucleating agents, γ-quinacridone and Ca-pimelate, are continuously receiving attention. The former, mostly applied as commercial pigment mixture Cinquasia red, has been in industrial use since the 1970s and has accordingly been studied before, including modifications of the composition [140, 141]. More recent work has been done in terms of cooling rate effect [71, 74] and other processing-related phenomena [142, 143]. The latter has the advantage of being both colorless and highly efficient in a wide range of iPP types [144, 145], but it showed some difficulties in market introduction. A recent study by Yue et al. [146] deals with the effect of thermal treatment on the β-nucleation efficiency of Ca-pimelate and sheds some light on the effects of drying and storage of this substance. A solution might be to facilitate handling and dispersion by supporting calcium carbonate micro- or nanoparticles [147].

In any case, the effect of impact strength improvement and a shift in the brittle-to-ductile transition temperature typical for β-nucleated iPP is limited to higher molecular weights at least for the case of homopolymers (see Figure 3.10) and random copolymers [34, 71]. In heterophasic copolymers (HECOs), it can be extended to types with lower molecular weight matrix and higher molecular weight elastomer phase [144, 145], but the typical application is still the pipe sector, because of the positive effect on crack propagation and the higher pressure resistance.

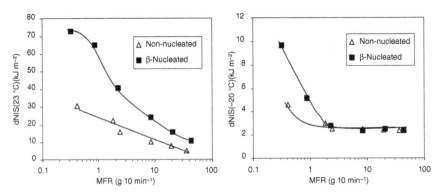

Figure 3.10 Double notched impact strength of non-nucleated and β-nucleated iPP homopolymers with different molar mass resp. MFR, results for (a) 23 °C and (b) −20 °C. Source: Reproduced with permission from Grein [35] © Springer Nature.

The recently dominant class of novel β-nucleating agent is WBG or WBG-II, mostly described as "rare earth complex" or "lanthanum complex," somewhat more precisely by the sum formula $Ca_xLa_{1-x}(LIG1)_m(LIG2)_n$ with LIG1 and LIG2 being organic ligands [148]. There is a long row of publications on the application of WBG, predominantly by Chinese groups, who evaluated the performance of this β-nucleating agent with different dispersion modes, cooling histories, and base polymer types [148–152]. While the WBG-complex is definitely suitable for high isotacticity iPP homopolymers, its activity seems to be affected by chain defects more than other types [39, 151].

As last aspect of nucleation, the deactivation of nucleating agents by other common additives should be discussed. A well-known example of deactivation is represented by the pair calcium stearate (CaSt), as acid scavenger or slip agent, and sodium benzoate (NaBz), as nucleating agent. The two additives can chemically interact with each other [153], resulting in the inhibition of the nucleating effect of NaBz toward iPP and a decrease in T_c and stiffness. CaSt has also been reported to hinder the self-assembly process of sorbitol-type nucleating agents [4], on which the nucleation efficiency is highly dependent. Moreover, glycerol monostearate (GMS), used as antistatic agent, is found to interact with NaBz in a similar way, resulting in the neutralization of the functions of both additives in PP [154].

3.4 Crystallization in Multiphase Copolymers, Blends, and Composites

There are several types of multiphase systems to be considered in this chapter, all developed and produced to extend the property range in terms of stiffness, impact strength, and other application-related properties. In terms of production volume, the biggest class are certainly iPP ICPs, also called HECOs of iPP or – previously and incorrectly – block copolymers [155], nearly exclusively based on ethylene and propylene as monomers and characterized by a crystalline iPP matrix with dispersed particles of mostly amorphous and elastomeric nature. The second class are iPP-based blends with other polymers, the ones with independently produced elastomers being the biggest sub-class here (and historically older than ICPs [156]). This class has an overlap with iPP blends with PE homo- and copolymers, ranging in density and crystallinity from high-density polyethylene (HDPE) [157] to polyolefins (PO)-plastomers and -elastomers from SSCs with higher α-olefins like 1-hexene or 1-octene [158]. Other and practically less relevant types are blends with other non-olefinic polymers like polyamide (PA) or polystyrene. Added to these blends is the large class of compounds and composites with non-polymeric modifiers, ranging from mineral fillers like talc [108–110] through GF [113–116] to organic reinforcements. Only selected aspects of this wide variety, all having relevance to crystallization and processing effects, will be discussed here.

Important structural parameters for the first class, iPP ICPs, are the crystalline phases of the iPP matrix as well as of the dispersed ethylene–propylene copolymer (EPC), the part of the composition that actually provides impact strength over a wide temperature range. Next to the EPC amount, which is one key factor for the stiffness/toughness balance [159], also its composition (like C_2 or ethylene content) and molecular weight – or rather the relation of molecular weight to the matrix, defining the viscosity ratio in the melt state – is being varied. Standard parameters for defining these two parameters are the C_2 content and the intrinsic viscosity of the xylene cold soluble (XCS) fraction (C_2(XCS) and IV(XCS), resp.), being industrial standards despite their inferiority in precision to results of modern fractionation techniques [160, 161].

Comonomer content variation in the EPC has been studied as a design parameter for ICPs by several groups over time, both for polymers based on Ti-based Ziegler–Natta type catalysts [162–165] and single-site (metallocene) catalysts [166]. When assuming the standard approach of sequential copolymerization of iPP homopolymer and EPC, increasing C_2(XCS) will first result in phase separation between 10 and 22 wt% [166], then to a gradual increase of EPC particle size due to reduced compatibility [164], accompanied by reduction of glass transition temperature [163]. At higher C_2(XCS), crystalline inclusions appear inside the EPC particles, basically composed of a linear LDPE with propylene (C_3) as comonomer [160].

One example of a systematic crystallization study comes from Zhejiang University, China, dealing with ICPs having a high EPC content from a two-stage bulk/gas-phase process [167–169]. A combination of DSC and polarizing heat-stage microscopy was applied on both full polymers and fractions, starting from a series of four materials including a homopolymer and three ICPs with 18–34 wt% EPC [167]. Regarding EPC composition, the data indicate a strong increase of the C_2(EPR) with the EPR content, meaning that more and more crystalline PE will be formed, as confirmed by PE crystallization peaks [168]. Next to possible PP/EPC, respectively, PP/PE interaction effects, this is an explanation for the increased overall crystallization rate observed in isothermal DSC. An earlier study by Yokoyama and Ricco [170], provides a more detailed explanation by differentiating between growth and nucleation effects, finding that growth is actually retarded (probably by particles getting in the way), but nucleation enhanced in presence of EPC (probably by particles acting as centers for nucleation). For systems with a low amount of EPC around 20 wt%, Doshev et al. found a distinct effect of the EPC composition on the spherulitic growth rate, being largest at intermediate contents of ethylene and propylene, as well as revealing ethylene co-unit crystallization well above 100 °C if the ethylene-co-unit content is larger than 70 wt% [164].

Still in quiescent mode, further characterization of the kinetics of crystallization can be achieved by employing fast scanning chip calorimetry (FSC), allowing application of process-relevant crystallization conditions. In short, iPP shows a bimodal temperature-dependence of the crystallization rate, with maxima at around 30 and 80 °C, associated with homogeneous and heterogeneous crystal nucleation, respectively [13]. In such a study, acceleration by EPC was obtained on a sample containing close to 30 wt% EPC with a C_2 content of around 40 wt%, but only for crystallization at rather high temperature. Results from FSC, which allows assessing crystallization at higher supercooling of the melt, suggest that the fast ethylene-co-unit crystallization process in the EPC, occurring at around 60–70 °C, slows down the iPP-matrix crystallization; at very low temperatures, it was proposed that EPC propylene-co-unit crystallization accelerates the iPP-matrix crystallization process again [171].

In a recent study, the crystallization kinetics of metallocene-catalyzed heterophasic iPP copolymers was analyzed as a function of the cooling rate and supercooling in non-isothermal and isothermal crystallization experiments, respectively [172]. FSC allowed assessing crystallization at processing-relevant conditions, and variation of the content (0–39 wt%) and composition (0–35 wt% propylene co-units) of the EPC-particles revealed qualitatively new insight about mechanisms of heterogeneous crystal nucleation (Figure 3.10). At high temperatures, the here studied C_2-rich EPC-particles accelerate crystallization of the iPP-matrix, with the acceleration or nucleation efficacy correlating to content. The crystallization time reduces by more than half in presence of 39 wt% EPC particles. At low temperature, homogeneous crystal nucleation in the iPP-matrix outpaces all heterogeneous nucleation effects, and the matrix-crystallization rate is independent of the sample composition. The obtained results lead to the conclusion that the crystallization kinetics of iPP can be affected significantly by the content and composition of EPC particles, even toward superfast crystallizing iPP grades (Figure 3.11).

For iPP blends with conventional amorphous EPR, spherulitic growth was found to be reduced only marginally in the concentration range up to 30 wt%, with no effects on nucleation density [173]. This situation changes when using an EPC based on a special catalyst and characterized by special high isotacticity in the C_3-sequences as modifier [174]. These were synthesized in a wide range of compositions (7.8–47.5 wt% C_2) and blended at 20 wt% with a PP homopolymer. The most C_3-rich of these modifiers were found to be miscible even in the solid phase, where co-crystallization was evident and a single glass transition was achieved, meaning that these resemble bimodal homo-random copolymers. In contrast, the C_2-rich types appear to nucleate the compositions.

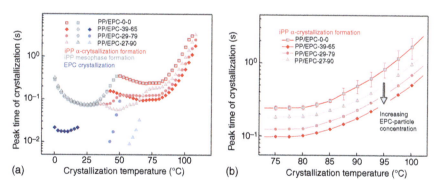

Figure 3.11 Peak-time of crystallization of iPP homopolymer and ICP as a function of the crystallization temperature (a) in the entire analyzed temperature range between 0 and 110 °C, and (b) in the high-temperature region above 75 °C. Source: Reproduced from Mileva et al. [172] © MDPI/Public Domain CC BY 4.0.

While numerous papers deal with morphology and mechanics of iPP blends with C_2/C_n plastomers [31, 158], only few of these study crystallization effects systematically. For an SSC-based copolymer with octene (EOC) having a density of 895 kg m^{-3}, blended with iPP homopolymer up to 40 wt%, the DSC results imply a stronger interaction at lower modifier content [175]. The T_c of both components is lower in this range, where also the particle size, from scanning electron microscopy (SEM) on cryo-fractured surfaces, appears to be smallest. In a more extensive study by Svoboda et al. [176], different plastomers with an octene content from 8.9 to 21.3 wt% were studied in 1 : 1 (wt/wt) blends with iPP. Spherulitic growth was studied in microscopy and DSC here, with RuO_4-contrasted transmission electron micrographs providing the respective phase morphology. The results (see Figure 3.12) imply an influence of the disperse phase on both nucleation and growth rate. Occlusion of PP for high molecular weight of the SSC-PE causes a reduction of G_c by confinement, while small SSC-PE particles appear capable of nucleating crystallization.

For conventional HDPEs, high-pressure-based LDPEs or normal linear low density polyethylenes (LLDPEs), the situation is clearer, although also here morphology can vary between particulate and co-continuous. As described by Jose et al. [177], crystallinity does not vary significantly in droplet-matrix morphology, while in co-continuous systems the crystallinity of each component might change by up to 20% due to the addition of the other component. Most authors agree that the presence of PE in a blend with iPP reduces the nuclei density of the latter, which is commonly explained by migration of nucleating heterogeneities from PP to PE domains during the melt-mixing process. The driving force for the migration of heterogeneous nuclei across the interface is the interfacial free energy difference of those nuclei when surrounded by iPP and PE melt, where it was observed that spherulites growing from PP melt do not reject PE inclusions [178]. The reason is that, since PP crystals have similar interfacial energy values as heterogeneous nuclei of PP, the contact between PE melt and PP crystals is preferred. Thus, the interfacial free energy of a nucleus in contact with PE melt is smaller than when it is surrounded by PP melt [179], meaning that the contact of a heterogeneous nucleus with PE melt is energetically advantageous.

In contrast to that, growth rate effects are reported diversely, with no changes in immiscible blends with HDPE and LDPE [179–182], but a significant decrease in growth rate for blends of PP and LLDPE [182]. Conversely, Blackadder et al. [183] found no changes in the melting behavior of the components in PP/HDPE blends, while the crystallization kinetics of iPP were enhanced when blended with HDPE, even if there was no variation of crystallinity. For the PE part of such blends, Rybnikár [181] reported a higher crystallization rate of linear PE blended with PP, explained by changes in the concentration of nuclei of different types or, more recently, by the nucleating action of the previously crystallized PP phase upon cooling from the melt [184].

This epitaxy between iPP and HDPE is a very relevant subject from both theoretical and practical perspective and has been studied extensively [184–189]. It occurs due to the dimensional matching of the PE interchain distance (4.94 Å) with the PP (10″1″) interplanar spacing (5.05 Å) [185, 187–189]. Therefore, the PE chains exactly

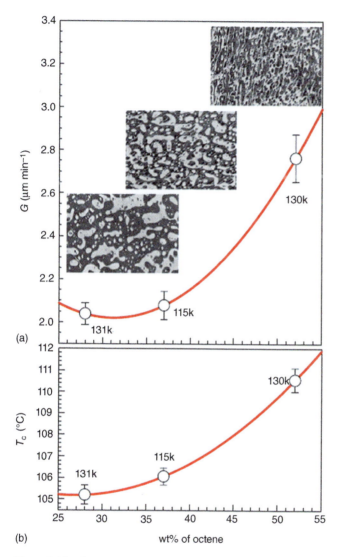

Figure 3.12 Crystallization kinetics from optical microscopy (a) and from DSC (b) as a function of octene content in EOC for iPP/EOC 50/50 blends; the structure of the three corresponding blends is shown by three inserted TEM pictures. Source: Reproduced with permission from Svoboda et al. [176]/Elsevier. .

fit into the valleys formed by the PP methyl groups (see Figure 3.13a). Furthermore, epitaxial growth occurs when the chain axes of the two phases are about 50° apart. Thus, the crystal dimension in the substrate must be greater than the critical nucleus size of the depositing phase. As well explained by Greso and Phillips [187] and experimentally proved by Yan et al. for thin-layered films [188], for the deposition of the first PE stem onto a PP substrate and the generation of a critical nucleus, the length of the PE stem must be smaller than the lamellar thickness of the PP substrate in the matching direction (see Figure 3.13b).

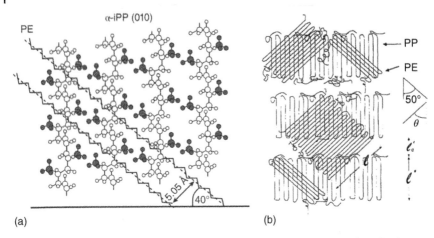

Figure 3.13 Epitaxy between iPP and (HD)PE. (a) Schematic representation of epitaxy between PE and iPP. Source: Adapted from Carmeli et al. [190]. (b) Sketch of the geometrical relationship for epitaxial growth of PE onto iPP crystals. Source: Reproduced with permission from Greso and Phillips [187] © Elsevier.

Recently, epitaxy was thoroughly demonstrated to occur also in PE/PP blends [190, 191]. The increase of iPP lamellar thickness that was achieved via self-nucleating and annealing iPP crystals was shown to directly correlate with the enhancement of PE crystallization kinetics via isothermal or non-isothermal crystallization of PE. The interfacial interactions between iPP and PE domains are affected by the crystallization order of the phases during cooling from the melt because this determines which component can act as a template for the epitaxial growth of the second component.

The cooling rate experienced by the phases during cooling from the melt strongly affects the order of crystallization. This is due to the fact that, despite the typically higher crystallization temperature of PP with respect to PE, the latter shows an intrinsically 10 times faster crystallization kinetics than PP [98]. Thus, an "inversion point" of the crystallization order can be detected. In a recent study [192], it was demonstrated via continuous cooling curve (CCC) diagrams that mutual nucleating effects between both polymers in a blend correlate well with the inversion point of the crystallization order (see Figure 3.14). Already before it has been shown that processing involving high shear can also cause PE crystallization to happen before iPP, thus also nucleating the matrix of an HDPE-modified ICP [193].

Among iPP blends with other polymers, naturally those with crystalline ones are of interest. An example is blends of PA (normally polyamide-6), which are targeted at combining the high thermomechanical performance of PA with the insensitivity to moisture of PP. Similarly to PE and iPP, PA-6 and iPP mutually influence each other's crystallization when blended. It was reported that iPP acts as a diluent at the crystallization temperature of PA-6, which is way higher than the one of PP, resulting in a decrease in the PA crystallization rate. On the contrary, the crystallization rate of iPP in the blends is increased due to the nucleating effect of the already-formed PA-6 crystals [193–196].

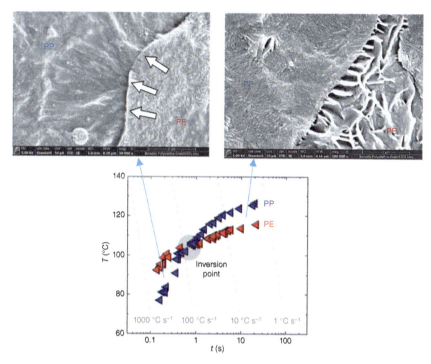

Figure 3.14 Morphological evidence of the correlation between inversion point and mutual nucleating effects between the phases. PP spherulites are nucleated on top of the already formed PE crystals (left SEM image), and PE lamellar stacks are nucleated at the interface with the already formed PP crystals (right SEM image). Source: Materials as in Carmeli et al. [192]/American Chemical Society/CC BY 4.0.

For reducing the interfacial tension and improving the adhesion between PA-6 and iPP, to get acceptable impact performance and ductility, normally compatibilizing agents have to be applied. The most used compatibilizers are maleic- and acrylic acid-grafted PP [193, 197]. It was also noted that the addition of a compatibilizer reduces this nucleating effect, which is a confirmation of the nucleating activity of the PA component at the interface between the phases. One alternative considered for compatibilization is inorganic nanoparticles, for which surface-modified nano-silica [198] and organophilic nanoclays [199] have been tested. In the latter case, neither a PP-affine nor a PA-affine clay platelet modification resulted in a preferential positioning at the interface of the two polymers. Aggregate-free dispersion and exfoliation could only be achieved in PA-6, and the nanoclay-modified blends presented improved mechanical performance over the non-modified references but did not reach the PA-6 nanocomposites, while the also tested micro-talc proved to be superior in nucleation performance toward both polymers.

Regarding compounds and composites, the effect of talc on iPP crystallization has been discussed above [107–112]. In practice, the combination of elastomer phases from using an ICP base or adding external impact modifiers needs to be considered, creating a heterogeneous morphology at multiple length scales [107].

Fibers of inorganic or organic nature are another type of reinforcement for iPP to improve its thermal and mechanical properties. Some reinforcing fibers have a significant effect on the crystallization behavior from the quiescent melt already, like aramid or carbon fibers [200, 201], but not the already discussed GF [114–116]. On nucleating fibers, a transcrystalline layer (TCL) is formed when a large number of nuclei are generated at the surface of the fiber, and, due to impingement among adjacent lamellae, the only direction of crystals growth is perpendicular to the fiber axis. The growth rate of the TCL does not depend on the fiber type, but it is determined only by the type of crystallizing polymer [200].

In any processing situation or generally when applying shear or extensional stress to a system comprising iPP and fibers (or highly anisotropic particles), the formation of a TCL is rather independent of fiber type or crystallization temperature [202]. Under flow, the shear stresses developed at the interface between GF and iPP melt induce an orientation of iPP molecules closer to the fiber along the fiber axis. The pre-oriented molecules can act as self-nuclei, like in other processes involving FIC [76, 78, 81, 83, 114, 115], drastically reducing the energy barrier to nucleation. In the "classical" experiments to study this, TCL was shown to form only when pulling the fiber in resp. through the melt [114, 116, 203, 204]. Looijmans et al. [116] recently reported that the nucleation efficiency in quiescent crystallization conditions correlates well with the anisotropy of TCL around the fiber obtained under shear flow crystallization, showing that the fiber/iPP interaction is still affected by polymer type.

Usually, α-phase is nucleated at the interface between fibers and iPP under shear flow. The observation of apparent β-phase TCLs grown on top of GF or other polymeric fibers was clarified by Varga and Karger-Kocsis [205]. They demonstrated that a two-layer transcrystalline structure is obtained on a GF under isothermal crystallization in a sheared melt of PP within the range of PP β-phase formation (130–140 °C). First, α-phase nuclei are formed on the GF surface due to the residual stresses, but β-phase crystals are nucleated and favored to grow on top of the α-nuclei, due to their higher growth rate in that temperature range. This effect can be strengthened by combining with selective nucleation [117, 118].

3.5 Processing Effects and Resulting Properties

In addition to the macromolecular structure, the processing parameters play a crucial role in controlling the final material properties of iPP and its copolymers. Each processing/conversion technology is characterized by a cooling and shear rate at which the melt is solidified. The former in combination with the temperature of solidification affects the final crystal structure, morphology, and material properties like stiffness/impact and transparency. The insufficient time for crystallization at large supercoolings or high cooling rates results in generation of transparent film with low stiffness due to the decreased crystallinity and smaller size of the crystals. Other post-processing effects like shrinkage and warpage (mostly in injection-molded articles), but also physical aging and long-term stability, have also been found to be strongly affected by crystallinity.

3.5 Processing Effects and Resulting Properties

Processing effects can be categorized roughly into thermal history, pressure, and flow phenomena. The former are mostly reflected by cooling rate effects, which have, especially since the development of suitable equipment for achieving high cooling rates in a controlled manner, been studied for many polymers [13, 53, 54, 71, 72, 74, 172, 192, 206–210]. Depending on polymer structure, the effect of cooling rate can range from simple crystallinity reductions to the formation of different crystal modifications, practically always involving massive changes in mechanics and optics. The effect on the polymorphism of iPP homopolymer has been discussed in Chapter 2: Higher cooling rates cause a transition from α-crystalline to the mesomorphic phase (with γ- and β-polymorphs disappearing even before), accompanied by reduced density and change of crystal structure from lamellar to nodular [13, 71]. At cooling rates higher than $1000\,K\,s^{-1}$, quenching into fully amorphous (glassy) structure can occur when the quench is performed below the glass transition temperature [206]. Both transitions being affected by the presence of heterogeneous nuclei [210] and high molecular weight fractions for homopolymers, meaning that highly pure iPP with a narrow MWD will be most easy to quench [207]. This is in line with the practical preference for "controlled rheology" (peroxide-degraded) grades for cast film extrusion and fiber spinning.

Copolymerization can shift the critical cooling rate for such transformations significantly, as studied, for example, for ethylene, butene, hexene, and octene random copolymers of iPP [26, 53, 54, 209, 211]. The study of Cavallo et al. on a series of ethylene–propylene (EP) random copolymers with increasing ethylene content has been discussed above (see Figure 3.5) [53, 54]. The critical rate of cooling has been confirmed in experiments where combination of FSC, wide angle X-ray scattering (WAXS) analysis, and atomic force microscopy (AFM) has been applied [26, 209, 211]. In Figure 3.15, a schematic representation of the crystal structure formation as a function of the cooling rate and comonomer type (ethylene [Eth], butene [But], hexene [Hex], and octene [Oct]) is presented. In the right-hand side, AFM images represent the crystal morphology of samples cooled slowly or rapidly from the quiescent melt [209, 211]. The critical rate of cooling for suppression of crystallization and formation of monoclinic lamellae is shifted toward lower cooling rate with increasing the length of the comonomer branch. In case of iPP-Oct, cooling at rates of $1–5\,K\,s^{-1}$ triggers mesophase formation, which for iPP-But is shifted to almost $50–80\,K\,s^{-1}$ depending on the comonomer content.

Phase transition changes and properties directly on cast films of iPP produced from non-isotropic melt have mostly been examined on films produced at low undercoolings or chill roll temperatures higher than 60 °C. In biaxially stretched (biaxially oriented PP [BOPP]) film, it has been found that mostly polymer rheology and die temperature affect the film morphology [212]. In regard to the effect of processing parameters on the structure formation, it has been demonstrated that fibrillary crystalline structures are generated in a drawing process conducted at 110 and 140 °C in cast films based on blends of iPP and 20 wt% of random propylene–ethylene copolymer. The highly oriented crystalline structure resulted in increase of the final melting temperature by 10 °C [213]. The highest tensile strength along transverse direction (TD) was observed for films prepared at 140 °C, and the effect of drawing temperature along machine direction (MD) was less significant. Another study

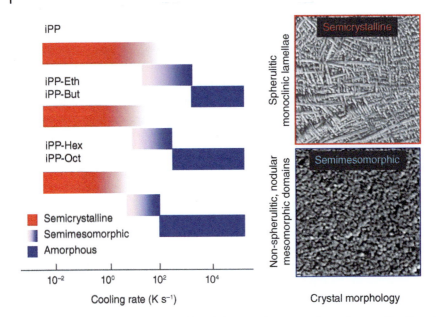

Figure 3.15 Schematic representation of the crystal structure formation as a function of the cooling rate and comonomer type (ethylene [Eth], butene [But], hexene [Hex], and octene [Oct]). In the right-hand side, AFM images represent the crystal morphology of samples cooled slowly or rapidly from the quiescent melt (materials as in [211]). Source: Mileva et al. [26]/Elsevier; Mileva et al. [209]/Springer Nature.

by the same group confirmed that films containing solely lamellar crystals showed higher crystallinity and tensile strength but lower barrier to oxygen [214]. A more recent work on conventional BOPP process variations shows that even seemingly small variations in isotacticity and polydispersity affect the processing behavior of iPP homopolymers, but the pre-cast film structure is equally decisive [215].

The few studies dedicated to cast film extrusion of iPP homopolymers and propylene–ethylene or propylene-1-butene random copolymers confirm that at lower chill roll temperatures (T_{CR}), like in the range from 10 to 30 °C, mesophase forms in case of the EP random copolymers and monoclinic α-crystals in iPP homopolymers, respectively [216–220]. In cases where low T_{CR} triggers formation of the mesomorphic form of iPP, an increase in the chill roll temperature results in formation of the stable monoclinic form. This "quenching" effect [51] makes films more susceptible to post-crystallization and physical aging, especially when being exposed to elevated temperatures, like in a heat sterilization process [219].

In two recent studies [72, 221], it has been possible to define the main parameters governing the generation of specific crystal structures to reach targeted mechanical and optical properties (see Figure 3.16). In particular, the degree of undercooling or the chill roll temperature plays the major role in the formation of the different polymorphic forms of iPP in the analyzed cast films, while the initial melt temperature is of much less relevance. Large undercooling (chill roll temperature of 15 °C) triggered the predominant formation of the mesomorphic phase with low degree

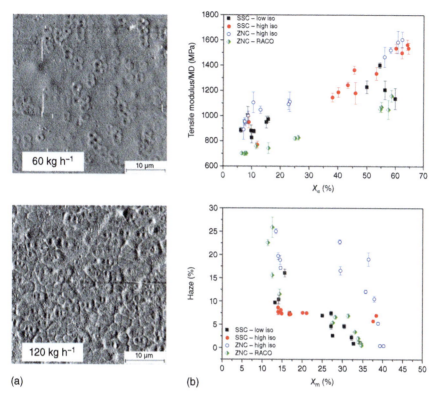

Figure 3.16 Cast film processing effects on structure and properties for different iPP homo- and copolymers: (a) AFM images from the surface of iPP homopolymer films produced with identical T_{CR} and different throughput. Source: Materials as in Mileva et al. [221]/Springer Nature. (b) correlation between modulus and α-phase content (top) and haze and β-phase content (bottom). Source: Reproduced from Di Sacco et al. [72] © MDPI/Public Domain CC BY 4.0.

of crystallinity (below 40%), while at low undercooling (chill roll temperature of 70 °C) monoclinic phase develops at high crystallinity (above 50%). Even minor variations in the chain structure, like isotacticity in iPP homopolymers based on ZNC or SSC [72] change the relation between crystal structure and performance. The tensile modulus of such films is more strongly related to the α-phase content, and haze rather than to the mesophase (Figure 3.16b). Variation in the processing speed contributes mainly to a slight change in crystallinity, and surface roughness invoked by FIC [221, 222]. The surface impressions of isolated spherulites (Figure 3.16a) have been described before in the literature but are frequently overshadowed by the presence of processing aids like antiblocking agents [218].

In injection molding, the formation of flow-induced structures in crystalline polymers was found to be important rather early. The common term "skin-core morphology" [223] was attributed to the combination of highly oriented surface layers made up primarily of the "shish-kebab" structures christened by Keller and coworkers [224] with a bulk of largely unoriented spherulitic structures. An analysis of the actual structure and genesis of shish-kebab layers was achieved with a combination of X-ray scattering and transmission electron microscopy (TEM) [225]: The "shish" core of highly oriented molecules forms the primary nucleus, while the "kebab" shell is formed by disk-like lamellar crystals. Moving from the skin to the more isotropic core of an injection-molded part, one passes through an intermediate zone commonly called "fine-grained layer," where nucleus density has been increased by flow without formation of an oriented superstructure. Ideally, the core is least affected by processing and has an isotropic spherulitic structure, as already shown in Figure 3.1.

Numerous FIC studies have been done for iPP, because of the strong effect of processing variations on final performance. Over time, the problem has been approached from two sides. First, using real-life injection molding followed by a more or less comprehensive analysis of the resulting parts or specimens, phenomenological work regarding the effects of polymer structure, nucleation, and processing parameters was performed [76, 77, 225–228]. While this approach means accepting a complex flow and thermal history, it was definitely helpful in obtaining important correlations, some of which were also found in the more fundamental studies discussed below, like the enhancing effects of high polydispersity [76, 81, 83] and nucleation [77, 78, 105] on skin layer thickness. The latter explains differences in performance combinations between nucleating agents. While α-nucleation generally gives higher stiffness, particulate organophosphate types with their high particle anisotropy maximize the modulus as well as the relative thickness of highly oriented skin layers. In contrast, optical performance is better optimized by soluble nucleating agents of the sorbitol- or trisamide-type, where more homogeneous crystal structures (i.e. less skin layers) are achieved in molding. Another specific finding is the formation of β-modification at the transition from skin to fine-grained layer in absence of selective nucleation [227, 228], which has likely the same root cause as the one close to GF [205].

The second approach involved more elementary work regarding the origin of flow-induced nuclei and their development into more or less oriented superstructures [77, 226, 229, 230]. For the latter part of FIC research, the necessity of developing specific setups and experimental protocols became quickly obvious [231–233], allowing fully control of the flow and thermal history. Collecting respective data must be followed up by structural modeling considering molecular orientation processes, heat transfer, and flow fields [83, 234–239]. The various groups have presented quite different ideas regarding the details of the formation and growth of nuclei, but also the resulting overall crystallization process under realistic conditions. Maybe the best understanding has been developed by the group of Peters at TU Eindhoven (NL), resulting in a fully quantitative model [83, 237]. It was found that cumulative flow effects produce a combination of nucleation density increase at lower stress levels (similar to the original concept of Janeschitz-Kriegl et al. [229]), leading to oriented structure formation at higher levels, as shown in Figure 3.17a. The finally developed model allows understanding the effect of MWD resp. high molecular weight fraction and nucleation, including the result on crystallinity, polymorphism, and orientation. It also improved the understanding

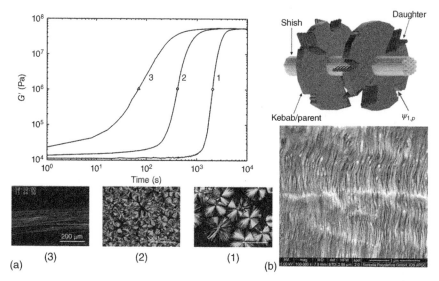

Figure 3.17 Principles of flow-induced crystallization and shish formation: (a) evolution of the storage modulus during crystallization of iPP homopolymer HD120MO (MFR 8 g 10 min^{-1}) at $T_{exp} = 135\,°C$, measured under quiescent conditions (1) and after shearing at $\gamma = 60\,s^{-1}$ for $t_s = 1$ second (2) and $t_s = 6$ seconds (3); optical micrographs indicate the characteristic morphology. Source: Housmans et al. [236]/American Chemical Society. (b) Principle of parent–daughter lamellar order in a shish-kebab structure. Source: (Top) Reproduced with permission from Roozemond et al. [237], © 2016, Elsevier. SEM image of actual shish-kebab structures in the skin layer of an injection molded part. Source: (Bottom) With Permission of Borealis AG.

of daughter lamella formation both in spherulitic and shish-kebab structures, as shown in Figure 3.17b: Space filling between lamellae is happening by this process, which clearly contributes to polymorphism in molding processes, like the β-phase formation discussed above.

Going back to actual processing, an additional level of complexity needs to be considered with multi-phase systems like ICPs, blends, and compounds. Here, the crystalline morphology is combined with the phase morphology and particle orientation, as shown for iPP/EOC blends [240] and also for ICPs [241]. Elastomer or copolymer particles orient strongly in the skin layer, are nearly spherical in the core, and can have a complex structure in the transition zone. Especially the latter study [241], where SEM-based phase morphology has been combined with spatially resolved WAXS, suggests an interaction between the orientation of the EPC particles and the matrix crystallinity. This effect is certainly higher in case of crystalline inclusions in the dispersion, as seen for an HDPE-modified ICP [193]. The complexity is increased further for compounds with additional inorganic reinforcements like talc [242], where a two-dimensional (disk-like) orientation of the dispersed elastomer phase has been described, affecting shrinkage and thermal expansion on top of mechanics.

One more conversion type to be mentioned is fiber spinning, where significant work has also been done on iPP (fibers account for ~28% of the applications globally). Quite different fiber spinning condition sets allow variation of properties like modulus and elongation at break even for the predominantly used iPP homopolymers. The parameters, however, have to be seen in relation to the polymer structure, especially its MWD [243, 244]. A clear correlation between stiffness and crystal orientation as seen, e.g. by birefringence, but at least partial quenching into mesophase is also possible [245]. Only rather recently, systematic work on melt-blown fiber production, where the process does not involve mechanical pull-off or stretching and where effectively only webs can be tested, has been done [246, 247]. In order to get optimal web properties like softness and filtration efficiency, also here the parameter set needs to be adapted to the (rather low) molecular weight of the respective polymers.

3.6 Investigation Methods for PP Crystallization and Morphology

As shown from the extensive and yet not even complete review of morphologies of iPP-based compositions in the previous paragraphs, the study of morphology and crystallinity of polyolefins is a vast field, and several different experimental methods are used for this purpose. On the one hand, methods like microscopies (optical, electron, scanning probe microscopies, etc.) provide a great insight into the finest features of the morphology developed. However other techniques that provide more representative information on bulk properties (X-ray diffraction [XRD], thermal analysis, etc.) are also of fundamental importance.

When discussing the use of microscopy techniques, the importance of specimen preparation is often overlooked. Specimens from polyolefins are always prepared by cryo-ultramicrotomy, probably with the only exception of the last, room temperature cut in the case of RuO_4-stained specimens.

Artifacts can easily be introduced in this process, such that the preparation of high-quality specimens for further analysis is of fundamental importance and often linked to the mastery and skills of well-trained operators. Typical artifacts like scratches, compression, folds, chattering, and thickness differences within sections introduce contrast or features that are not real and can lead to misinterpretations. In particular, in the case of ultra-cryocutting for AFM analysis, very high-quality cuts on typically small regions are required.

For SEM analysis, the specimens need to be further treated in order to induce topography, and thus contrast in the images, based on the morphology developed. Historically, the specimens were exposed to acids in order to remove the amorphous phase. Palmer and Cobbold [248] first used fuming nitric acid (95%) at 80 °C to expose the lamellar morphology of PE. With time, milder and milder recipes have been used in order not to overetch the surface, therefore revealing more elusive details. Very well-known is the recipe of Bassett and Olley [249], using 2 : 1 concentrated sulfuric acid/phosphoric acid with 0.7 w/v% potassium permanganate. The use of more diluted sulfuric acid only however provides a more gentle etching and it solves issues related to artifacts from the limited solubility of the etched material.

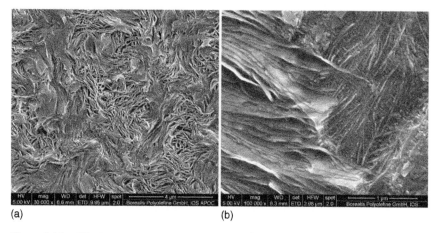

Figure 3.18 SEM images collected after etching PE (a) and iPP (b) samples with diluted sulfuric acid (75%) and potassium permanganate 1 w/v% for 15 minutes.

Figure 3.18 shows examples of morphologies of PE and iPP imaged after etching with diluted sulfuric acid (75%) and potassium permanganate 1 w/v% for 15 minutes.

Moreover, the specimens are often stained with OsO_4 or RuO_4 to introduce contrast between the amorphous and crystalline phases, or to stain the EPC or other elastomer inclusions [250]. AFM provides an interesting alternative in many cases, since the contrast is based on factors like the difference in mechanical properties of the phases, and thus no sample etching or staining is required. Additionally, samples can be easily placed in the microscope and imaged directly on the unmodified surface. Figure 3.19 shows an example of an iPP film, where small rubber particles can be observed at the surface (left). Care must be taken of the thermal history here: Upon thermal treatment, the rubber phase undergoes extended diffusion to the surface, creating new structures which, for example, change the optical properties of the film.

Figures 3.18 and 3.19 show the level of detail that can be learned through microscopy about morphology, but a complete understanding of the phenomena taking place often requires the support of techniques that can provide more general information on a larger volume of the sample. XRD is very important for the study of polyolefins as it leads to a quantification of the degree of crystallinity [251], content of different polymorphs [252], and an average evaluation of the lamellar thickness, obviously on a more representative scale than a micrograph. For example, Figure 3.20 shows patterns collected on a β nucleated iPP, upon increasing the temperature [253]. Diffractions belonging to the α phase appear at ~145 °C, allowing one to study quantitatively the β–α transition. For e.g. subtracting an amorphous halo, it is possible to quantify the total amount of crystallinity at each temperature, and the (changing) amount of β phase through the comparison of the heights of the reflections of the α and β phases. This crucial information can then be nicely complemented by microscopy, as the AFM image collected at 149 °C on the same sample shows melting β-phase lamellae as well as patches of cross-hatched α phase.

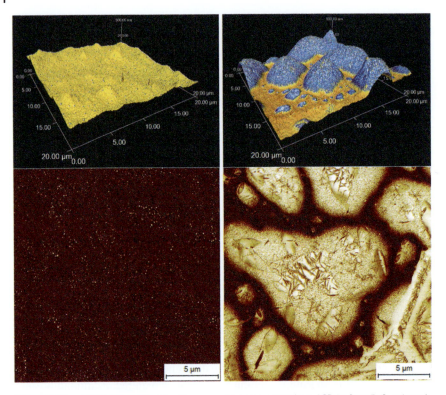

Figure 3.19 AFM images of a film from an ethylene–propylene ICP, before (left column) and after (right column) temperature treatment. Note the extensive diffusion of EPC to the surface. The top images are topography colored with the phase signal. The bottom two images are standard phase images.

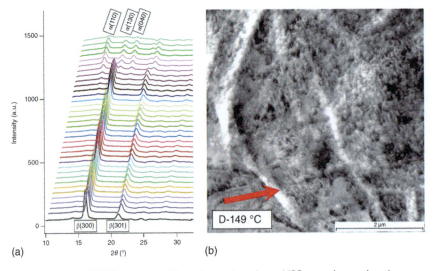

Figure 3.20 (a) WAXS patterns collected on a β nucleated iPP, upon increasing the temperature. (b) The AFM image of the same sample was collected in a heating stage, at 149 °C. Source: Reproduced with permission from Tranchida et al. [253]/John Wiley & Sons.

3.6 Investigation Methods for PP Crystallization and Morphology

Interesting alternatives to XRD are provided by spectroscopy-based techniques. Nielsen et al. [254] clearly showed that bands related to chains in helical conformation can be used, together with others assigned to chains in non-helical conformation, to measure the degree of crystallinity of iPP. Coupling the technique with heating scans, Khafagy [255] was able to show that up to 10 °C above the melting temperature, short-range order in the order of 12 monomer units can still be observed. Similar information was also obtained by Fourier-transform infrared (spectroscopy) (FTIR), for example by Zhu et al. [256] showing the existence of ordered structures at temperature much higher than the melting temperature, as evidenced by the persistence of the 998 cm^{-1} band at up to 210 °C. Again, the combination of different techniques is key to the profound understanding of complex phenomena. Di Sacco et al. [257] combined Raman spectroscopy with small-angle X-ray scattering (SAXS)/WAXS to gain insight into the meso–α transition of iPP, and the details of comonomer inclusion in the lattice together with its role to hinder the reorganization of chains in helical conformation.

Thermal analysis also plays a crucial role in the study of morphology and crystallization. The possibility to study the crystallization process with DSC during cooling from the melt obviously provides fundamental information, as does the study of the morphology developed through the final melting scan of a DSC experiment. It must be stressed that the first heating scan, normally used only to erase the previous thermal history, can actually be extremely informative when studying morphologies developed under processing conditions or with controlled cooling, as it will be discussed in the next paragraphs. The limit of DSC, however, is related to the low cooling and heating rates that can be attained. For the former, cooling rates of up to c. 50 K min^{-1} can be obtained, which should be compared to tens up to thousand K s^{-1} experienced during the solidification in processing conditions [206, 207]. For the latter, low heating rates have been recognized as an issue due to the possibility of rearrangements in the morphology during heating, hindering the possibility of knowing in detail the starting morphology. Early attempts to overcome these limits led, for example, to the development of HyperDSC [208, 258], with improvements of 1 order of magnitude for cooling and heating rates.

More progress was achieved by further decreasing the cell and sample sizes, which led to FSC [13, 172]. Extremely high cooling and heating rates can be obtained with these systems, albeit at the price of low sample mass, i.e. in the order of tens or few hundreds nanograms. The study of filled systems with, e.g. GF or talc is therefore difficult, and caution must be paid when studying systems like blends that could phase separate on a length scale of the size of the sample. Additionally, it has been recognized that coupling studies done with FSC together with other techniques such as optical microscopy [259], TEM [260], AFM [261], or WAXS [262] provides exceptional insight into the structure development even at high cooling rates.

As an alternative to these combined in situ approaches, home-made instruments have been developed in order to prepare samples cooled at high cooling rates [207] and even under high pressure [263]. The advantage of these instruments is that relatively large samples can be prepared, typically with thickness of c. 100 μm, as required in order to avoid thermal gradients inside the sample and to achieve a

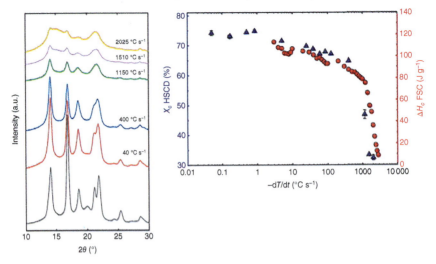

Figure 3.21 Results from a proprietary fast-cooling device: (a) WAXS patterns measured on specimens of iPP homopolymer crystallized at the cooling rates indicated in the figure; (b) comparison of the degree of crystallinity measured on these specimens and crystallization enthalpy measured during FSC experiments. Source: Reproduced from Carmeli et al. [264] © Elsevier/Public Domain CC BY 4.0.

homogenous morphology, but with length and width of several centimeters. The cooling rate is not controlled, however, and it is typically very high at high temperatures, while it flattens out at low temperatures. A possible approach is to define the cooling rate of the sample preparation overall, as the cooling rate obtained at the temperature where the material shows the maximum crystal growth rate. On the other hand, cooling during a real process is also ballistic in nature, without control of the cooling rate, and therefore specimens prepared with these instruments can be considered as good replicas of real, fast processes. Figure 3.21 shows WAXS patterns as measured on samples prepared at different cooling rates, up to 2000 K min^{-1} [264]. The comparison between the degree of crystallinity measured by WAXS on samples prepared in one such home-made instrument and the crystallization enthalpy (as proxy to degree of crystallization) measured by FSC on the same material shows indeed that the two methods can be considered as valid alternatives to each other.

Acknowledgments

The authors want to thank all colleagues assisting in the preparation of this work by supplying material and being available for discussions. Specific thanks go to Juliane Braun, Martin Obadal, Gottfried Kandioller, and Walter Schaffer for providing unpublished data and images. We further would like to thank our scientific partners Prof. Gerrit Peters (Eindhoven, NL), Prof. Béla Pukánszky (Budapest, HU), and Prof. René Androsch (Halle, DE) for their many years of coaching and collaboration.

References

1 Natta, G., Pino, P., Corradini, P. et al. (1955). Crystalline high polymers of α-olefins. *Journal of the American Chemical Society* 77: 1708–1710. https://doi.org/10.1021/ja01611a109.

2 Taniike, T. and Terano, M. (2013). The use of donors to increase the isotacticity of polypropylene. In: *Polyolefins: 50 Years After Ziegler and Natta I*, Advances in Polymer Science, vol. 257 (ed. W. Kaminsky). Berlin, Heidelberg: Springer https://doi.org/10.1007/12_2013_224.

3 Gahleitner, M. and Severn, J.R. (2008). Designing polymer properties. In: *Tailor-Made Polymers Via Immobilization of Alpha-Olefin Polymerization Catalysts* (ed. J.R. Severn and J.C. Chadwick). Weinheim: Wiley-VCH https://doi.org/10.1002/9783527621668.ch1.

4 Keller, A. (1957). A note on single crystals in polymers: evidence for a folded chain configuration. *Philosophical Magazine* 2: 1171–1175. https://doi.org/10.1080/14786435708242746.

5 Phillips, R.A. and Wolkowicz, M.D. (2005). Polypropylene morphology. In: *Polypropylene Handbook* (ed. N. Pasquini), 147–264. Munich: Hanser Publishers.

6 Bernstein, J. (2011). Polymorphism – a perspective. *Crystal Growth & Design* 11: 632–650. https://doi.org/10.1021/cg1013335.

7 Cavallo, D. and Alfonso, G.C. (2015). Concomitant crystallization and cross-nucleation in polymorphic polymers. In: *Polymer Crystallization II*, Advances in Polymer Science, vol. 277 (ed. F. Auriemma, G.C. Alfonso, and C. de Rosa). Cham: Springer https://doi.org/10.1007/12_2015_330.

8 Xin, R., Zhang, J., Sun, X. et al. (2018). Polymorphic behavior and phase transition of poly(1-butene) and its copolymers. *Polymers* 10: 556. https://doi.org/10.3390/polym10050556.

9 Brückner, S., Meille, S.V., Petraccone, V., and Pirozzi, B. (1991). Polymorphism in isotactic polypropylene. *Progress in Polymer Science* 16: 361–404. https://doi.org/10.1016/0079-6700(91)90023-E.

10 De Rosa, C., Auriemma, F., Vollaro, P. et al. (2011). Crystallization behavior of propylene–butene copolymers: the trigonal form of isotactic polypropylene and Form I of isotactic poly(1-butene). *Macromolecules* 44: 540–549. https://doi.org/10.1021/ma102534f.

11 Lotz, B. (2014). A new ε crystal modification found in stereodefective isotactic polypropylene samples. *Macromolecules* 47: 7612–7624. https://doi.org/10.1021/ma5009868.

12 Natta, G., Peraldo, M., and Corradini, P. (1967). Smectic mesomorphous modification of isotactic polypropylene. In: *Stereoregular Polymers and Stereospecific Polymerizations: The Contributions of Giulio Natta and His School to Polymer Chemistry*, vol. 2 (ed. G. Natta and F. Danusso). Pergamon/Elsevier https://doi.org/10.1016/B978-1-4831-9882-8.50030-0.

13 Androsch, R., Di Lorenzo, M.L., Schick, C., and Wunderlich, B. (2010). Mesophases in polyethylene, polypropylene, and poly(1-butene). *Polymer* 51: 4639–4662. https://doi.org/10.1016/j.polymer.2010.07.033.

14 Karger-Kocsis, J. (1999). Amorphous or atactic polypropylene. In: *Polypropylene, Polymer Science and Technology Series*, vol. 2 (ed. J. Karger-Kocsis). Dordrecht: Springer. https://doi.org/10.1007/978-94-011-4421-6_2.

15 Resconi, L., Jones, R.L., Rheingold, A.L., and Yap, G.P.A. (1996). High-molecular-weight atactic polypropylene from metallocene catalysts. 1. $Me_2Si(9\text{-}Flu)_2ZrX_2$ (X = Cl, Me). *Organometallics* 15: 998–1005. https://doi.org/10.1021/om950197h.

16 Chen, J.-H. and Chang, Y.-L. (2007). Isothermal crystallization kinetics and morphology development of isotactic polypropylene blends with atactic polypropylene. *Journal of Applied Polymer Science* 103: 1093–1104. https://doi.org/10.1002/app.25354.

17 Nam, B.-K., Park, O.O., and Kim, S.-C. (2015). Properties of isotactic polypropylene/atactic polypropylene blends. *Macromolecular Research* 23: 809–813. https://doi.org/10.1007/s13233-015-3106-1.

18 Minami, Y., Takebe, T., Kanamaru, M., and Okamoto, T. (2015). Development of low isotactic polyolefin. *Polymer Journal* 47: 227–234. https://doi.org/10.1038/pj.2014.110.

19 Razavi, A. (2013). Syndiotactic polypropylene: discovery, development, and industrialization via bridged metallocene catalysts. *Advances in Polymer Science* 258: 43–116. https://doi.org/10.1007/12-2013-220.

20 Schardl, J., Sun, L., Kimura, S., and Sugimoto, R. (1996). Potential film applications for syndiotactic polypropylene. *Journal of Plastic Film & Sheeting* 12: 157–164. https://doi.org/10.1177/875608799601200208.

21 Rastogi, S., La Camera, D., van der Burgt, F. et al. (2001). Polymorphism in syndiotactic polypropylene: thermodynamic stable regions for Form I and Form II in pressure–temperature phase diagram. *Macromolecules* 34: 7730–7736. https://doi.org/10.1021/ma0109119.

22 De Rosa, C., Auriemma, F., and Vinti, V. (1998). On the Form II of syndiotactic polypropylene. *Macromolecules* 31: 7430–7435. https://doi.org/10.1021/ma980789m.

23 Guadagno, L., D'Antello, C., Naddeo, C., and Vittoria, V. (2000). Polymorphism of oriented syndiotactic polypropylene. *Macromolecules* 33: 6023–6030. https://doi.org/10.1021/ma0005973.

24 Zhang, J., Yang, D., Thierry, A. et al. (2001). Isochiral Form II of syndiotactic polypropylene produced by epitaxial crystallization. *Macromolecules* 34: 6261–6267. https://doi.org/10.1021/ma010758i.

25 de Ballesteros, O.R., De Stefano, F., Auriemma, F. et al. (2021). Evidence of nodular morphology in syndiotactic polypropylene from the quenched state. *Macromolecules* 54: 7540–7551. https://doi.org/10.1021/acs.macromol.1c01011.

26 Mileva, D., Androsch, R., Cavallo, D., and Alfonso, G.C. (2012). Structure formation of random isotactic copolymers of propylene and 1-hexene or 1-octene at rapid cooling. *European Polymer Journal* 48: 1082–1092. https://doi.org/10.1016/j.eurpolymj.2012.03.009.

27 Zhang, X., Zhao, Y., Wang, Z. et al. (2005). Morphology and mechanical behavior of isotactic polypropylene (iPP)/syndiotactic polypropylene (sPP) blends and fibers. *Polymer* 46: 5956–5965. https://doi.org/10.1016/j.polymer.2005.05.004.

28 Zou, X.-X., Yang, W., Zheng, G.-Q. et al. (2007). Crystallization and phase morphology of injection-molded isotactic polypropylene (iPP)/syndiotactic polypropylene (sPP) blends. *Journal of Polymer Science Part B: Polymer Physics* 45: 2948–2955.

29 Garnier, L., Duquesne, S., Bourbigot, S., and Delobel, R. (2009). Non-isothermal crystallization kinetics of iPP/sPP blends. *Thermochimica Acta* 481: 32–45. https://doi.org/10.1016/j.tca.2008.10.006.

30 Bourbigot, S., Garnier, L., Revel, B., and Duquesne, S. (2013). Characterization of the morphology of iPP/sPP blends with various compositions. *eXPRESS Polymer Letters* 7: 224–237. https://doi.org/10.3144/expresspolymlett.2013.21.

31 Truong, L.T., Larsen, Å.G., and Roots, J. (2017). Morphology, crystalline features, and tensile properties of syndiotactic polypropylene blends. *Journal of Applied Polymer Science* 134: 44611. https://doi.org/10.1002/app.44611.

32 Natta, G. (1960). Progress in the stereospecific polymerization. *Die Makromolekulare Chemie* 35: 94–131. https://doi.org/10.1002/macp.1960.020350105.

33 Norton, D.R. and Keller, A. (1985). The spherulitic and lamellar morphology of melt-crystallized isotactic polypropylene. *Polymer* 26: 704–716. https://doi.org/10.1016/0032-3861(85)90108-9.

34 Varga, J. (2002). β-Modification of isotactic polypropylene: preparation, structure, processing, properties and application. *Journal of Macromolecular Science, Part B Physics* 41: 1121–1171. https://doi.org/10.1081/MB-120013089.

35 Grein, C. (2005). Toughness of neat, rubber modified and filled β-nucleated polypropylene: from fundamentals to applications. *Advances in Polymer Science* 188: 43–104. https://doi.org/10.1007/b136972.

36 Meille, V.S., Ferro, D.R., Brückner, S. et al. (1994). Structure of beta-Isotactic polypropylene: a long-standing structural puzzle. *Macromolecules* 27: 2615–2622. https://doi.org/10.1021/ma00087a034.

37 Keith, H.D., Padden, F.J., Walter, M.N., and Wyckoff, H.W. (1959). Evidence for a second crystal form of polypropylene. *Journal of Applied Physics* 30: 1485–1489. https://doi.org/10.1063/1.1734986.

38 Busse, K., Kressler, J., Maier, R.-D., and Scherble, J. (2000). Tailoring of the α-, β- and γ-modification in isotactic polypropene and propene/ethene random copolymers. *Macromolecules* 33: 8775–8780. https://doi.org/10.1021/ma000719.

39 Wang, J., Gahleitner, M., Gloger, D., and Bernreitner, K. (2020). β-Nucleation of isotactic polypropylene: chain structure effects on the effectiveness of two different nucleating agents. *eXPRESS Polymer Letters* 14: 491–502. https://doi.org/10.3144/expresspolymlett.2020.39.

40 Addink, E.J. and Beintema, J. (1961). Polymorphism of crystalline polypropylene. *Polymer* 2: 185–193. https://doi.org/10.1016/0032-3861(61)90021-0.

41 Turner-Jones, A. (1971). Development of the γ-crystal form in random copolymers and their analysis by DSC and X-ray methods. *Polymer* 12: 487–508. https://doi.org/10.1016/0032-3861(71)90031-0.

42 Meile, V.S. and Brückner, S. (1989). Non-parallel chains in crystalline γ-isotactic polypropylene. *Nature* 340: 455–457. https://doi.org/10.1038/340455a0.

43 Hosier, I.L., Alamo, R.G., Esteso, P. et al. (2007). Formation of the α and γ polymorphs in random metallocene – propylene copolymers. Effect of concentration and type of comonomer. *Macromolecules* 49: 6600–6616. https://doi.org/10.1021/ma030157m.

44 Krache, R., Benavente, R., López-Majada, J.M. et al. (2007). Competition between α, β, and γ polymorphs in a α-nucleated metallocenic isotactic polypropylene. *Macromolecules* 40: 6871–6878. https://doi.org/10.1021/ma0710636.

45 Poon, B., Rogunova, M., Hiltner, A. et al. (2004). Structure and properties of homogeneous copolymers of propylene and 1-hexene. *Macromolecules* 38: 1232–1243. https://doi.org/10.1021/ma000584p.

46 De Rosa, C., Auriemma, F., Talarico, G., and De Ballesteros, O.R. (2007). Structure of isotactic propylene–pentene copolymers. *Macromolecules* 40: 8531–8532. https://doi.org/10.1021/ma701985m.

47 De Rosa, C., Dello Iacono, S., Auriemma, F. et al. (2006). Crystal structure of isotactic propylene–hexene copolymers: the trigonal form of isotactic polypropylene. *Macromolecules* 39: 6098–6196. https://doi.org/10.1021/ma0606354.

48 Lotz, B. (2014). A new ε crystal modification found in stereodefective isotactic polypropylene samples. *Macromolecules* 47: 7612–7624. https://doi.org/10.1021/ma5009868.

49 Corradini, P., Petraccone, V., De Rosa, C., and Guerra, G. (1986). On the structure of the quenched mesomorphic phase of isotactic polypropylene. *Macromolecules* 19: 2699–2703. https://doi.org/10.1021/ma00165a006.

50 Caldas, V., Brown, G.R., Nohr, R.S. et al. (1994). The structure of the mesomorphic phase of quenched isotactic polypropylene. *Polymer* 35: 899–907. https://doi.org/10.1016/0032-3861(94)90931-8.

51 Gesovich, D.M. and Geil, P.H. (1968). Morphology of quenched polypropylene. *Polymer Engineering and Science* 8: 202–209. https://doi.org/10.1002/pen.760080305.

52 McAllister, P.B., Carter, T.J., and Hinde, R.M. (1978). Structure of the quenched form of polypropylene. *Journal of Polymer Science, Polymer Physics Edition* 16: 49–57. https://doi.org/10.1002/pol.1978.180160105m.

53 Cavallo, D., Azzurri, F., Floris, R. et al. (2010). Continuous cooling curves diagrams of propene/ethylene random copolymers. The role of ethylene counits in mesophase development. *Macromolecules* 43: 2890–2896. https://doi.org/10.1021/ma902865e.

54 Cavallo, D., Portale, G., Balzano, L. et al. (2010). Real-time WAXD detection of mesophase development during quenching of propene/ethylene copolymers. *Macromolecules* 43: 10208–10212. https://doi.org/10.1021/ma1022499.

55 Tranchida, D., Mileva, D., Resconi, L. et al. (2015). Molecular and thermal characterization of a nearly perfect isotactic poly(propylene). *Macromolecular Chemistry and Physics* 216: 2171–2178. https://doi.org/10.1002/macp.201500189.

56 Nakamura, K., Shimizu, S., Umemoto, S. et al. (2008). Temperature dependence of crystal growth rate for α and β forms of isotactic polypropylene. *Polymer Journal* 40: 915–922. https://doi.org/10.1295/polymj.PJ2007231.

57 van Schooten, J., van Hoorn, H., and Boerma, J. (1961). Physical and mechanical properties of polypropylene fractions. *Polymer* 2: 161–184. https://doi.org/10.1016/0032-3861(61)90020-9.

58 Janimak, J.J., Cheng, S.Z.D., Giusti, P.A., and Hsieh, E.T. (1991). Isotacticity effect on crystallization and melting in poly(propylene) fractions. 2. Linear crystal growth rate and morphology study. *Macromolecules* 24: 2253–2260. https://doi.org/10.1021/ma00009a020.

59 Paukkeri, R. and Lehtinen, A. (1993). Thermal behaviour of polypropylene fractions: 1. Influence of tacticity and molecular weight on crystallization and melting behaviour. *Polymer* 34: 4075–4082. https://doi.org/10.1016/0032-3861(93)90669-2.

60 Gahleitner, M., Bachner, C., Ratajski, E. et al. (1999). Effects of the catalyst system on the crystallization of polypropylene. *Journal of Applied Polymer Science* 73: 2507–2515.

61 Fujiyama, M. and Inata, H. (2002). Crystallization and melting characteristics of metallocene isotactic polypropylenes. *Journal of Applied Polymer Science* 85: 1851–1857. https://doi.org/10.1002/app.10797.

62 Gahleitner, M., Jääskeläinen, P., Ratajski, E. et al. (2005). Propylene–ethylene random copolymers: comonomer effects on crystallinity and application properties. *Journal of Applied Polymer Science* 95: 1073–1081. https://doi.org/10.1002/app.21308.

63 De Rosa, C., Auriemma, F., de Ballesteros, O.R. et al. (2007). Crystallization behavior of isotactic propylene–ethylene and propylene–butene copolymers: effect of comonomers versus stereodefects on crystallization properties of isotactic polypropylene. *Macromolecules* 40: 6600–6616. https://doi.org/10.1021/ma070409.

64 De Rosa, C., Auriemma, F., de Ballesteros, O.R. et al. (2007). Tailoring the physical properties of isotactic polypropylene through incorporation of comonomers and the precise control of stereo- and regioregularity by metallocene catalysts. *Chemistry of Materials* 19: 5122–5130. https://doi.org/10.1021/cm071502f.

65 Jeon, K., Chiari, Y.L., and Alamo, R.G. (2008). Maximum rate of crystallization and morphology of random propylene ethylene copolymers as a function of comonomer content up to 21 mol%. *Macromolecules* 41: 95–108. https://doi.org/10.1021/ma070757b.

66 Vestberg, T., Parkinson, M., Fonseca, I., and Wilén, C.-E. (2012). Poly(propylene-*co*-ethylene) produced with a conventional and a self-supported Ziegler–Natta catalyst: effect of ethylene and hydrogen concentration on activity and polymer structure. *Journal of Applied Polymer Science* 124: 4889–4896. https://doi.org/10.1002/app.35586.

67 Tynys, A., Fonseca, I., Parkinson, M., and Resconi, L. (2012). Quantitative ^{13}C NMR analysis of isotactic ethylene–propylene copolymers prepared with

metallocene catalyst: effect of ethylene on polymerization mechanisms. *Macromolecules* 45: 7704–7710. https://doi.org/10.1021/ma301102p.

68 Agarwal, V., van Erp, T.B., Balzano, L. et al. (2014). The chemical structure of the amorphous phase of propylene–ethylene random copolymers in relation to their stress–strain properties. *Polymer* 55: 896–905. https://doi.org/10.1016/j.polymer.2013.12.051.

69 Guan, X., Ding, L., Tong, W., and Xiang, F.Y.M. (2021). Effect of molecular weight and isotacticity distribution on hard elastic polypropylene cast films and membranes. *Polymer International* 70: 212–221. https://doi.org/10.1002/pi.6117.

70 Foresta, T., Piccarolo, S., and Goldbeck-Wood, G. (2001). Competition between α and γ phases in isotactic polypropylene: effects of ethylene content and nucleating agents at different cooling rates. *Polymer* 42: 1167–1176. https://doi.org/10.1016/S0032-3861(00)00404-3.

71 Gahleitner, M., Mileva, D., Androsch, R. et al. (2016). Crystallinity-based product design: utilizing the polymorphism of isotactic PP homo- and copolymers. *International Polymer Processing* 31: 618–627. https://doi.org/10.3139/217.3242.

72 Di Sacco, F., Gahleitner, M., Wang, J., and Portale, G. (2020). Systematic investigation on the structure–property relationship in isotactic polypropylene films processed via cast film extrusion. *Polymers* 12: 1636. https://doi.org/10.3390/polym12081636.

73 Paulik, C., Tranninger, C., Wang, J. et al. (2021). Catalyst type effects on structure/property relations of polypropylene random copolymers. *Macromolecular Chemistry and Physics* 222: 2100302. https://doi.org/10.1002/macp.202100302.

74 Mollova, A., Androsch, R., Mileva, D. et al. (2013). Crystallization of isotactic polypropylene containing beta-phase nucleating agent at rapid cooling. *European Polymer Journal* 49: 1057–1065. https://doi.org/10.1016/j.europolymj.2013.01.15.

75 Tzoganakis, C., Vlachopoulos, J., Hamielec, A.E., and Shinozaki, D.M. (1989). Effect of molecular weight distribution on the rheological and mechanical properties of polypropylene. *Polymer Engineering and Science* 29: 390–396. https://doi.org/10.1002/pen.760290607.

76 Phillips, R., Herbert, G., News, J., and Wolkowicz, M. (1994). High modulus polypropylene: effect of polymer and processing variables on morphology and properties. *Polymer Engineering and Science* 34: 1731–1743. https://doi.org/10.1002/pen.760342304.

77 Gahleitner, M., Wolfschwenger, J., Bernreitner, K. et al. (1996). Crystallinity and mechanical properties of PP-homopolymers as influenced by molecular structure and nucleation. *Journal of Applied Polymer Science* 61: 649–657. https://doi.org/10.1002/(SICI)1097-4628(19960725)61:4<649::AID-APP8>3.0.CO;2-L.

78 Jerschow, P. and Janeschitz-Kriegl, H. (1997). The role of long molecules and nucleating agents in shear induced crystallization of isotactic polypropylenes. *International Polymer Processing* 12: 72–77. https://doi.org/10.3139/217.970072.

79 Chvátalová, L., Navrátilová, J., Čermák, R. et al. (2009). Joint effects of molecular structure and processing history on specific nucleation of isotactic polypropylene. *Macromolecules* 42: 7413–7417. https://doi.org/10.1021/ma9005878.

80 Horváth, Z., Sajó, I.E., Stoll, K. et al. (2010). The effect of molecular mass on the polymorphism and crystalline structure of isotactic polypropylene. *eXPRESS Polymer Letters* 4: 101–114. https://doi.org/10.3144/expresspolymlett.2010.15.

81 Pantani, R., Balzano, L., and Peters, G.W.M. (2012). Flow-induced morphology of iPP solidified in a shear device. *Macromolecular Materials and Engineering* 297: 60–67. https://doi.org/10.1002/mame.201100158.

82 Horváth, Z., Menyhárd, A., Doshev, P. et al. (2013). Effect of molecular architecture on the crystalline structure and stiffness of iPP homopolymers: modeling based on annealing experiments. *Journal of Applied Polymer Science* 130: 3365–3373. https://doi.org/10.1002/APP.39585.

83 Troisi, E.M., Arntz, S.A.J.J., Roozemond, P.C. et al. (2017). Application of a multi-phase multi-morphology crystallization model to isotactic polypropylenes with different molecular weight distributions. *European Polymer Journal* 97: 397–408. https://doi.org/10.1016/j.europolymj.2017.09.042.

84 Tranninger, C. (2012). Transparent thin wall packaging material with improved stiffness and flowability. EP 2935343 B1, priority 21.12.2012, grant 28.03.2018, Austria.

85 DeMaio, V.V. and Dong, D. (1997). The effect of chain structure on melt strength of polypropylene and polyethylene. In: *Proceedings of the 43rd SPE ANTEC Conference*, 1512–1516. Society of Plastics Engineers.

86 Gotsis, A.D., Zeevenhoven, B.L.F., and Hogt, A.H. (2004). The effect of long chain branching on the processability of polypropylene in thermoforming. *Polymer Engineering and Science* 44: 973–982. https://doi.org/10.1002/pen.20089.

87 Weng, W., Markel, E.J., and Dekmezian, A.H. (2001). Synthesis of long-chain branched propylene polymers via macromonomer incorporation. *Macromolecular Rapid Communications* 22: 1488–1499. https://doi.org/10.1002/1521-3927(20011201)22:18<1488::AID-MARC1488>3.0.CO;2-I.

88 Paavola, S., Saarinen, T., Löfgren, B., and Pikänen, P. (2004). Propylene copolymerization with non-conjugated dienes and α-olefins using supported metallocene catalyst. *Polymer* 45: 2099–2110. https://doi.org/10.1016/j.polymer.2004.01.053.

89 Rätzsch, M., Arnold, M., Borsig, E. et al. (2002). Radical reactions on polypropylene in the solid state. *Progress in Polymer Science* 27: 1195–1282. https://doi.org/10.1016/S0079-6700(02)00006-0.

90 Jørgensen, J.K., Redford, K., Ommundsen, E., and Stor, A. (2007). Molecular structure and shear rheology of long chain branched polypropylene formed by light cross-linking of a linear precursor with 1,3-benzenedisulfonyl azide. *Journal of Applied Polymer Science* 106: 950–960. https://doi.org/10.1002/app.26674.

91 Auhl, D., Stange, J., Münstedt, H. et al. (2004). Long-chain branched polypropylenes by electron beam irradiation and their rheological properties. *Macromolecules* 37: 9465–9472. https://doi.org/10.1021/ma030579w.

92 Lugão, A.B., Artel, B.W.H., Yoshiga, A. et al. (2007). Production of high melt strength polypropylene by gamma irradiation. *Radiation Physics and Chemistry* 76: 1691–1695. https://doi.org/10.1016/j.radphyschem.2007.03.013.

93 Reichelt, N., Stadlbauer, M., Folland, R., and Park, C.B. (2003). PP-blends with tailored foamability and mechanical properties. *Cellular Polymers* 22: 315–327. https://doi.org/10.1177/026248930302200503.

94 Rätzsch, M., Hesse, A., Reichelt, N., et al. (1998). Polyolefin molded parts of improved stiffness and heat resistance. EP 890612 B1, Priority 09.07.1997, grant 04.05.2005, Austria.

95 Tian, J., Yu, W., and Zhou, C. (2006). Crystallization kinetics of linear and long-chain branched polypropylene. *Journal of Macromolecular Science, Part B Physics* 45: 969–985. https://doi.org/10.1080/00222340600870507.

96 Yu, F., Zhang, H., Liao, R. et al. (2009). Flow induced crystallization of long chain branched polypropylenes under weak shear flow. *European Polymer Journal* 45: 2110–2118. https://doi.org/10.1016/j.eurpolymj.2009.03.011.

97 Gajzlerová, L., Navrátilová, J., Ryzí, A. et al. (2020). Joint effects of long-chain branching and specific nucleation on morphology and thermal properties of polypropylene blends. *eXPRESS Polymer Letters* 14: 952–961. https://doi.org/10.3144/expresspolymlett.2020.77.

98 Gahleitner, M., Grein, C., Kheirandish, S., and Wolfschwenger, J. (2011). Nucleation of polypropylene homo- and copolymers. *International Polymer Processing* 26: 2–20. https://doi.org/10.3139/217.2411.

99 Blomenhofer, M., Ganzleben, S., Hanft, D. et al. (2005). "Designer" nucleating agents for polypropylene. *Macromolecules* 38: 3688–3695. https://doi.org/10.1021/ma0473317.

100 Menyhárd, A., Gahleitner, M., Varga, J. et al. (2009). The influence of nucleus density on optical properties in nucleated isotactic polypropylene. *European Polymer Journal* 45: 3138–3148. https://doi.org/10.1016/j.eurpolymj.2009.08.006.

101 Horváth, Z., Menyhárd, A., Doshev, P. et al. (2016). Improvement of the impact strength of ethylene–propylene random copolymers by nucleation. *Journal of Applied Polymer Science* 133: 43823. https://doi.org/10.1002/app.43823.

102 Liu, X., Liu, X., Li, Y. et al. (2020). Nanoengineering of transparent polypropylene containing sorbitol-based clarifier. *Journal of Polymer Research* 27: 198. https://doi.org/10.1007/s10965-020-02169-3.

103 Jang, G.-S., Cho, W.-J., and Ha, C.-S. (2001). Crystallization behavior of polypropylene with or without sodium benzoate as a nucleating agent. *Journal of Polymer Science Part B: Polymer Physics* 39: 1001–1016. https://doi.org/10.1002/polb.1077.

104 Libster, D., Aserin, A., and Garti, N. (2007). Advanced nucleating agents for polypropylene. *Advances in Polymer Technology* 18: 685–695. https://doi.org/10.1002/pat.970.

105 Housmans, J.-W., Gahleitner, M., Peters, G.W.M., and Meijer, H.E.H. (2009). Structure–property relations in molded, nucleated isotactic polypropylene. *Polymer* 50: 2304–2319. https://doi.org/10.1016/j.polymer.2009.02.050.

106 Meijer-Vissers, T. and Goossens, H. (2012). The influence of the cooling rate on the nucleation efficiency of isotactic poly(propylene) with bis(3,4-dimethylbenzylidene)sorbitol. *Macromolecular Symposia* 330: 150–165. https://doi.org/10.1002/masy.201300027.

107 McGenity, P.M., Hooper, J.J., Paynter, C.D. et al. (1992). Nucleation and crystallization of polypropylene by mineral fillers: relationship to impact strength. *Polymer* 33: 5215–5224. https://doi.org/10.1016/0032-3861(92)90804-6.

108 Leong, Y.W., Abu Bakar, M.B., Mohd. Ishak, Z.A. et al. (2004). Comparison of the mechanical properties and interfacial interactions between talc, kaolin, and calcium carbonate filled polypropylene composites. *Journal of Applied Polymer Science* 91: 3315–3326. https://doi.org/10.1002/app.13542.

109 Castillo, L.A., Barbosa, S.E., and Capiati, N.J. (2013). Influence of talc morphology on the mechanical properties of talc filled polypropylene. *Journal of Polymer Research* 20: 152. https://doi.org/10.1007/s10965-013-0152-2.

110 Qiu, F., Wang, M., Hao, Y., and Guo, S. (2014). The effect of talc orientation and transcrystallization on mechanical properties and thermal stability of the polypropylene/talc composites. *Composites: Part A* 58: 7–15. https://doi.org/10.1016/j.compositesa.2013.11.011.

111 Layachi, A., Makhlouf, A., Frihi, D. et al. (2019). Non-isothermal crystallization kinetics and nucleation behavior of isotactic polypropylene composites with micro-talc. *Journal of Thermal Analysis and Calorimetry* 138: 1081–1095. https://doi.org/10.1007/s10973-019-08262-0.

112 Wittmann, J.-C. and Lotz, B. (1990). Epitaxial crystallization of polymers on organic and polymeric substrates. *Progress in Polymer Science* 15: 909–948. https://doi.org/10.1016/0079-6700(90)90025-V.

113 Gahleitner, M., Grein, C., and Bernreitner, K. (2012). Synergistic mechanical effects of calcite micro- and nanoparticles and β-nucleation in polypropylene copolymers. *European Polymer Journal* 48: 49–59. https://doi.org/10.1016/j.eurpolymj.2011.10.013.

114 Varga, J. and Karger-Kocsis, J. (1996). Rules of supermolecular structure formation in sheared isotactic polypropylene melts. *Journal of Polymer Science Part B: Polymer Physics* 34: 657–670. https://doi.org/10.1002/(SICI)1099-0488(199603)34:4<657::AID-POLB6>3.0.CO;2-N.

115 Duplay, C., Monasse, B., Haudin, J.-M., and Costa, J.-L. (2000). Shear-induced crystallization of polypropylene: influence of molecular weight. *Journal of Materials Science* 35: 6093–6103. https://doi.org/10.1023/A:1026731917188.

116 Looijmans, S.F.S.P., Carmeli, E., Puskar, L. et al. (2020). Polarization modulated infrared spectroscopy: a pragmatic tool for polymer science and engineering. *Polymer Crystallization* 3: e10138. https://doi.org/10.1002/pcr2.10138.

117 Assouline, E., Pohl, S., Fulchiron, R. et al. (2000). The kinetics of α and β transcrystallization in fibre-reinforced polypropylene. *Polymer* 41: 7843–7854. https://doi.org/10.1016/S0032-3861(00)00113-0.

118 Xie, H.-Q., Zhang, S., and Xie, D. (2005). An efficient way to improve the mechanical properties of polypropylene/short glass fiber composites. *Journal of Applied Polymer Science* 96: 1414–1420. https://doi.org/10.1002/app.21575.

119 Yoshimoto, S., Ueda, T., Yamanaka, K. et al. (2001). Epitaxial act of sodium 2,2′-methylene-bis-(4,6-di-*t*-butylphenylene)phosphate on isotactic polypropylene. *Polymer* 42: 9627–9631. https://doi.org/10.1016/S0032-3861(01)00510-9.

120 Patil, N., Invigorito, C., Gahleitner, M., and Rastogi, S. (2013). Influence of a particulate nucleating agent on the quiescent and flow-induced crystallization of isotactic polypropylene. *Polymer* 54: 5883–5891. https://doi.org/10.1016/j.polymer.2013.08.004.

121 Gao, J., Meng, X., Deng, Z. et al. (2022). Enhancement of "in-situ" dispersed NA11 for the mechanical and crystallization properties of polypropylene. *Journal of Polymer Research* 29: 168. https://doi.org/10.1007/s10965-022-02920-y.

122 Nagasawa, S., Fujimori, A., Masuko, T., and Iguchi, M. (2005). Crystallisation of polypropylene containing nucleators. *Polymer* 46: 5241–5250. https://doi.org/10.1016/j.polymer.2005.03.099.

123 Wang, B., Utzeri, R., Castellano, M. et al. (2020). Heterogeneous nucleation and self-nucleation of isotactic polypropylene microdroplets in immiscible blends: from nucleation to growth-dominated crystallization. *Macromolecules* 53: 5980–5991. https://doi.org/10.1021/acs.macromol.0c01167.

124 Zhang, Y.-F. and Xin, Z. (2006). Effects of substituted aromatic heterocyclic phosphate salts on properties, crystallization, and melting behaviors of isotactic polypropylene. *Journal of Applied Polymer Science* 100: 4868–4874. https://doi.org/10.1002/app.23209.

125 Zhang, Y.-F., He, B., Hou, H.-H., and Guo, L.-H. (2017). Isothermal crystallization of isotactic polypropylene nucleated with a novel aromatic heterocyclic phosphate nucleating agent. *Journal of Macromolecular Science, Part B Physics* 56: 811–820. https://doi.org/10.1080/00222348.2017.1385360.

126 Wang, J. and Dou, Q. (2008). Crystallization behavior and optical and mechanical properties of isotactic polypropylene nucleated with rosin-based nucleating agents. *Polymer International* 57: 233–239. https://doi.org/10.1002/pi.2329.

127 Zhang, Y.-F., Hou, H.-H., and Guo, L.-H. (2018). Effects of cyclic carboxylate nucleating agents on nucleus density and crystallization behavior of isotactic polypropylene. *Journal of Thermal Analysis and Calorimetry* 131: 1483–1490. https://doi.org/10.1007/s10973-017-6669-6.

128 Broda, J. (2003). Nucleating activity of the quinacridone and phthalocyanine pigments in polypropylene crystallization. *Journal of Applied Polymer Science* 90: 3957–3964. https://doi.org/10.1002/app.13083.

129 van der Meer, D.W., Milazzo, D., Sanguineti, A., and Julius Vancso, G. (2005). Oriented crystallization and mechanical properties of polypropylene nucleated on fibrillated polytetrafluoroethylene scaffolds. *Polymer Engineering and Science* 45: 458–468. https://doi.org/10.1002/pen.20297.

130 Alcazar, D., Ruan, J., Thierry, A., and Lotz, B. (2006). Structural matching between the polymeric nucleating agent isotactic poly(vinylcyclohexane) and isotactic polypropylene. *Macromolecules* 39: 2832–2840. https://doi.org/10.1021/ma052651r.

131 Mileva, D., Gahleitner, M., Shutov, P. et al. (2019). Effect of supercooling on crystal structure of nucleated isotactic polypropylene. *Thermochimica Acta* 677: 194–197. https://doi.org/10.1016/j.tca.2019.03.031.

132 Seiji, Y. and Hisao, T. (1994). The phase transition in the gel state of the 1,3:2,4-di-O-benzylidene-D-sorbitol/ethylene glycol system. *Bulletin of the Chemical Society of Japan* 67: 2053–2056. https://doi.org/10.1246/bcsj.67.2053.

133 Balzano, L., Rastogi, S., and Peters, G.W.M. (2008). Flow induced crystallization in isotactic polypropylene-1,3:2,4-bis(3,4-dimethylbenzylidene)sorbitol blends: implications on morphology of shear and phase separation. *Macromolecules* 41: 399–408. https://doi.org/10.1021/ma071460g.

134 Schawe, J.E.K., Budde, F., and Alig, I. (2018). Nucleation activity at high supercooling: sorbitol-type nucleating agents in polypropylene. *Polymer* 153: 587–596. https://doi.org/10.1016/j.polymer.2018.08.054.

135 Nguon, O.J., Charlton, Z., Kumar, M. et al. (2020). Interactions between sorbitol-type nucleator and additives for polypropylene. *Polymer Engineering and Science* 60: 3046–3055. https://doi.org/10.1002/pen.25535.

136 Sowinski, P., Piorkowska, E., Boyer, S.A.E., and Haudin, J.-M. (2021). High-pressure crystallization of iPP nucleated with 1,3:2,4-bis(3,4-dimethylbenzylidene)sorbitol. *Polymers* 13: 145. https://doi.org/10.3390/polym13010145.

137 Kristiansen, P.M., Gress, A., Smith, P. et al. (2006). Phase behavior, nucleation and optical properties of the binary system isotactic polypropylene/N,N',N''-tris-isopentyl-1,3,5-benzene-tricarboxamide. *Polymer* 47: 249–253. https://doi.org/10.1016/j.polymer.2005.08.053.

138 Varga, J., Stoll, K., Menyhárd, A., and Horváth, Z. (2011). Crystallization of isotactic polypropylene in the presence of a β-nucleating agent based on a trisamide of trimesic acid. *Journal of Applied Polymer Science* 121: 1469–1480. https://doi.org/10.1002/app.33685.

139 Abraham, F., Kress, R., Smith, P., and Schmidt, H.-W. (2013). A new class of ultra-efficient supramolecular nucleating agents for isotactic polypropylene. *Macromolecular Chemistry and Physics* 214: 17–24. https://doi.org/10.1002/macp.201200487.

140 Sterzynski, T., Lambla, M., Georgi, F., and Thomas, M. (1997). Studies of the *trans*-quinacridone nucleation of poly(ethylene-*b*-propylene). *International Polymer Processing* 12: 64–71. https://doi.org/10.3139/217.970064.

141 Sterzynski, T., Calo, P., Lambla, M., and Thomas, M. (1997). Trans- and dimethyl quinacridone nucleation of isotactic polypropylene. *Polymer Engineering and Science* 37: 1917–1927. https://doi.org/10.1002/pen.11842.

142 Barczewski, M., Matykiewicz, D., and Hoffmann, B. (2017). Effect of quinacridone pigments on properties and morphology of injection molded isotactic polypropylene. *International Journal of Polymer Science* 7043297. https://doi.org/10.1155/2017/7043297.

143 Gohn, A.M., Rhoades, A.M., Okonski, D., and Androsch, R. (2018). Effect of melt-memory on the crystal polymorphism in molded isotactic polypropylene. *Macromolecular Materials and Engineering* 303: 1800148. https://doi.org/10.1002/mame.201800148.

144 Chen, H.B., Karger-Kocsis, J., Wu, J.S., and Varga, J. (2002). Fracture toughness of α- and β-phase polypropylene homopolymers and random- and block-copolymers. *Polymer* 43: 6505–6514. https://doi.org/10.1016/S0032-3861(02)00590-6.

145 Grein, C. and Gahleitner, M. (2008). On the influence of nucleation of iPP/EPR blends with different molecular characteristics. *eXPRESS Polymer Letters* 2: 392–397. https://doi.org/10.3144/expresspolymlett.2008.47.

146 Yue, Y., Hu, D., Zhang, Q. et al. (2018). The effect of structure evolution upon heat treatment on the beta-nucleating ability of calcium pimelate in isotactic polypropylene. *Polymer* 149: 55–64. https://doi.org/10.1016/j.polymer.2018.06.060.

147 Zhang, Z., Tao, Y., Yang, Z., and Mai, K. (2008). Preparation and characteristics of nano-$CaCO_3$ supported β-nucleating agent of polypropylene. *European Polymer Journal* 44: 1955–1961. https://doi.org/10.1016/j.europolymj.2008.04.022.

148 Liu, Z., Liu, X., Li, L. et al. (2019). Crystalline structure and remarkably enhanced tensile property of β-isotactic polypropylene via overflow microinjection molding. *Polymer Testing* 76: 448–454. https://doi.org/10.1016/j.polymertesting.2019.04.002.

149 Yi, Q.-F., Wen, X.-J., Dong, J.-Y., and Han, C.C. (2008). A novel effective way of comprising a β-nucleating agent in isotactic polypropylene (i-PP): polymerized dispersion and polymer characterization. *Polymer* 49: 5053–5063. https://doi.org/10.1016/j.polymer.2008.09.037.

150 Luo, F., Geng, C., Wang, K. et al. (2009). New understanding in tuning toughness of β-polypropylene: the role of β-nucleated crystalline morphology. *Macromolecules* 42: 9325–9331. https://doi.org/10.1021/ma901651f.

151 Luo, F., Zhu, Y., Wang, K. et al. (2012). Enhancement of β-nucleated crystallization in polypropylene random copolymer via adding isotactic polypropylene. *Polymer* 53: 4861–4870. https://doi.org/10.1016/j.polymer.2012.08.037.

152 Zhang, Y., Sun, T., Wei Jiang, W., and Han, G. (2018). Crystalline modification of a rare earth nucleating agent for isotactic polypropylene based on its self-assembly. *Royal Society Open Science* 5: 180247. https://doi.org/10.1098/rsos.180247.

153 Dieckmann, D. (2001). Effect of various acid neutralizers on the crystallization temperature of nucleated polypropylene. *Journal of Vinyl and Additive Technology* 7: 51–55. https://doi.org/10.1002/vnl.10264.

154 Dieckmann, D., Nyberg, W., Lopez, D., and Barnes, P. (1996). The additive/antistat interaction in polypropylene. *Journal of Vinyl and Additive Technology* 2: 57–62. https://doi.org/10.1002/vnl.10095.

155 Gahleitner, M., Doshev, P., and Tranninger, C. (2013). Heterophasic copolymers of polypropylene – development, design principles and future challenges. *Journal of Applied Polymer Science* 130: 3028–3037. https://doi.org/10.1002/app.39626.

156 Liang, J.Z. and Li, R.K.Y. (2000). Rubber toughening in polypropylene: a review. *Journal of Applied Polymer Science* 77: 409–417. https://doi.org/10.1002/(SICI)1097-4628(20000711)77:2<409::AID-APP18>3.0.CO;2-N.

157 Teh, J., Rudin, A., and Keung, J.C. (1994). A review of polyethylene–polypropylene blends and their compatibilization. *Advances in Polymer Technology* 13: 1–23. https://doi.org/10.1002/adv.1994.060130101.

158 Kontopoulou, M., Wang, W., Gopakumar, T.G., and Cheung, C. (2003). Effect of composition and comonomer type on the rheology, morphology and properties of ethylene-α-olefin copolymer/polypropylene blends. *Polymer* 44: 7495–7504. https://doi.org/10.1016/j.polymer.2003.08.043.

159 Kotter, I., Grellmann, W., Koch, T., and Seidler, S. (2006). Morphology-toughness correlation of polypropylene/ethylene–propylene rubber blends. *Journal of Applied Polymer Science* 100: 3364–3371. https://doi.org/10.1002/app.23708.

160 Cheruthazhekatt, S., Pijpers, T.F.J., Harding, G.W. et al. (2012). Multidimensional analysis of the complex composition of impact polypropylene copolymers: combination of TREF, SEC-FTIR-HPer DSC, and high temperature 2D-LC. *Macromolecules* 45: 2025–2034. https://doi.org/10.1021/ma2026989.

161 Jeremic, L., Albrecht, A., Sandholzer, M., and Gahleitner, M. (2020). Rapid characterization of high-impact ethylene–propylene copolymer composition by crystallization extraction separation: comparability to standard separation methods. *International Journal of Polymer Analysis and Characterization* 25: 581–596. https://doi.org/10.1080/1023666X.2020.1821151.

162 Fan, Z.-Q., Zhang, Y.-Q., Xu, J.-T. et al. (2003). Structure and properties of polypropylene/poly(ethylene-*co*-propylene) in-situ blends synthesized by spherical Ziegler–Natta catalyst. *European Polymer Journal* 39: 795–804. https://doi.org/10.1016/S0014-3057(02)00287-2.

163 Doshev, P., Lach, R., Lohse, G. et al. (2005). Fracture characteristics and deformation behavior of heterophasic ethylene–propylene copolymers as a function of the dispersed phase composition. *Polymer* 46: 9411–9422. https://doi.org/10.1016/j.polymer.2005.07.029.

164 Doshev, P., Lohse, G., Henning, S. et al. (2006). Phase interactions and structure evolution of heterophasic ethylene–propylene copolymers as a function of system composition. *Journal of Applied Polymer Science* 101: 2825–2837.

165 Grein, C., Gahleitner, M., Knogler, B., and Nestelberger, S. (2007). Melt viscosity effects in ethylene–propylene copolymers. *Rheologica Acta* 46: 1083–1089. https://doi.org/10.1007/s00397-007-0200-0.

166 Aarnio-Winterhof, M., Doshev, P., Seppälä, J., and Gahleitner, M. (2017). Structure–property relations of heterophasic ethylene–propylene copolymers based on a single-site catalyst. *eXPRESS Polymer Letters* 11: 152–161. https://doi.org/10.3144/expresspolymlett.2017.16.

167 Lin, Z., Peng, M., and Zheng, Q. (2004). Isothermal crystallization behavior of polypropylene catalloys. *Journal of Applied Polymer Science* 93: 877–882. https://doi.org/10.1002/app.20501.

168 Xu, J.-T., Fu, S.-Z., Wang, X.-P. et al. (2005). Effect of the structure on the morphology and spherulitic growth kinetics of polyolefin in-reactor alloys. *Journal of Applied Polymer Science* 98: 632–638. https://doi.org/10.1002/app.22074.

169 Shangguan, Y., Song, Y., and Zheng, Q. (2007). Kinetics analysis on spherulite growth rate of polypropylene catalloys. *Polymer* 48: 4567–4577. https://doi.org/10.1016/j.polymer.2007.05.066.

170 Yokoyama, Y. and Ricco, T. (1997). Crystallization and morphology of reactor-made blends of isotactic polypropylene and ethylene–propylene rubber. *Journal of Applied Polymer Science* 66: 1007–1014. https://doi.org/10.1002/(SICI)1097-4628(19971107)66:6<1007::AID-APP1>3.0.CO;2-M.

171 Cai, J., Luo, R., Lv, R. et al. (2017). Crystallization kinetics of ethylene-*co*-propylene rubber/isotactic polypropylene blend investigated via chip-calorimeter measurement. *European Polymer Journal* 96: 79–86. https://doi.org/10.1016/j.eurpolymj.2017.09.003.

172 Mileva, D., Wang, J., Gahleitner, M. et al. (2020). New insights into crystallization of heterophasic isotactic polypropylene by fast scanning chip calorimetry. *Polymers* 12: 1683–1698. https://doi.org/10.3390/polym12081683.

173 Martuscelli, E., Silvestre, C., and Abate, G. (1982). Morphology, crystallization and melting behaviour of films of isotactic polypropylene blended with ethylene–propylene copolymers and polyisobutylene. *Polymer* 23: 229–237. https://doi.org/10.1016/0032-3861(82)90306-8.

174 Nitta, K.-H., Shin, Y.-W., Hashiguchi, H. et al. (2005). Morphology and mechanical properties in the binary blends of isotactic polypropylene and novel propylene-*co*-olefin random copolymers with isotactic propylene sequence 1. Ethylene–propylene copolymers. *Polymer* 46: 965–975. https://doi.org/10.1016/j.polymer.2004.11.033.

175 Gao, J., Wang, D., Yu, M., and Yao, Z. (2004). Nonisothermal crystallization, melting behavior, and morphology of polypropylene/metallocene-catalyzed polyethylene blends. *Journal of Applied Polymer Science* 93: 1203–1210. https://doi.org/10.1002/app.20505.

176 Svoboda, P., Svobodova, D., Slobodian, P. et al. (2009). Crystallization kinetics of polypropylene/ethylene–octene copolymer blends. *Polymer Testing* 28: 215–222. https://doi.org/10.1016/j.polymertesting.2008.12.007.

177 Jose, S., Aprem, A.S., Francis, B. et al. (2004). Phase morphology, crystallisation behaviour and mechanical properties of isotactic polypropylene/high density polyethylene blends. *European Polymer Journal* 40: 2105–2115. https://doi.org/10.1016/j.eurpolymj.2004.02.026.

178 Galeski, A., Pracella, M., and Martuscelli, E. (1984). Polypropylene spherulite morphology and growth rate changes in blends with low-density polyethylene. *Journal of Polymer Science, Polymer Physics Edition* 22: 739–747. https://doi.org/10.1002/pol.1984.180220415.

179 Bartczak, Z., Galeski, A., and Pracella, M. (1986). Spherulite nucleation in blends of isotactic polypropylene with high-density polyethylene. *Polymer* 27: 537–543. https://doi.org/10.1016/0032-3861(86)90239-9.

180 Wenig, W. and Meyer, K. (1980). Investigation of the crystallization behaviour of polypropylene–polyethylene blends by optical microscopy. *Colloid and Polymer Science* 258: 1009–1014. https://doi.org/10.1007/BF01382395.

181 Rybnikár, F. (1988). Crystallization and morphology in blends of isotactic polypropylene and linear polyethylene. *Journal of Macromolecular Science, Part B Physics* 27: 125–144. https://doi.org/10.1080/00222348808245759.

182 Shanks, R.A., Li, J., and Yu, L. (2000). Polypropylene–polyethylene blend morphology controlled by time–temperature–miscibility. *Polymer* 41: 2133–2139. https://doi.org/10.1016/S0032-3861(99)00399-7.

183 Blackadder, D.A., Richardson, M.J., and Savill, N.G. (1981). Characterization of blends of high density polyethylene with isotactic polypropylene. *Die Makromolekulare Chemie* 182: 1271–1282. https://doi.org/10.1002/macp.1981.021820426.

184 Rybnikar, F. and Kaszonyiova, M. (2014). Epitaxial crystallization of linear polyethylene in blends with isotactic polypropylene. *Journal of Macromolecular Science, Part B Physics* 53: 217–232. https://doi.org/10.1080/00222348.2013.808522.

185 Lotz, B. and Wittmann, J.C. (1987). Polyethylene–isotactic polypropylene epitaxy: analysis of the diffraction patterns of oriented biphasic blends. *Journal of Polymer Science Part B: Polymer Physics* 25: 1079–1087. https://doi.org/10.1002/polb.1987.090250509.

186 Petermann, J. and Xu, Y. (1991). The origin of heteroepitaxy in the system of uniaxially oriented isotactic polypropylene and polyethylene. *Journal of Materials Science* 26: 1211–1215. https://doi.org/10.1007/BF00544457.

187 Greso, A.J. and Phillips, P.J. (1994). The role of secondary nucleation in epitaxial growth: the template model. *Polymer* 35: 3373–3376. https://doi.org/10.1016/0032-3861(94)90897-4.

188 Yan, S., Petermann, J., and Yang, D. (1997). Effect of lamellar thickness on the epitaxial crystallization of PE on oriented iPP films. *Polymer Bulletin* 38: 87–94. https://doi.org/10.1007/s002890050023.

189 Yan, S., Yang, D., and Petermann, J. (1998). Controlling factors for the occurrence of heteroepitaxy of polyethylene on highly oriented isotactic polypropylene. *Polymer* 39: 4569–4578. https://doi.org/10.1016/S0032-3861(97)10137-9.

190 Carmeli, E., Fenni, S.E., Caputo, M.R. et al. (2021). Surface nucleation of dispersed polyethylene droplets in immiscible blends revealed by polypropylene matrix self-nucleation. *Macromolecules* 54: 9100–9112. https://doi.org/10.1021/acs.macromol.1c01430.

191 Yan, S., Katzenberg, F., and Petermann, J. (1999). Epitaxial and graphoepitaxial growth of isotactic polypropylene (iPP) from the melt on highly oriented high density polyethylene (HDPE) substrates. *Journal of Polymer Science Part B: Polymer Physics* 37: 1893–1898. https://doi.org/10.1002/(SICI)1099-0488(19990801)37:15<1893::AID-POLB13>3.0.CO;2-L.

192 Carmeli, E., Kandioller, G., Gahleitner, M. et al. (2021). Continuous cooling curve diagrams of isotactic-polypropylene/polyethylene blends: mutual nucleating effects under fast cooling conditions. *Macromolecules* 54: 4834–4846. https://doi.org/10.1021/acs.macromol.1c00699.

193 Van Drongelen, M., Gahleitner, M., Spoelstra, A.B. et al. (2015). Flow-induced solidification of high-impact polypropylene copolymer compositions: morphological and mechanical effects. *Journal of Applied Polymer Science* 132: 42040. https://doi.org/10.1002/app.42040.

194 Campoy, I., Arribas, J.M., Zaporta, M.A.M. et al. (1995). Crystallization kinetics of polypropylene–polyamide compatibilized blends. *European Polymer Journal* 31: 475–480. https://doi.org/10.1016/0014-3057(94)00185-5.

195 Marco, C., Ellis, G., Gómez, M.A. et al. (1997). Rheological properties, crystallization, and morphology of compatibilized blends of isotactic polypropylene and polyamide. *Journal of Applied Polymer Science* 65: 2665–2677. https://doi.org/10.1002/(SICI)1097-4628(19970926)65:13<2665::AID-APP8>3.0.CO;2-9.

196 Pigłowski, J., Gancarz, I., Wlaźlak, M., and Kammer, H.W. (2000). Crystallization in modified blends of polyamide and polypropylene. *Polymer* 41: 6813–6824. https://doi.org/10.1016/S0032-3861(00)00034-3.

197 Holsti-Miettinen, R.M., Seppälä, J.V., Ikkala, O.T., and Reima, I.T. (1994). Functionalized elastomeric compatibilizer in PA 6/PP blends and binary interactions between compatibilizer and polymer. *Polymer Engineering and Science* 34: 395–404. https://doi.org/10.1002/pen.760340504.

198 Caro, A.S., Parpaite, T., Otazhagine, B. et al. (2017). Viscoelastic properties of polystyrene/polyamide-6 blend compatibilized with silica/polystyrene Janus hybrid nanoparticles. *Journal of Rheology* 61: 305. https://doi.org/10.1122/1.4975334.

199 Gahleitner, M., Kretzschmar, B., Pospiech, D. et al. (2006). Morphology and mechanical properties of polypropylene/polyamide 6 nanocomposites prepared by a two-step melt-compounding process. *Journal of Applied Polymer Science* 100: 283–291. https://doi.org/10.1002/app.23102.

200 Thomason, J.L. and Van Rooyen, A.A. (1990). Investigation of the transcrystallised interphase in fibre-reinforced thermoplastic composites. In: *Integration of Fundamental Polymer Science and Technology-4* (ed. P.J. Lemstra and L.A. Kleintjens), 335–342. Dordrecht: Springer https://doi.org/10.1007/978-94-009-0767-6_39.

201 Schoolenberg, G.E. and Van Rooyen, A.A. (1993). Transcrystallinity in fiber-reinforced thermoplastic composites. *Composite Interfaces* 1: 243–252. https://doi.org/10.1163/156855493X00103.

202 Thomason, J.L. and Van Rooyen, A.A. (1992). Transcrystallized interphase in thermoplastic composites, Part II Influence of interfacial stress, cooling rate, fibre. *Journal of Materials Science* 27: 897–907. https://doi.org/10.1007/BF01197639.

203 Gray, D.G. (1974). "Transcrystallization" induced by mechanical stress on a polypropylene melt. *Journal of Polymer Science, Polymer Letters Edition* 12: 645–650. https://doi.org/10.1002/pol.1974.130121107.

204 Devaux, E. and Caze, C. (2000). Evolution of the interfacial stress transfer ability between a glass fibre and a polypropylene matrix during polymer crystallization. *Journal of Adhesion Science and Technology* 14: 965–974. https://doi.org/10.1163/156856100743004.

205 Varga, J. and Karger-Kocsis, J. (1993). The occurence of transcrystallization or row-nucleated cylindritic crystallization as a result of shearing in a glass-fiber-reinforced polypropylene. *Composites Science and Technology* 48: 191–198. https://doi.org/10.1016/0266-3538(93)90136-5.

206 Piccarolo, S., Saiu, M., Brucato, V., and Titomanlio, G. (1992). Crystallization of polymer melts under fast cooling. II. High-purity iPP. *Journal of Applied Polymer Science* 46: 1097–4628. https://doi.org/10.1002/app.1992.070460409.

207 Brucato, V., Piccarolo, S., and La Carrubba, V. (2002). An experimental methodology to study polymer crystallization under processing conditions. The influence of high cooling rates. *Chemical Engineering Science* 57: 4129–4143. https://doi.org/10.1016/S0009-2509(02)00360-3.

208 Pijpers, M. and Mathot, V. (2008). Optimization of instrument response and resolution of standard- and high-speed power compensation DSC benefits for the study of crystallization, melting and thermal fractionation. *Journal of Thermal Analysis and Calorimetry* 93: 319–327. https://doi.org/10.1007/s10973-007-8924-8.

209 Mileva, D., Androsch, R., and Radusch, H.-J. (2009). Effect of structure on light transmission in isotactic polypropylene and random propylene-1-butene copolymers. *Polymer Bulletin* 62: 561–571. https://doi.org/10.1007/s00289-008-0034-7.

210 Mileva, D., Androsch, R., Zhuravlev, E. et al. (2012). Homogeneous nucleation and mesophase formation in glassy isotactic polypropylene. *Polymer* 53: 277–282. https://doi.org/10.1016/j.polymer.2011.11.064.

211 Mileva, D., Wang, J., Androsch, R. et al. (2021). Crystallization of random metallocene-catalyzed propylene-based copolymers with ethylene and 1-hexene on rapid cooling. *Polymers* 13: 2091. https://doi.org/10.3390/polym13132091.

212 Aniunoh, K. and Harrison, G. (2010). The processing of polypropylene cast films. I. Impact of material properties and processing conditions on film formation. *Polymer Engineering and Science* 50: 1151–1160. https://doi.org/10.1002/pen.21637.

213 Sadeghi, F., Tabatabaei, S.H., Ajji, A., and Carreau, P.J. (2010). Properties of uniaxial stretched polypropylene films: effect of drawing temperature and random copolymer content. *The Canadian Journal of Chemical Engineering* 88: 1091–1098. https://doi.org/10.1002/cjce.20372.

214 Tabatabaei, S. and Abdellah, A. (2011). Effect of initial crystalline morphology on properties of polypropylene cast films. *Journal of Plastic Film & Sheeting* 27: 223–233. https://doi.org/10.1177/8756087911407921.

215 Gloger, D., Rossegger, E., Gahleitner, M., and Wagner, C. (2020). Plastic drawing response in the biaxially oriented polypropylene (BOPP) process: polymer structure and film casting effects. *Journal of Polymer Engineering* 40: 743–752. https://doi.org/10.1515/polyeng-2019-0220.

216 Cimmino, S., Martuscelli, E., Nicolais, L., and Silvestre, C. (1978). Thermal and mechanical properties of isotactic random propylene–butene-1 copolymer. *Polymer* 19: 1222–1223. https://doi.org/10.1016/0032-3861(78)90075-7.

217 Marega, C., Marigo, A., Saini, R., and Ferrari, P. (2001). The influence of thermal treatment and processing on the structure and morphology of poly(propylene-*ran*-1-butene) copolymers. *Polymer International* 50: 442–448. https://doi.org/10.1002/pi.655.

218 Resch, K., Wallner, G., Teichert, C. et al. (2006). Optical properties of highly transparent polypropylene cast films: influence of material structure, additives, and processing conditions. *Polymer Engineering and Science* 46: 520–531. https://doi.org/10.1002/pen.20503.

219 Gahleitner, M., Grein, C., Blell, R. et al. (2011). Sterilization of propylene/ethylene random copolymers: annealing effects on crystalline structure and transparency as influenced by polymer structure and nucleation. *eXPRESS Polymer Letters* 5: 788–798. https://doi.org/10.3144/expresspolymlett.2011.77.

220 Meng, X., Shijun, Z., Jieying, L. et al. (2014). Influences of processing on the phase transition and crystallization of polypropylene cast films. *Journal of Applied Polymer Science* 131: 41100. https://doi.org/10.1002/APP.41100.

221 Mileva, D., Gahleitner, M., Gloger, D., and Tranchida, D. (2018). Crystal structure: a way to control properties in cast films of polypropylene. *Polymer Bulletin* 75: 5587–5598. https://doi.org/10.1007/s00289-018-2343-9.

222 Lamberti, G. (2011). Flow-induced crystallization during isotactic polypropylene film casting. *Polymer Engineering and Science* 51: 851–861. https://doi.org/10.1002/pen.21891.

223 Kantz, M.R., Newman, H.D. Jr., and Stigale, F.H. (1972). The skin-core morphology and structure–property relationships in injection-molded polypropylene. *Journal of Applied Polymer Science* 16: 1249–1260. https://doi.org/10.1002/app.1972.070160516.

224 Bashir, Z., Odell, J.A., and Keller, A. (1986). Stiff and strong polyethylene with shish kebab morphology by continuous melt extrusion. *Journal of Materials Science* 21: 3993–4002. https://doi.org/10.1007/PL00020271.

225 Kimata, S., Sakurai, T., Nozue, Y. et al. (2007). Molecular basis of the shish-kebab morphology in polymer crystallization. *Science* 316: 1014–1017. https://doi.org/10.1126/science.1140132.

226 Fujiyama, M., Wakino, T., and Kawasaki, Y. (1988). Structure of skin layer in injection-molded polypropylene. *Journal of Applied Polymer Science* 35: 29–49. https://doi.org/10.1002/app.1988.070350104.

227 Jiang, J., Wang, S., Sun, B. et al. (2015). Effect of mold temperature on the structures and mechanical properties of micro-injection molded polypropylene. *Materials & Design* 88: 245–251. https://doi.org/10.1016/j.matdes.2015.09.003.

228 Pan, Y., Guo, X., Zheng, G. et al. (2018). Shear-induced skin-core structure of molten isotactic polypropylene and the formation of β-crystal. *Macromolecular Materials and Engineering* 303: 1800083. https://doi.org/10.1002/mame.201800083.

229 Janeschitz-Kriegl, H., Ratajski, E., and Stadlbauer, M. (2003). Flow as an effective promotor of nucleation in polymer melts: a quantitative evaluation. *Rheologica Acta* 42: 355–364. https://doi.org/10.1007/s00397-002-0247-x.

230 Eder, G., Janeschitz-Kriegl, H., and Liedauer, S. (1990). Crystallization processes in quiescent and moving polymer melts under heat transfer conditions. *Progress in Polymer Science* 15: 629–714. https://doi.org/10.1007/s00397-002-0247-x.

231 Moitzi, J. and Skalicky, P. (1990). Shear-induced crystallization of isotactic polypropylene melts: isothermal WAXS experiments with synchrotron radiation. *Polymer* 34: 3168–3172. https://doi.org/10.1016/0032-3861(93)90385-N.

232 Housmans, J.-W., Balzano, L., Santoro, D. et al. (2009). A design to study flow induced crystallization in a multipass rheometer. *International Polymer Processing* 24: 185–197. https://doi.org/10.3139/217.2230.

233 van Erp, T.B., Balzano, L., Spoelstra, A.B. et al. (2012). Quantification of non-isothermal, multi-phase crystallization of isotactic polypropylene: the influence of shear and pressure. *Polymer* 53: 5896–5908. https://doi.org/10.1016/j.polymer.2012.10.027.

234 Zuidema, H., Peters, G.W.M., and Meijer, H.E.H. (2001). Development and validation of a recoverable strain-based model for flow-induced crystallization of polymers. *Macromolecular Theory and Simulations* 10: 447–460. https://doi.org/10.1002/1521-3919(20010601)10:5<447::AID-MATS447>3.0.CO;2-C.

235 Pantani, R., Coccorullo, I., Speranza, V., and Titomanlio, G. (2005). Modeling of morphology evolution in the injection molding process of thermoplastic polymers. *Progress in Polymer Science* 30: 1185–1222. https://doi.org/10.1016/j.progpolymsci.2005.09.001.

236 Housmans, J.-W., Steenbakkers, R.J.A., Roozemond, P.C. et al. (2009). Saturation of pointlike nuclei and the transition to oriented structures in flow-induced crystallization of isotactic polypropylene. *Macromolecules* 42: 5728–5740. https://doi.org/10.1021/ma802479c.

237 Roozemond, P.C., van Erp, T.B., and Peters, G.W.M. (2016). Flow-induced crystallization of isotactic polypropylene: modeling formation of multiple crystal phases and morphologies. *Polymer* 89: 69–80. https://doi.org/10.1016/j.polymer.2016.01.032.

238 Grosso, G., Troisi, E.M., Jaensson, N.O. et al. (2019). Modelling flow induced crystallization of IPP: multiple crystal phases and morphologies. *Polymer* 182: 121806. https://doi.org/10.1016/j.polymer.2019.121806.

239 Caelers, H.J.M., de Cock, A., Looijmans, S.F.S.P. et al. (2022). An experimentally validated model for quiescent multiphase primary and secondary crystallization phenomena in PP with low content of ethylene comonomer. *Polymer* 253: 124901. https://doi.org/10.1016/j.polymer.2022.124901.

240 Li, J., Zhang, X., Qu, C. et al. (2007). Hierarchy structure in injection molded polypropylene/ethylene–octane copolymer blends. *Journal of Applied Polymer Science* 105: 2252–2259. https://doi.org/10.1002/app.26225.

241 Liu, X., Miao, X., Cai, X. et al. (2020). The orientation of the dispersed phase and crystals in an injection-molded impact polypropylene copolymer. *Polymer Testing* 90: 106658. https://doi.org/10.1016/j.polymertesting.2020.106658.

242 Grestenberger, G., Potter, G.D., and Grein, C. (2014). Polypropylene/ethylene–propylene rubber (PP/EPR) blends for the automotive industry: basic correlations between EPR-design and shrinkage. *eXPRESS Polymer Letters* 8: 282–292. https://doi.org/10.3144/expresspolymlett.2014.31.

243 Andreassen, E., Myrhe, O.J., Hinrichsen, E.L., and Grøstad, K. (1994). Effects of processing parameters and molecular weight distribution on the tensile properties of polypropylene fibers. *Journal of Applied Polymer Science* 52: 1505–1517. https://doi.org/10.1002/app.1994.070521015.

244 Misra, S., Lu, F.-M., Spruiell, J.E., and Richeson, G.C. (1995). Influence of molecular weight distribution on the structure and properties of melt-spun polypropylene filaments. *Journal of Applied Polymer Science* 56: 1761–1779. https://doi.org/10.1002/app.1995.070561307.

245 Arvidson, S.A., Khan, S.A., and Gorga, R.E. (2010). Mesomorphic – α-monoclinic phase transition in isotactic polypropylene: a study of processing effects on structure and mechanical properties. *Macromolecules* 43: 2916–2924. https://doi.org/10.1021/ma1001645.

246 Ozturk, M.K., Venkataraman, M., and Mishra, R. (2018). Influence of structural parameters on thermal performance of polypropylene nonwovens. *Polymers for Advanced Technologies* 29: 3027–3034. https://doi.org/10.1002/pat.4423.

247 Drabek, J. and Zatloukal, M. (2020). Influence of molecular weight, temperature, and extensional rheology on melt blowing process stability for linear isotactic polypropylene. *Physics of Fluids* 32: 083110. https://doi.org/10.1063/5.0020773.

248 Palmer, R.P. and Cobbold, A.J. (1964). The texture of melt crystallised polythene as revealed by selective oxidation. *Die Makromolekulare Chemie* 74: 174–189. https://doi.org/10.1002/macp.1964.020740114.

249 Olley, R.H. and Bassett, D.C. (1982). An improved permanganic etchant for polyolefines. *Polymer* 23: 1707–1710. https://doi.org/10.1016/0032-3861(82)90110-0.

250 Pölt, P., Ingolic, E., Gahleitner, M. et al. (2000). Characterization of modified polypropylene by scanning electron microscopy. *Journal of Applied Polymer Science* 78: 1152–1161. https://doi.org/10.1002/1097-4628(20001031)78:5<1152::AID-APP250>3.0.CO;2-7.

251 Martinez Salazar, J., Gonzalez Ortega, J.C., and Balta Calleja, F.J. (1977). On the separation of crystalline and diffuse X-ray scattering in semicrystalline polymers. *Anales de Física* 73: 244–247.

252 Turner-Jones, A., Aizlewood, J.M., and Beckett, D.R. (1964). Crystalline forms of isotactic polypropylene. *Die Makromolekulare Chemie* 75: 134–157. https://doi.org/10.1002/macp.1964.020750113.

253 Tranchida, D., Kandioller, G., Schwarz, P. et al. (2020). Novel characterization of the β → α crystalline phase transition of isotactic polypropylene through high-temperature atomic force microscopy imaging and nanoindentation. *Polymer Crystallization* 3: e10140. https://doi.org/10.1002/pcr2.10140.

254 Nielsen, A.S., Batchelder, D.N., and Pyrz, R. (2002). Estimation of crystallinity of isotactic polypropylene using Raman spectroscopy. *Polymer* 43: 2671–2676. https://doi.org/10.1016/S0032-3861(02)00053-8.

255 Khafagy, R.M. (2006). In situ FT-Raman spectroscopic study of the conformational changes occurring in isotactic polypropylene during its melting and crystallization processes. *Journal of Polymer Science Part B: Polymer Physics* 44: 2173–2182. https://doi.org/10.1002/polb.20891v.

256 Zhu, X., Yan, D., Yao, H., and Zhu, P. (2000). In situ FTIR spectroscopic study of the regularity bands and partial-order melts of isotactic poly(propylene). *Macromolecular Rapid Communications* 21: 354–357. https://doi.org/10.1002/(SICI)1521-3927(20000401)21:7<354::AID-MARC354>3.0.CO;2-B.

257 Di Sacco, F., Saidi, S., Hermida-Merino, D., and Portale, G. (2021). Revisiting the mechanism of the meso-to-α transition of isotactic polypropylene and ethylene–propylene random copolymers. *Macromolecules* 54: 9681–9691. https://doi.org/10.1021/acs.macromol.1c01904.

258 Pijpers, M.F.J., Mathot, V.B.F., Goderis, B. et al. (2002). High-speed calorimetry for the study of the kinetics of (de)vitrification, crystallization, and melting of macromolecules. *Macromolecules* 35: 3601–3613. https://doi.org/10.1021/ma011122u.

259 Androsch, R. and Schick, C. (2015). Crystal nucleation of polymers at high supercooling of the melt. *Advances in Polymer Science* 276: 257–288. https://doi.org/10.1007/12_2015_325.

260 Grapes, M.D., LaGrange, T., Friedman, L.H. et al. (2014). Combining nanocalorimetry and dynamic transmission electron microscopy for in situ characterization of materials processes under rapid heating and cooling. *Review of Scientific Instruments* 85: 084902. https://doi.org/10.1063/1.4892537.

261 Zhang, R., Zhuravlev, E., Androsch, R., and Schick, C. (2019). Visualization of polymer crystallization by in situ combination of atomic force microscopy and fast scanning calorimetry. *Polymers* 11: 890. https://doi.org/10.3390/polym11050890.

262 Van Drongelen, M., Meijer-Vissers, T., Cavallo, D. et al. (2013). Microfocus wide-angle X-ray scattering of polymers crystallized in a fast scanning chip calorimeter. *Thermochimica Acta* 563: 33–37. https://doi.org/10.1016/j.tca.2013.04.007.

263 La Carrubba, V., Brucato, V., and Piccarolo, S. (2000). Isotactic polypropylene solidification under pressure and high cooling rates. A master curve approach. *Polymer Engineering and Science* 40: 2430–2441. https://doi.org/10.1002/pen.11375.

264 Carmeli, E., Cavallo, D., and Tranchida, D. (2021). Instrument for mimicking fast cooling conditions of polymers: design and case studies on polypropylene. *Polymer Testing* 97: 107164. https://doi.org/10.1016/j.polymertesting.2021.107164.

4

Crystallization of PE and PE-Based Blends, and Composites

Amirhosein Sarafpour[1], Gholamreza Pircheraghi[1], Farzad Gholami[2], Rouhollah Shami-Zadeh[1], and Farzad Jani[3]

[1] *Sharif University of Technology, Department of Materials Science and Engineering, Polymeric Materials Research Group (PMRG), Azadi St., Tehran, Tehran, Iran*
[2] *Georgia Institute of Technology, School of Materials Science and Engineering, Atlanta, GA 30332, USA*
[3] *Pars Special Economic Energy Zone, JAM Petrochemical Co., Research and Development, Asalouyeh, 75118-11368 Boushehr, Iran*

4.1 An Introduction to Polyethylene, Its Crystallization, and Kinetics

4.1.1 Basics of Structure and Morphology

Product, process, and fabrication advances in polyethylene (PE) have evolved over the past 75 years to suit an expanding market's needs. PE is the most common type of plastic because it can be made in a wide range of ethylene homopolymers and copolymers [1].

All kinds of PE have roughly the same backbone of covalently bonded carbon and hydrogen atoms with dangling hydrogens; the diversity comes from the branches that alter the material's properties [2]. Branches might be simple alkyl groups or acid and ester functions. Meagerly, flaws in the polymer backbone, particularly vinyl groups at the chain ends, cause molecular variations. Branches and other faults in the regular chain structure impair crystallinity, in which chains with few flaws have better crystallinity than chains with many.

The total density of PE increases with crystallinity because crystalline portions pack better than noncrystalline sections. However, there are different ethylene homopolymers and copolymers. Scientific and industrial communities commonly divide PE resins into three main categories: high density polyethylene (HDPE), low density polyethylene (LDPE), and linear low density polyethylene (LLDPE) [3]. The chain structures for HDPE, LDPE, and LLDPE are illustrated in Figure 4.1a.

HDPE has a linear structure with fewer than one side branch per 200 CH_2 units [6]. It has high crystallinity and, thus, high tensile strength and hardness. HDPE resins with no comonomer are brittle and susceptible to environmental stress cracking. Adding modest quantities of α-olefins (1-butene, 1-hexene, or 1-octene)

Polymer Crystallization: Methods, Characterization, and Applications, First Edition.
Edited by Jyotishkumar Parameswaranpillai, Jenny Jacob, Senthilkumar Krishnasamy, Aswathy Jayakumar, and Nishar Hameed.
© 2023 WILEY-VCH GmbH. Published 2023 by WILEY-VCH GmbH.

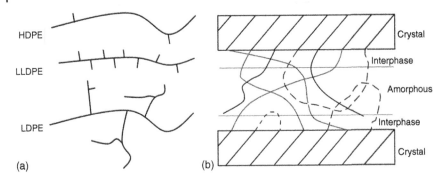

Figure 4.1 (a) The chain structures of HDPE, LLDPE, and LDPE. Source: Ragaert et al. [4]/John Wiley & Sons. (b) Crystalline, interphase, and amorphous arrangement in semi-crystalline polyethylene. Source: Allegra [5].

to the PE backbone enhances tie-chain concentration while decreasing density and crystallinity. This phenomenon improves environmental stress crack resistance (ESCR) [7].

Later developments produced bimodal HDPE resins with a very low molecular weight homopolymer portion and a very high molecular weight portion with copolymers using 1-hexene or 1-butene. Bimodal HDPE resins have better stiffness, impact resistance, toughness, and ESCR balance than unimodal HDPE resins [6]. Additionally, HDPE can have a molecular weight of up to 10 million, which is branded as ultra high molecular weight polyethylenes (UHMWPEs). It has excellent abrasion and impact resistance compared to its lower molecular weight counterparts [8, 9].

LDPE has more branches than HDPE, which reduces its crystallinity. The presence of a high level of "tree-like" long-chain branching (LCB), typically two to three LCB/1000 carbon atoms, is a notable molecular characteristic of LDPE resins. Intermolecular chain transfer reactions produce long branches as long as the primary chains, while intramolecular chain transfer reactions produce small branches of roughly four CH_2 units. A high amount of LCB in LDPE resins results in outstanding extrusion processability and melt strength, but a lower abrasion and stiffness properties [2, 10].

The third form of PE is LLDPE with short branches. Because of the absence of LCB, these resins were referred to as "linear" LDPE. Ethylene is copolymerized with butene, hexene, or octene to create LLDPE. The branches reduce density compared to HDPE because they interfere with crystallization. The number of branches affects crystallinity, enabling LLDPE to be made into flexible films or rigid structures [10].

The broader classification of PE based on density is represented in Table 4.1.

As a semi-crystalline polymer, PE has two or more solid phases. At least one of which has molecular chain segments structured in a regular three-dimensional array and one or more phases in which chains are disordered. Semi-crystallinity is crucial because PE may be considered a composite material comprising crystalline and noncrystalline areas, which is why it is called a semi-crystalline polymer [2]. Hence, crystalline and noncrystalline phases, as well as their sizes, shapes, orientations, and connections, all have a role in determining the physical characteristics of PE.

Table 4.1 Commercial classification of polyethylene.

PE type	Symbol	α-Olefin content (mol%)	Crystallinity (%)	Density (g cm^{-3})
High density	HDPE	0	65–70	>0.960
		0.2–0.5	60–65	0.941–0.959
Ultrahigh molecular weight	UHMWPE	0	30–40	0.930–0.935
Medium density	MDPE	1–2	45–55	0.926–0.940
Linear low density	LLDPE	2.5–3.5	30–45	0.915–0.925
Low density	LDPE	20–30CH$_3$/1000C	45–55	0.910–0.940
Very low density	VLDPE	>4	<25	<0.915

Source: Adapted from Kissin [6].

For instance, a PE sample's physical and mechanical behaviors are determined by the arrangement of the phases in relation to one another, their relative quantities, and the degree to which they are connected. Making a solid PE specimen that is not semi-crystalline is almost impossible. A completely amorphous PE would be a soft material, whereas a PE sample entirely made up of a crystalline structure would be brittle. Indeed, PE has a good balance of resilience and toughness.

The crystalline portions are embedded in a continuous matrix of noncrystalline phases. The noncrystalline phase comprises two sections; a partially ordered section next to the crystallites and disordered chains in the remaining spaces. The three-phase morphology of PE is shown schematically in Figure 4.1b.

The structural feature known as a lamella is formed when PE molecules crystallize and fold on themselves several times. The lamella thickness is the average length of a PE chain's straight segment in a lamella. The spacing between adjacent chains in the lamellae is 4.4–4.6 A. In HDPE resins, the folds joining neighboring straight parts of the chain comprise a few ethylene units and produce amorphous areas. Each lamella's sandwich structure comprises a crystalline core of folded chains and two amorphous areas on both sides of the core.

All PE resins are semi-crystalline polymers, except those very low density polyethylene (VLDPE) resins with a high-α olefin content (ethylene elastomers). The degree of branching in chains determines the crystalline structure of PE. Different morphological characteristics of PE with different branching degrees are depicted in Figure 4.2.

Disordered molecular segments are generally continuous with those in crystallites in noncrystalline areas. The three kinds of segments illustrated in Figure 4.3 are made up of these segments. "Tie chains" link two neighboring crystalline domains,

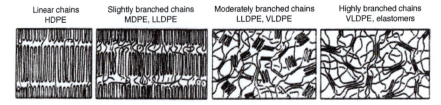

Figure 4.2 Effect of branching on semi-crystalline morphology of polyethylene. Source: Kissin [6]/Carl Hanser Verlag.

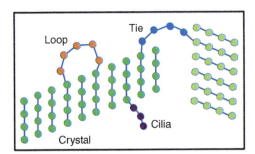

Figure 4.3 Polymer chain segments. Source: Zhai et al. [11]/American Chemical Society.

"loops" that can return to the crystallite from which they started, and "cilia" that terminate in a chain end.

The connection between adjacent crystallites has an important influence on a sample's physical properties. The sample's molecular weight, branching level, and crystallization conditions all influence the ratio of chains returning to the same crystallite vs. those traversing the disordered areas.

Spherulites are the most prevalent large-scale morphological structure made up of crystalline and noncrystalline areas. The spherulites' major structural components are rod-like fibrils (rays) radiating from the spherulite's core to the periphery. Spherulites collide with one another as they expand, forming irregular polyhedrons. The fibrils comprise stacked lamellae that twist, branch, and occupy the whole spherulite space. Figure 4.4 demonstrates the overall structure of a typical spherulite in a semi-crystalline polymer such as PE and microscopic images of the spherulites seen in an isothermally crystallized HDPE at 120 °C.

PE has orthorhombic, monoclinic, and hexagonal unit cells. Orthorhombic PE is the most common crystalline form of linear PE. The orthorhombic cell has the following dimensions: $a = 0.740$ nm, $b = 0.493$ nm, and $c = 0.253$ nm [6]. Short-chain branches in PE chains cause the orthorhombic cell to expand somewhat; a length rises to 0.77 nm, and b rises to 0.5 nm. The density of the crystals decreases as a result of this expansion [6].

Pseudo-monoclinic PE is the second most common crystalline form. PE's pseudo-monoclinic crystal form (also known as the triclinic) is a metastable phase that forms under elongation conditions.

PE's hexagonal crystal structure is created through crystallization at extraordinarily high pressures. This structure is interesting for researchers in laboratories. It is not produced under any commercially viable circumstances at this time. Because

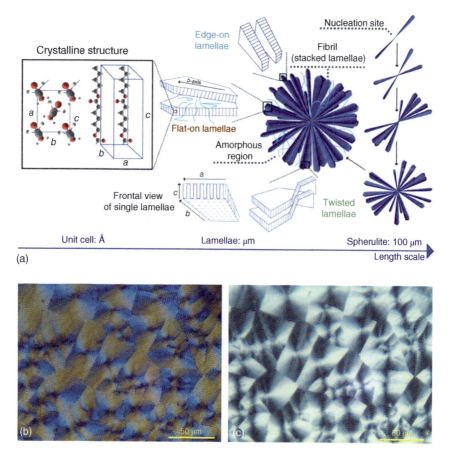

Figure 4.4 (a) A schematic representing the hierarchical structure of the polymer spherulite. Source: Zhai et al. [11]/American Chemical Society/CC BY 4.0. Microscopic images of an HDPE sample crystallized under isothermal condition at 120 °C in (b) polarized and (c) phase contrast modes. Source: Koziol et al. [12]/American Chemical Society/CC BY 4.0.

individual chain stems spin at random phase angles with regard to their neighbors, the hexagonal phase is also known as the "rotator" phase [2]. Figure 4.5 shows different types of PE crystal unit cells that are possible to be formed under different circumstances described above.

To study basics of PE crystallization, it is necessary to know common values for thermal properties of the material, including melting temperature (T_m), crystallization temperature (T_c), and degree of crystallinity (X_c). Table 4.2 summarizes some reported values in the literature for each thermal property, although it should be considered that these values may change considerably with different variables, most importantly the specific PE grade and conditions of the testing. Another property is the glass transition temperature (T_g), which has various reported values for PE, but most commonly, it is accepted to be in the range of −130 to −110 °C [2, 15].

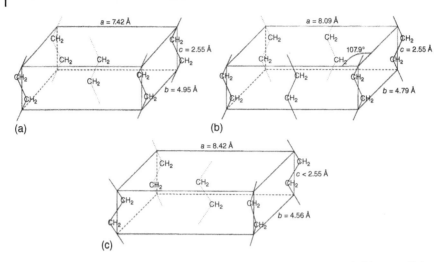

Figure 4.5 Different types of unit cells in polyethylene: (a) orthorhombic (b) monoclinic, and (c) hexagonal. Source: Peacock [2]/Taylor & Francis Group.

Table 4.2 Typical thermal properties of some PE grades [13, 14].

Property PE grade	T_m (°C)	T_c (°C)	X_c (%)
HDPE	132	117	61
LLDPE	122	109	36
LDPE	111	97	34
UHMWPE	135	115	52

4.1.2 Theory of Crystallization and Its Kinetics

In semicrystalline polymers, there is a significant fraction of amorphous phase in-between the crystalline lamellae. Crystallization in semicrystalline polymers can happen in different forms, such as single crystals, polycrystalline aggregates, and highly oriented structures, formation of which depends on the crystallization condition and thermomechanical history of the polymer. The kinetics of the crystallization can be studied under isothermal and non-isothermal conditions. In traditional studies for polymer crystallization, the conditions are idealized and all external factors are considered to be constant. Isothermal crystallization is often described by the Avrami equation [16, 17]. The Avrami equation makes it possible to calculate the crystallinity fraction as a function of time:

$$1 - X(t) = \exp(-kt^n) \quad (4.1)$$

where $X(t)$ is the fraction of transformed material at time t, n is the Avrami index, depending on dimensionality and time dependence of crystal growth, and k is the overall crystallization rate constant [16–18]. The value of Avrami index in case of polymers usually reveals the dimension of crystals, e.g. a value of 2 suggests a two-dimensional lamellar growth while three indicates three-dimensional spherulitic growth. Generally, studies in which crystallization is considered as an isothermal process, are limited to idealized conditions, in which external conditions are constant. In real situations, however, the external conditions change continuously, which makes the treatment of non-isothermal crystallization more complex. However, the study of crystallization in a continuously changing environment is of greater interest since industrial processes proceed generally under non-isothermal conditions. Several methods have been proposed to study the kinetics of non-isothermal crystallization, most of which are developed based on the Avrami equation. A list of some of the proposed equations in the literature is provided in Table 4.3.

4.2 Experimental Study on Crystallization Kinetics of Polyethylene

4.2.1 Isothermal Crystallization

Various studies have been conducted on crystallization behavior of different grades of PE [29–32], their blends [33–36], and composites [37–41], employing the isothermal crystallization technique. It has proven to be a reliable and straightforward method for accurate monitoring of polymer crystallization kinetics and an indicator for predicting the morphology and properties of the polymer after the crystallization from the melt.

Sarafpour et al. [42] implemented the Avrami equation parameters along with the rheological properties to predict the behavior of PE compounded with carbon black masterbatch. In this study, the correlation between the Avrami index n and rheological properties of the blends shows that when the molecular weight of PE-based masterbatch carrier decreases, the tendency of the polymer to form structures containing more spherulites increases significantly. Figure 4.6a,b shows the scanning electron microscopy (SEM) image as well as schematical representation of microstructure of the blend of HDPE with low and high molecular weight masterbatch carriers, respectively. It has been shown that the Avrami index shows linear relation with consistency constant, zero shear viscosity, and Carreau–Yasuda parameters (Figure 4.6c,d). It can be concluded from this research that the isothermal crystallization parameters are helpful for predicting the rheological and structural properties of the semicrystalline polymers.

In another study to correlate the isothermal crystallization and long-term performance of the PE, Gholami et al. [43] have provided information about utilization of isothermal crystallization to predict long-term resistance to slow crack growth (SCG), an important parameter for PE pipes. The SCG is an important phenomenon in many thermoplastics, especially in PE and its application in pipes, as it involves a usually undetectable crack growth in the polymer under stresses that much

Table 4.3 List of some of proposed equations for the non-isothermal crystallization kinetics.

Author(s)	Proposed equation	Parameters
Ziabicki [19]	$E(t) = \ln\left(\int_0^t \frac{ds}{\tau_{\frac{1}{2}}}\right)^n \left[1 + a_1 \int_0^t \frac{ds}{\tau_{\frac{1}{2}}} + a_2 \left(\int_0^t \frac{ds}{\tau_{\frac{1}{2}}}\right)^2 + a_3 \cdots\right]$	$a_1, a_2, \ldots, a_n$ are the coefficients of the series, n is the Avrami exponent, $\tau_{\frac{1}{2}}$ is half-time of crystallization, and s is the time required
Nakamura et al. [20, 21]	$X(t) = 1 - \exp\left[-\left(\int_0^t K'(T)d\tau\right)^n\right]$	n is the Avrami index determined from isothermal crystallization data, $X(t)$ is the fraction of transformed material at time t, and K' is related to the Avrami constant K: $K' = K^{\frac{1}{n}}$
Patel et al. [22]	$\frac{dX(t)}{dt} = nK'(T)[1 - X(t)]\left\{\ln\left[\frac{1}{1-X(t)}\right]\right\}^{\frac{1}{n}}$	n is the Avrami index determined from isothermal crystallization data, $X(t)$ is the fraction of transformed material at time t, and K' is related to the Avrami constant K: $K' = K^{\frac{1}{n}}$
Ozawa [23]	$1 - X(t) = \exp[-K(t)/\lambda^m]$	λ is the cooling rate, $X(t)$ is the relative crystallinity, and $K(t)$ and m are the Ozawa crystallization rate constant and Ozawa exponent, respectively
Liu et al. [24] Also referred to as Mo's method(	$\log(\lambda) = \log F(t) - a \log t$ where: $F(t) = \left[\frac{K(t)}{k}\right]^{\frac{1}{m}}$	a is the ratio between the Avrami and Ozawa exponents (n/m). $F(T)$ refers to the value of the cooling rate, chosen at unit crystallization time when the system has a certain degree of crystallization. k is the Avrami crystallization rate constant, and $K(t)$ is the Ozawa crystallization rate constant
Dietz [25]	$\frac{dX}{dt} = nX(T)(1-X)\,t^{(n-1)}\exp\left(-\frac{\gamma X}{1-X}\right)$	γ lies between 0 and 1, n is the Avrami index determined from isothermal crystallization data, $X(t)$ is the fraction of transformed material at time t
Lim et al. [26]	$\ln G + \frac{U^*}{R(T_b - z^i - T_\infty)} = \ln G_0 - \frac{K_g}{(T_c - z^i)[T_m^0 - (T_b - z^i)]f}$	T_b is the temperature at which the first measurable data is recorded, χ is the cooling rate, K_g is the energy required for the formation of a nucleus of critical size and T_m^0 is the equilibrium melting temperature, T_c is crystallization temperature, U^* is a universal constant characteristic of the activation energy of chain motion (reptation) in the melt and is equal to 1500 cal mol^{-1}, G is the growth rate to be obtained from thermal optical microscopy
Kissinger [27]	$\dfrac{d\left(\ln\frac{\chi}{T_p^2}\right)}{d\left(\frac{1}{T_p}\right)} = -\dfrac{\Delta E}{R}$	R is the gas constant, χ is heating rate, ΔE is activation energy, and T_p is the temperature rate of crystallization is maximum
Nadkarni et al. [28]	$\Delta T_c = P\chi + \Delta T_c^0$	where ΔT_c^0 is the degree of undercooling required, and P is a process sensitivity factor

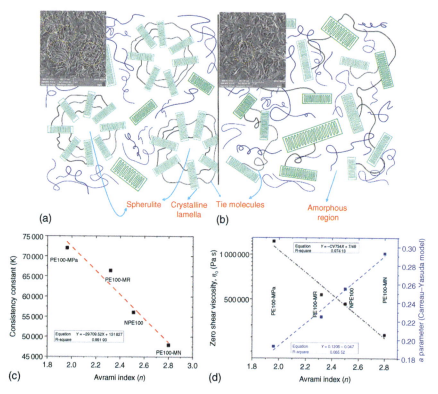

Figure 4.6 Schematic of a structure (a) with three-dimensional spherulitic structure and (b) with two-dimensional lamellae. Values of Avrami index vs. (c) consistency constant in power law model and (d) zero shear viscosity and parameter a obtained from Carreau–Yasuda model. Source: Sarafpour et al. [42]/John Wiley & Sons.

lower the yield strength of the material, leading to premature brittle failure in time scales much shorter than expected lifetime of the material. In this study, the correlation between the Avrami index n and the failure time in full notched creep test (FNCT, details of which can be seen in ISO 16770 standard) indicates that the smaller Avrami index parameter is corresponding with longer failure time in FNCT experiment (Figure 4.7). This behavior is because of the microstructural differences between the samples, in which samples with larger Avrami index show higher tendency to form more spherulites in the structure, while in the samples with lower Avrami index, the lamellae are spread randomly in the structure. Since the boundary of the spherulites acts as stress concentration regions in the samples under constant load, the crack can propagate in this area, and thus the samples with more 3D structure show shorter lifetime in FNCT test than the sample with structure comprising random lamellae. Results of this study indicate that isothermal crystallization provides unique results, and can be a much more straightforward and less time-consuming procedure for indirect determination of creep resistance than the standard SCG test, making it a useful and fast complementary test method for evaluation of creep performance in semi-crystalline polymers.

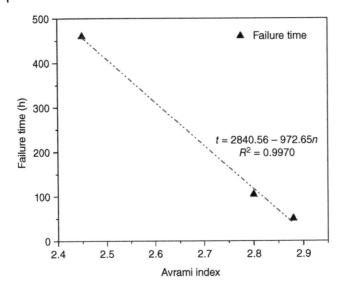

Figure 4.7 The relation of creep failure time of PE samples against Avrami index n. Source: Gholami et al. [43]/with permission of Elsevier.

4.2.2 Non-isothermal Crystallization

Nakamura et al. [20, 21] performed the first extensive and detailed research on the non-isothermal crystallization of PE. In this study, the non-isothermal kinetics of HDPE was investigated via X-ray scattering technique to identify changes in crystallinity and evaluate the integrated intensity of the 2θ range in which the (110) reflection is located under various cooling conditions. They developed the theoretical treatments to be used for analyzing the change in crystallinity of semicrystalline polymers, which also showed good agreement with the experimental data.

Supaphol and Spruiell [44] implemented modified light-depolarizing microscopy technique in order to investigate the non-isothermal crystallization kinetics of HDPE with two high and low molecular weight values (101 000 and 77 000 g mol^{-1}). They used different cooling rates in the range of 0–2500 °C min^{-1}. It has been observed that non-isothermal crystallization in PE significantly depends on the cooling rate. By increasing the cooling rate, the crystallization temperature decreased from ~122 to ~112 °C for both low and high molecular weight polymers (Figure 4.8a). The crystallization rate vs. cooling rate for both high and low molecular weight PE show similar trend; however, the lower molecular weight polymer crystallized in a relatively shorter time over all ranges of undercooling studied (Figure 4.8b). This behavior indicates that in the low molecular weight PE, the mobility of the polymeric chains in the melt structure is higher, which makes it easier for the chains to transfer themselves from entangled melt to the crystalline phase. To investigate the effect of cooling rate on the nucleation process, Avrami analysis was applied to the experimental data on the non-isothermal crystallization of PE and, the results indicated that the Avrami index for all samples, which crystallized under different cooling conditions, is close to 3, indicating a 3D spherulitic structure. Moreover, after the completion

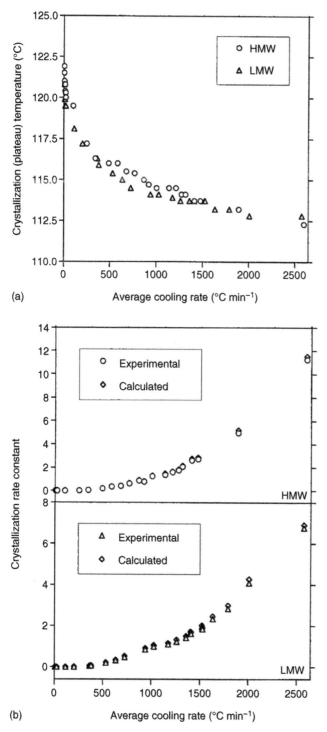

Figure 4.8 (a) Typical plateau temperature and (b) crystallization rate constant vs. average cooling rate. Source: Supaphol et al. [44]/John Wiley & Sons.

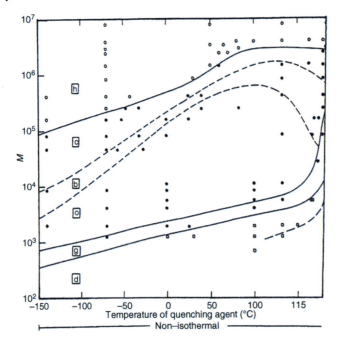

Figure 4.9 Morphological map for molecular weight fractions of linear polyethylene. Source: Mandelkern et al. [45]/American Chemical Society. Plot of molecular weight against either quenching or isothermal crystallization temperatures. Morphological forms are indicated by letters as defined in text.

of crystallization, the spherulite dimensions were almost the same. These findings demonstrated that number of spherulites and nucleation density did not change with cooling rate or crystallization temperature.

The morphology or supermolecular structure of linear and branched PE crystallized under non-isothermal conditions has been investigated by Mandelkern et al. [45]. It has been observed that depending on molecular mass, concentration of branch groups, and quenching temperature, wide range of morphological structures could obtain. In this study, a morphological map for the non-isothermal crystallization of linear PE is given in Figure 4.9, which indicates the possible morphology of the PE in the specific range of molecular mass and quenching temperature. In this map regions marked "a," "b," and "c" denote spherulites; region "d" represents thin rods; region "g" corresponds with rods and sheet-like crystals; in the region "h" randomly oriented lamellae forms. The random-type morphology can be formed for molecular masses of 100k or higher for crystallization after relatively rapid cooling. The map shown in Figure 4.9 makes it possible to obtain various morphologies in PE structures from samples with the same molecular mass by choosing the appropriate crystallization conditions.

Two separate studies by Krumme et al. [46] and Shen et al. [47] on bimodal PE materials revealed that by increasing the molar mass in non-isothermal crystallization, the rate of crystallization first increases, then reaches a peak at the broadest molecular weight distribution, and eventually decreases.

4.3 Nucleation Theory

Crystallization of the PE, like other semicrystalline polymers, follows a nucleation-growth mechanism from polymer melt. The early stage of crystallization process is nucleation, which determines the physical and mechanical properties of the polymers remarkably. There are several descriptions for the nucleation theory, which can be categorized into classical nucleation theory (CNT) and nonclassical nucleation theory. A thorough review of both theories has been conducted by [48].

CNT assumes the similarity of macroscopic properties of the nucleus and the bulk crystal and then goes on to obtain a critical nucleus size in a homogenous nucleation based on the Gibbs free energy variations.

The accuracy of the CNT is low and it fails to quantitatively predict the nucleation theory, especially in the high molecular weight polymers, due to several assumptions, including: (i) existence of the energy barrier in the system, (ii) bulk properties of the nuclei, even if they are significantly small, and (iii) presence of a sharp density interface between the nuclei and surrounding liquid. As a result, several other theories have been proposed to predict the nucleation process, known as the nonclassical nucleation theories, namely diffuse interface theory (DIT) and density function theory (DFT). DIT assumes a density profile in the interface of nucleus and the liquid and also considers the free energy of the interface as a function of nuclei radius (Figure 4.10).

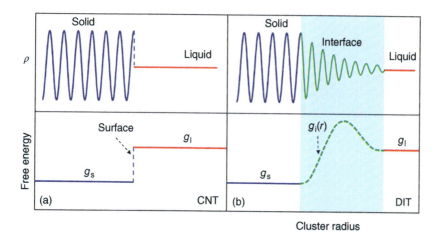

Figure 4.10 The distribution of density ρ and free energy density of interface is a function of radius for (a) classical nucleation theory and (b) diffuse nucleation theory. Source: Reproduced with permission from Tang et al. [48], © 2019, American Chemical Society.

4.4 Crystal Growth

Determination of the crystal growth rate in the semi-crystalline polymers is a complicated issue, which can be overcome by some simplifying assumptions. The most popular and accurate assumption is considering same growth rate in crystal lamellae and radius of the spherulite r. The growth rate of the spherulites can be observed by optical microscope by measuring the propagation of the growth in the front of the spherulite. It has been shown that at constant temperature, the radius of the spherulite grows linearly with temperature. Diagram of growth rate and an example of an isotactic polypropylene (iPP) spherulite growth with time are shown in Figure 4.11.

In semicrystalline polymers, the crystal growth rate vs. temperature graph exhibits a bell-shaped diagram between glass transition temperature (T_g) and melting point (T_m). This graph shows a maximum that occurs at $T_{c,max}$. Same trend for the nucleation rate graph can be observed in the polymers, but in this graph the maximum nucleation rate (I_{max}) occurs in lower temperatures (Figure 4.12). From Figure 4.12, it can be observed that at temperatures close to the T_m, the rate of nucleation is significantly low, and for formation of the nuclei, high supercooling is necessary. By reducing temperature toward T_g, the nucleation facilitates; however, the crystal growth rate encounters barrier of chain immobility, which limits the crystallization. Generally, at $T < T_g$, there is no chain mobility and as a result, growth cannot happen. On the other hand, at $T > T_m$, nuclei are not stable and in the absence of the nuclei, all polymer chains are in the molten state with high degree of mobility [49–51].

Side chains, branched molecular structure, polarity, and chain inflexibility are the common parameters, which can reduce the maximum crystal growth rate significantly. Among all semi-crystalline polymers, PE (non-branched HDPE type), thanks

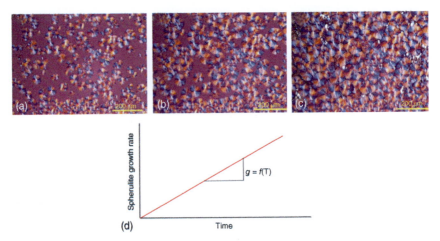

Figure 4.11 A demonstration of spherulite growth rate of an isotactic polypropylene sample seen using POM at (a) 30, (b) 60, and (c) 90 seconds under isothermal conditions at 130 °C. (d) Spherulite radius (r) growth curve at a constant rate ($g = f(T)$).

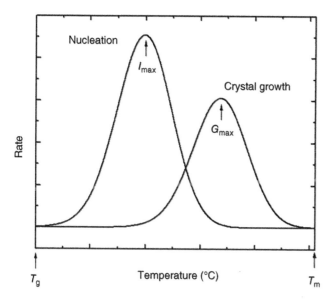

Figure 4.12 Schematic plots for the primary nucleation rate (*I*) and crystal growth rate (*G*) as a function of the isothermal crystallization (or nucleation) temperature. Source: Lorenzo et al. [49]/John Wiley & Sons.

to its linear structure, flexible chains, and lack of side branches, has the highest crystallization rate given in Table 4.4. On the other hand, polycarbonate with the most rigid chain as well as *para*-phenylene groups show the lowest crystal growth rate. Due to extremely low crystal growth rate of polycarbonate, this polymer is commonly marketed as an amorphous polymer. In comparison with PE, the growth rate of the polypropylene (PP) lamellae is 100 times slower. This is due to the presence of

Table 4.4 Characteristic values of the crystallization parameters for some polymers.

Polymer	Maximum crystal growth rate ($\mu m\ s^{-1}$)
Polyethylene (HDPE)	33
Nylon 6,6 (PA 6,6)	20
Nylon 6 (PA 6)	3
Polypropylene (PP)	0.33
PET	0.12
Polystyrene (PS)	0.0042
Polycarbonate (PC)	0.0002

Polyethylene has one of the fastest rates of crystallization among many common thermoplastic materials.
Source: Adapted from Canevarolo [52].

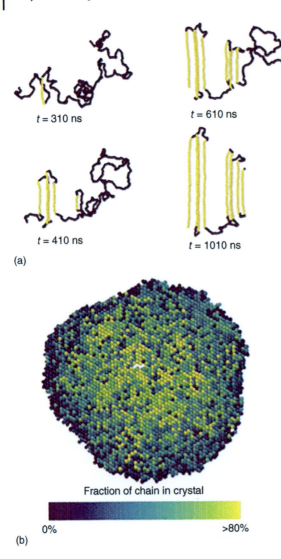

Figure 4.13 (a) Crystallization of chains, snapshots from the trajectory of a chain that eventually becomes completely embedded in a crystal. (b) Stems in a crystal are colored by the crystalline fraction of the chain they belong to. Source: Reproduced from Verho et al. [56]/American Chemical Society/CC BY 4.0.

the methyl side group, which makes its crystallization conformation helical, a more difficult conformation compared to the planar zig-zag type for PE [53–55].

Verho et al. [56] performed large-scale atomistic molecular dynamics simulations in order to investigate the growth of lamellar crystals of PE. In this study, the growth of the PE lamellae was evaluated in two different regions: the growth at the front or edge of lamellae and at the center of lamellae. Individual chain trajectories can provide insight into the formation and growth of lamellae. Figure 4.13a depicts how a chain progressively gets involved in the formation of the crystals. Eventually, the entire polymer chain gets involved in the formation of the crystal. Further stem lengthening is not possible at this point without decreasing the number of stems. As a result, the thickening of the crystal lamella inside the chain's region will be hindered. The crystallinity of chains within a lamella is depicted in Figure 4.13b. The illustration shows that stems toward the edge are more likely to belong to chains with low crystallinity. However, many chains in the core area are practically entirely crystalline and hence have no accessible slack length. The findings of this study help to understand the growth of the PE crystals in the initial stage of the growth at the edge of crystals and subsequent stage at the center of the crystals. It has been observed that at the edge of the crystals, where two crystalline and amorphous phases are exposed to each other, number of the available chains inside amorphous region, which can diffuse into the lamellae is high; as a result, the growth rate is remarkably high. However, in the central region, the crystal's growth is delayed due to a lack of available new chains. Also, it has been shown that the crystallinity is different in these two regions. In the center of the lamella, where there are no accessible free chains, the crystallinity of the chains can be as high as 90% (Figure 4.13b) [56].

4.5 PE Blends and Co-crystallization

Blending is one of the common techniques in developing polymers with high performances for various applications [57]. In blending of semi-crystalline polymers, co-crystallization is a rare phenomenon, which needs to be considered as a secondary crystallization process [58] and refers to crystallization of the blend component in same lamella [59]. The co-crystallization phenomenon requires several conditions, including high miscibility between the blend components in molten state, similarity in the crystallization rates, similarity in the molecular structures, and similarity in the crystalline lattice structures [58–60]. Among these, first two conditions are related to the kinetic accessibility required to build co-crystals, while the other two are thermodynamically required for stability of the formed co-crystals in the structure. Due to these relatively hard-to-meet requirements, co-crystallization is not a common phenomenon in polymer blends. Co-crystallization has been observed only for a few blends of polymers, including blends of PE [13, 61], fluorocarbon polymers [62, 63], poly(3-alkylthiophene)s [64], and poly(aryl ether ketone)s [65, 66], among which the copolymerization in the blends of different grades of PE is the most common case.

Researchers have utilized a variety of techniques to investigate the co-crystallization behavior of different PE blends [67–71]. The effect of side chain branch content and distribution on the co-crystallization in polymer blends was investigated by Rana [72] and Zhao et al. [73] via small-angle X-ray scattering (SAXS), differential scanning calorimetry (DSC), and wide-angle X-ray scattering (WAXS) characterization methods. Thermal fractionation technology has provided a novel approach to examine co-crystallization; these techniques include temperature-rising elution fractionation (TREF) [74, 75], crystallization analysis fractionation (CRYSTAF) [74–76], step crystallization (SC) [75, 77], and successive self-nucleation annealing (SSA) [75, 78, 79]. The TREF process separates the part of material that has been previously crystallized from solution on an inert substrate during very slow cooling or multiple steps by elution of polymer fractions at successively rising temperatures. The principle of CRYSTAF is similar to that of TREF, but with shorter experimental time by applying different procedures [80]. Anantawaraskul et al. [69] used CRYSTAF to investigate the crystallization behavior of four different types of PE with different comonomers and discovered that the cooling rate and chain crystallizability resemblance were the most significant factors regulating co-crystallization during CRYSTAF analysis.

SC and SSA techniques were established on the basis of DSC, as contrasted to TREF and CRYSTAF. The methylene sequence length (MSL) is used to determine the crystallization temperature of the chain segment in SC and SSA. Longer methylene chain segments will readily arrange themselves into a crystal lattice to build thicker lamellae at higher temperatures [77, 79, 81]. As a result, MSL is the most important factor affecting these fractionation techniques. Both intermolecular and intramolecular structural inhomogeneity can be monitored by these technologies. Furthermore, as compared to TREF and CRYSTAF, DSC-based thermal fractionation techniques offer remarkable benefits in terms of not using solvents, greater resolution, and shorter experimental times [79, 81]. The SSA technique may fractionate a polymer sample in substantially less time and with much higher resolution than other well-known thermal fractionation methods, such as SC. Lorenzo et al. [82] have demonstrated that using high scanning rates (e.g. 50 °C min^{-1}) may significantly minimize the experimental duration of the SSA technique. Co-crystallization behavior has been studied using both SC and SSA. Arnal et al. [68] examined the crystallization behavior of various HDPE/LLDPE blends in depth and concluded that the effects of nucleation, re-organization, and melting point depression (dilution effects) caused by one type of PE on the other tend to produce an overlap of the crystallization and melting endotherms in conventional DSC cooling and heating scans, which may lead to partial miscibility interpretations. Even if the nucleation and dilution effects remain after SSA, the thermal fractionation process can separate them from co-crystallization by comparing the relative quantities of the thermal fractions formed by SSA.

In separate studies conducted by Tashiro et al. [83–85] and Cho et al. [86], it has been observed that the blends of HDPE and LLDPE showed miscibility in both the melt and the crystalline phase, which indicates that these two polymers can form co-crystallization. However, Lee et al. [87] and Cho et al. [86] showed that in LDPE/LLDPE blend systems, the polymers are miscible in melt phase and immiscible in crystalline phase; as a consequence, no co-crystallization has been reported in these blends [86]. Rana and coworkers [71] have pointed out that the crystalline phase is very ordered and selective, where only linear polymeric chain can be involved in its formation. Therefore, any small variation in the chemical structure of the PE chains may result in rejection of that chain from participating in lamellae formation. Co-crystallization occurs when the chains of both components of the PE blend are linear and long enough to be able to contribute to the formation of a single lamella.

Eslamian et al. [13] adopted DSC, SAXS, WAXS, and SEM characterization techniques to investigate the co-polymerization in ternary polymeric systems. In this study, HDPE, LDPE, and LLDPE blends with various ratios have been used. It has been shown that LDPE and LLDPE form tiny co-crystals, a so-called tie-crystal, which is located in the amorphous region between the large lamellae of the HDPE and LLDPE polymers. Figure 4.14a,b confirms the presence of these tiny co-crystals. Moreover, in the SEM image of the HDPE and ternary blends samples (after etching to remove the amorphous region), which are illustrated in Figure 4.14c and d, respectively, it is shown that in the ternary blend of the PE samples, amorphous region contains tiny crystals that are removed in etching process. Another indicator of the presence of the tiny co-crystals between the large crystals is the mechanical properties of the samples. It is shown that in the ternary blend samples, the tiny co-crystals strengthen the amorphous region and play a major role in the mechanical performance of the blend. According to that hypothesis, tiny co-crystals are dispersed within the amorphous region between two adjacent coarse HD-rich co-crystals (Figure 4.14e).

Sun et al. [77] adopted SSA technique via DSC to investigate the co-crystallization in the blend of HDPE and LLDPE. In this study, blends with different compositions were subjected to SSA cycles. The diagram for final heating cycle of the samples after the SSA process is shown in Figure 4.15. In Figure 4.15, it can be observed that because of the negligible amount of short chain branches (SCB) on the HDPE sample, the polymer chains cannot be fractionated in the SSA cycles. On the other hand, the LLDPE contains large amount of SCB in its molecular structure. In the DSC graph of the melting of the samples after SSA analysis, each peak that shows up at a certain temperature is an indicator for a fraction of materials with same methylene chain segments length and same SCB content in their molecular structure [43, 78]. The highest melting temperature relates to the longest successive methylene chain segments with negligible branching, and the lowest represents segments in which the regularity of methylene sequences was destroyed and the crystals have less perfection caused by SCB. In this study, when HDPE was added to the blend system in sufficient content to increase the number of melting peaks, co-crystallization most likely occurred to a large extent because the interaction between HDPE and LLDPE molecules resulted in the formation of a new fraction with an intermediate lamellar thickness. As the composition of the blend changed, the thickness of the original fraction decreased slightly as a new thermal fraction with much thicker lamella appeared, indicating that some long linear methylene segments were affected by HDPE and transferred from the original fraction to the newly appeared fraction.

Krishnaswamy et al. [88] evaluated the effect of 1-hexene short-chain branches in blends of PE comprising short and long chain fractions and the positioning of the branches either on short or long chains. The results indicated that when the SCB is located on the long chains, their growth in crystal domain becomes significantly hindered compared to rapid motion of shorter chains, which occupy nearby crystal sites on the growth front. This forces the long chains to either enter somewhere else in the same lamella, or enter a new growing lamella, effectively establishing tie molecules and consequently improving the mechanical properties and resistance to SCG significantly. The above phenomenon is schematically represented in Figure 4.16a.

The establishment of more tie molecules due to placement of SCBs on longer chains (sample L_b/S) exhibits itself in force–displacement curves of the two samples shown in Figure 4.16b, where the sample with SCBs placed on its longer chains

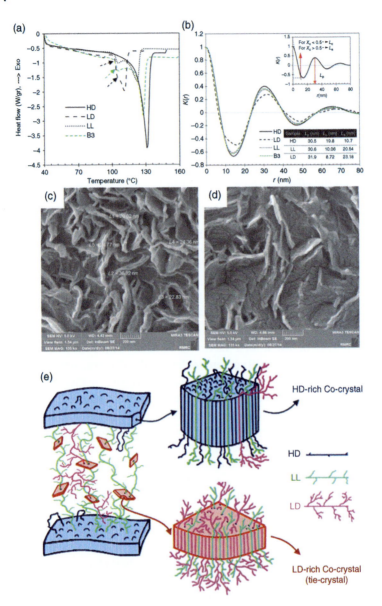

Figure 4.14 (a) DSC thermogram of the LDPE/LLDPE/HDPE blend on the pure sample curves. (b) The one-dimensional correlation function graphs of the pure and ternary blend samples. The top inset represents the method by which the long period (L_p), the lamellar thickness (L_c), and the thickness of the amorphous layer (L_a) are calculated. Also, the values of L_p, L_c, and L_a of the pure samples are given in the tabular inset. FE-SEM images of the cross-section of cryogenically fractured and etched samples of (c) pure HDPE and (d) ternary blend. (e) Schematic diagram of the crystalline morphology of the blends. The large crystals are HD-rich lamellae, and the tiny crystals, which are dispersed within the amorphous region between the large lamellae, are LD-rich co-crystals. Source: Eslamian et al. [13]/John Wiley & Sons.

Figure 4.15 DSC heating curves of all melt-mixed samples after SSA thermal fractionation. Source: Reproduced with permission from Sun et al. [77], © 2013, Taylor & Francis Group.

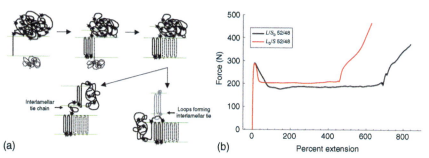

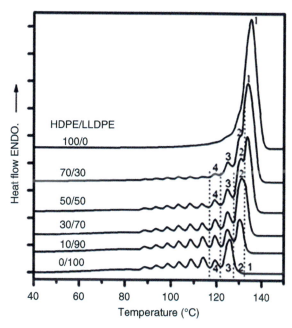

Figure 4.16 (a) Schematics of a polyethylene blend in which SCBs are located on the long chains. The figure demonstrates that easier crystallization of shorter chains (dotted curves) forces the long chains to either form loop chain in the same lamella or interlamellar tie chains between two different lamellae. (b) Load–displacement curve to compare the placement of SCBs on short chains or long chains. Source: Reproduced with permission from [88].

shows both a higher plateau load and a lower natural draw ratio, meaning the force required to yield the sample beyond a certain point is higher for the L_b/S sample. This makes the natural draw ratio and tie molecule concentration an effective indicator of long-term performance of PE material.

A study by Song et al. [33] found that addition of UHMWPE to bimodal pipe grade PE100 causes enhancement in crystallization kinetics of the HDPE material, in which UHMWPE segments effectively become nucleation sites, leading to both thicker lamellae and enhanced crystallization kinetics in isothermal crystallization. The fold surface energy of neat HDPE was shown to be reduced with addition of UHMWPE to the matrix. On the lower side of molecular weight spectrum, a study by [89] in commercial bimodal PE100, with addition of a lower molecular weight PE, thickness and amount of thick crystals in the structure increase, while overall crystallinity remains almost the same, which in turn improves the fracture toughness of the blends.

On the issue of blending PE, it is famously known that PP and PE are mostly considered thermodynamically immiscible unless a proper compatibilizer is employed [90, 91], thus their blend usually shows inferior properties compared to their respective homopolymers. Compared to PP, PE has a much higher crystallization rate, but a study by Carmeli et al. [92] has shown that depending on the cooling rate and type of PP used in PE/PP blends, i.e. nucleated or non-nucleated PP, an *inversion point* can be determined at different temperatures in continuous cooling curves (CCCs) diagrams of different PE/PP blends. Before the inversion point, i.e. at lower cooling rates, a nucleated PP can crystallize before the PE, while after it, PE crystallizes faster. Depending on which component of the blend crystallizes faster, the interface of the first one to crystallize can act as nucleating site for epitaxial growth of the second, slower component. Figure 4.17a shows the CCC diagram of the PE/PP blend and the corresponding morphological SEM images before and after inversion point.

It is clear that before the inversion point, at which PP crystallizes faster than PE, a transcrystalline layer of PE lamellae with preferred direction of growth has

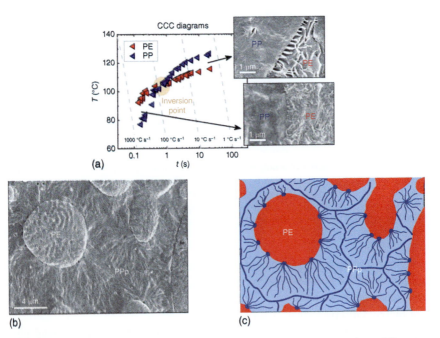

Figure 4.17 (a) A continuous cooling curve (CCC) diagram for PE/PP blends at different cooling rates and corresponding SEM images for samples prepared before and after inversion point. (b) SEM image of PE blended with non-nucleated PP (PPp), showing PE domains and surrounding PP areas and (c) Schematic representation of (b), where nucleation sites of spherulites are shown on the interface of crystallized PE domain. Source: Carmeli et al. [92]/American Chemical Society/CC BY 4.0.

nucleated from the already crystallized PP domain. Figure 4.17b shows an SEM image of the morphology of immiscible PP/PE blend after the inversion point, where the spherulites of PP have nucleated at the interface of solid PE domains, while the lamella bundles of the nucleated spherulites grow until they impinge on neighboring spherulites. Schematics of this phenomenon are represented in Figure 4.17c.

4.6 PE Nanocomposites

While PE is one of the most versatile mass-production polymers on the market with some excellent characteristics, there are still some properties that may have room for improvement, e.g. mechanical properties, electrical properties, optical properties, magnetic properties, medical properties, etc.

Making use of PE nanocomposites is one of the popular approaches in the research for enhancement of properties of PE in various areas. PE is one of the most notable matrices used in manufacturing and processing of thermoplastic-based nanocomposites. As discussed earlier in the chapter, the crystallization behavior of PE and subsequent semi-crystalline structure of the material can have a significant effect on final properties of the polymer. Herein, incorporating nano-sized fillers into the PE can affect the crystallization behavior and crystalline structure, as it has been thoroughly investigated in the literature [37, 40, 93–96]. Various types of nanofillers have been used in different studies for different applications, including, but not limited to: graphene and graphene oxide, carbon nanotubes, cellulose whiskers, copper nanoparticles, nanosilica, nanoclay, and starch.

In a thorough study of crystallization kinetics of nanocomposites with semi-crystalline polymer matrix by [97], authors sum up the behavior of growing polymer crystals and nanoparticles by dividing the nanoparticles into two types: mobile and immobile nanoparticles. Generally, one-dimensional particles like carbon nanotubes, two-dimensional plates like graphene, and sheet-like particles like clay are considered "immobile" relative to the growth rate of the crystal front in polymer matrix, while tiny spherical particles in order of 10–100 nm can be considered mobile depending on the crystallization growth rate. This means that if the average ratio of timescale for motion of nanofillers to that of the growing crystal (Péclet number) is much bigger than 1 (Figure 4.18a), the nanoparticles are considered immobile and they will naturally get trapped in the fast-moving crystal domains. But if the ratio is close to or less than 1, particles are considered mobile, meaning they can be pushed by the growth front of the crystal (Figure 4.18b).

Based on review of various studies, if the nanofiller content is less than a critical value, the particles can effectively act as heterogenous nucleating sites, while above this critical value, the particles can actually confine the polymer chains and reduce their mobility, effectively reducing T_c. This phenomenon is shown in Figure 4.18c.

Figure 4.18d shows that for mobile nanofillers, depending on rate of crystallization, nanofillers can either remain well dispersed in case of very fast crystallization, while getting assembled in sheet formation in interlamellar space at slow crystallization and finally, they get pushed into inter-spherulitic spaces in the case of very slow crystallization rates.

It is a well-established observation that various nano-sized fillers in PE can have a considerable nucleating effect in the crystallization process, which in turn can affect the final structure of the nanocomposite material [98–101]. Although there is a near consensus on nucleating effect of nanofillers in the PE, various studies report different results on final crystalline structure, overall crystallinity, and crystal size depending on type and size of the filler used. One study [99] confirms that although graphene nanoplatelets do act as nucleation sites in the HDPE matrix, they can slightly decrease overall crystallinity and make crystallization more difficult, especially at its final stages, due to nanoparticles hindering the chain mobility, and this phenomenon is intensified with increasing the diameter and size of the nanoplatelets.

Another study by Yuan et al. [98] on PE/Nanoclay found that although adding nanoclay causes the nucleating effect by nanoparticles through reducing free energy barrier and also crystallization half-time, it can, on the other hand, cause crystal growth retardation by increasing the activation energy compared to neat PE matrix.

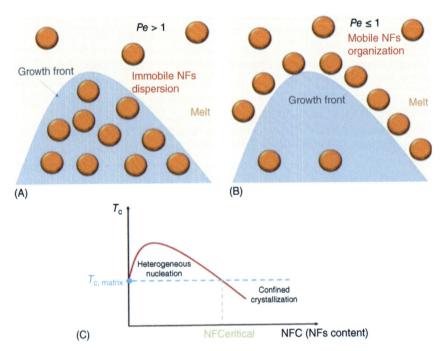

Figure 4.18 (a) Assembly situation of immobile nanofillers getting trapped at the crystal domain. (b) Arrangement of mobile nanofillers in the growth front of the crystal. (c) Behavior of crystallization temperature as a function of nanofiller content. (d) The distribution of mobile nanofillers after the crystallization of the polymer matrix depending on the rate of crystallization. Source: Altorbaq et al. [97]/with permission of Elsevier.

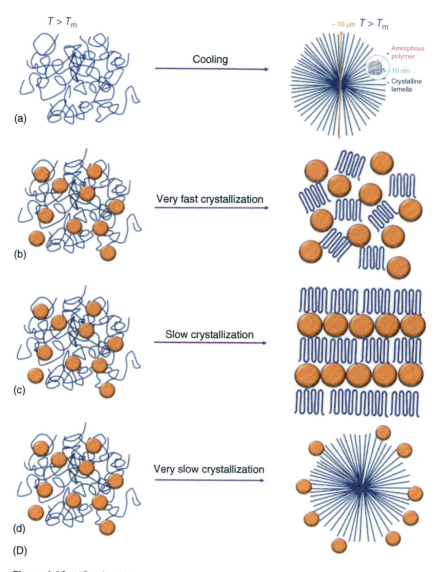

Figure 4.18 (Continued)

A study by Depan et al. [102] on PE/carbon nanotubes confirmed yet again the nucleating agent role of CNT nanofiller, as immobile particles, in the PE matrix. The nanotubes also changed the crystal morphology from spherulitic to disk-shaped, causing the crystals to grow laterally along the nanotube axis, effectively forming a nanohybrid shish-kebab structure. Figure 4.19 shows SEM images of the shish-kebab structure formed along the CNT axis for both LDPE and HDPE nanocomposites.

Figure 4.19 SEM micrographs of LDPE/CNT (a, b) and HDPE/CNT (c, d) nanocomposites and the shish–kebab crystal structure grown along carbon nanotube axis. Source: Depan et al. [102]/John Wiley & Sons.

4.7 Summary

This chapter provides an overview of the fundamental aspects of polyethylene crystallization. The various types of polyethylene structures, their crystallization kinetics and nucleation theory, and the factors affecting their crystallization behavior are discussed. The co-crystallization behavior of HDPE and its blends with other polyethylene types, as well as the influence of co-monomer distribution and molecular weight distribution on the mechanical properties of polyethylene are also explored. Additionally, the chapter presents a brief overview of some characterization techniques employed by different studies which are suitable for studying the crystalline structure and morphology of polyethylene. Finally, the effect of nanoparticles on nucleation theory and crystallization kinetics of polyethylene-based nanocomposites is discussed, providing an early understanding of how nanoparticles can affect the crystallization behavior of polyethylene.

References

1 Spalding, M.A. and Chatterjee, A. (2017). *Handbook of Industrial Polyethylene and Technology: Definitive Guide to Manufacturing, Properties, Processing, Applications and Markets Set*. Wiley.

2 Peacock, A. (2000). *Handbook of Polyethylene: Structures, Properties and Applications*. New York: Taylor & Francis.
3 Satkowski, M.M. (1990). The crystallization and morphology of polyethylene and its blends. Doctoral dissertations. University of Massachusetts Amherst.
4 Ragaert, K., Delva, L., van Damme, N. et al. (2016). Microstructural foundations of the strength and resilience of LLDPE artificial turf yarn. *Journal of Applied Polymer Science* 133 (43): 1–12.
5 Allegra, G. (ed.) (2005). *Interphases and Mesophases in Polymer Crystallization I*, vol. 180.
6 Kissin, Y.V. (2012). *Polyethylene: End-Use Properties and Their Physical Meaning*. Hanser.
7 Patel, R.M., Jain, P., Story, B., and Chum, S. (2008). Polyethylene: an account of scientific discovery and industrial innovations. In: *Innovations in Industrial and Engineering Chemistry* (ed. W.H. Flank, M.A. Abraham, and M.A. Matthews), 71–102. ACS Publications.
8 Zhang, H., Zhao, S., Xin, Z. et al. (2019). Wear resistance mechanism of ultrahigh-molecular-weight polyethylene determined from its structure–property relationships. *Industrial and Engineering Chemistry Research* 58 (42): 19519–19530.
9 Zhang, Z. and Ren, S. (2019). Functional gradient ultrahigh molecular weight polyethylene for impact-resistant armor. *ACS Applied Polymer Materials* 1 (8): 2197–2203.
10 Nowlin, T.E. (2014). *Business and Technology of the Global Polyethylene Industry*. Hoboken, NJ: Wiley.
11 Zhai, Z., Morthomas, J., Fusco, C. et al. (2019). Crystallization and molecular topology of linear semicrystalline polymers: simulation of uni- and bimodal molecular weight distribution systems. *Macromolecules* 52 (11): 4196–4208.
12 Koziol, P., Kosowska, K., Liberda, D. et al. (2022). Super-resolved 3D mapping of molecular orientation with vibrational techniques. *Journal of the American Chemical Society* 144 (31): 14278–14287.
13 Eslamian, M., Bagheri, R., and Pircheraghi, G. (2016). Co-crystallization in ternary polyethylene blends: tie crystal formation and mechanical properties improvement. *Polymer International* 65 (12): 1405–1416.
14 Melk, L. and Emami, N. (2018). Mechanical and thermal performances of UHMWPE blended vitamin E reinforced carbon nanoparticle composites. *Composites Part B: Engineering* 146: 20–27.
15 Stehling, F.C. and Mandelkern, L. (1970). The glass temperature of linear polyethylene. *Macromolecules* 3 (2): 242–252.
16 Avrami, M. (2004). Kinetics of phase change. II Transformation–time relations for random distribution of nuclei. *Journal of Chemical Physics* 8 (2): 212.
17 Avrami, M. (1939). Kinetics of phase change. I General theory. *Journal of Chemical Physics* 7 (12): 1103–1112.
18 Lorenzo, A.T., Arnal, M.L., Albuerne, J., and Müller, A.J. (2007). DSC isothermal polymer crystallization kinetics measurements and the use of the Avrami equation to fit the data: guidelines to avoid common problems. *Polymer Testing* 26 (2): 222–231.

19 Ziabicki, A. (1967). Kinetics of polymer crystallization and molecular orientation in the course of melt spinning. *Applied Polymer Symposia*, 6 (1), 1–18.

20 Nakamura, K., Watanabe, T., Katayama, K., and Amano, T. (1972). Some aspects of nonisothermal crystallization of polymers. I. Relationship between crystallization temperature, crystallinity, and cooling conditions. *Journal of Applied Polymer Science* 16 (5): 1077–1091.

21 Nakamura, K., Katayama, K., and Amano, T. (1973). Some aspects of non-isothermal crystallization of polymers. II. Consideration of the isokinetic condition. *Journal of Applied Polymer Science* 17 (4): 1031–1041.

22 Patel, R.M., Bheda, J.H., and Spruiell, J.E. (1991). Dynamics and structure development during high-speed melt spinning of nylon 6. II. Mathematical modeling. *Journal of Applied Polymer Science* 42 (6): 1671–1682.

23 Ozawa, T. (1971). Kinetics of non-isothermal crystallization. *Polymer (Guildford)* 12 (3): 150–158.

24 Liu, T., Mo, Z., Wang, S., and Zhang, H. (1997). Nonisothermal melt and cold crystallization kinetics of poly(aryl ether ether ketone ketone). *Polymer Engineering and Science* 37 (3): 568–575.

25 Dietz, W. (1981). Sphärolithwachstum in Polymeren. *Colloid and Polymer Science* 259 (4): 413–429.

26 Lim, G.B.A., McGuire, K.S., and Lloyd, D.R. (1993). Non-isothermal crystallization of isotactic polypropylene in dotriacontane. II: Effects of dilution, cooling rate, and nucleating agent addition on growth rate. *Polymer Engineering and Science* 33 (9): 537–542.

27 Kissinger, H.E. (1956). Variation of peak temperature with heating rate in differential thermal analysis. *Journal of Research of the National Bureau of Standards* 57 (4): 217–221.

28 Nadkarni, V.M., Bulakh, N.N., and Jog, J.P. (1993). Assessing polymer crystallizability from nonisothermal crystallization behavior. *Advances in Polymer Technology* 12 (1): 73–79.

29 Gao, R., He, X., Zhang, H. et al. (2016). Molecular dynamics study of the isothermal crystallization mechanism of polyethylene chain: the combined effects of chain length and temperature. *Journal of Molecular Modeling* 22 (3): 1–12.

30 Hubert, L., David, L., Seguela, R. et al. (2001). Physical and mechanical properties of polyethylene for pipes in relation to molecular architecture. I. Microstructure and crystallisation kinetics. *Polymer (Guildford)* 42 (20): 8425–8434.

31 Matsui, K., Seno, S., Nozue, Y. et al. (2013). Influence of branch incorporation into the lamella crystal on the crystallization behavior of polyethylene with precisely spaced branches. *Macromolecules* 46 (11): 4438–4446.

32 Kamal, M.R. and Chu, E. (1983). Isothermal and nonisothermal crystallization of polyethylene. *Polymer Engineering and Science* 23 (1): 27–31.

33 Song, S., Wu, P., Ye, M. et al. (2008). Effect of small amount of ultra high molecular weight component on the crystallization behaviors of bimodal high density polyethylene. *Polymer (Guildford)* 49 (12): 2964–2973.

34 Puig, C.C. (2001). Enhanced crystallisation in branched polyethylenes when blended with linear polyethylene. *Polymer (Guildford)* 42 (15): 6579–6585.

35 Chen, F., Shanks, R.A., and Amarasinghe, G. (2001). Crystallisation of single-site polyethylene blends investigated by thermal fractionation techniques. *Polymer (Guildford)* 42 (10): 4579–4587.

36 Hill, M.J., Barham, P.J., and Keller, A. (1992). Phase segregation in blends of linear with branched polyethylene: the effect of varying the molecular weight of the linear polymer. *Polymer (Guildford)* 33 (12): 2530–2541.

37 Trujillo, M., Arnal, M.L., Mu, A.J. et al. (2008). Thermal fractionation and isothermal crystallization of polyethylene nanocomposites prepared by in situ polymerization. *Macromolecules* 41 (6): 2087–2095.

38 Zou, P., Tang, S., Fu, Z., and Xiong, H. (2009). Isothermal and non-isothermal crystallization kinetics of modified rape straw flour/high-density polyethylene composites. *International Journal of Thermal Sciences* 48 (4): 837–846.

39 Vega, J.F., da Silva, Y., Vicente-Alique, E. et al. (2014). Influence of chain branching and molecular weight on melt rheology and crystallization of polyethylene/carbon nanotube nanocomposites. *Macromolecules* 47 (16): 5668–5681.

40 Amigo, N., Palza, H., Canales, D. et al. (2019). Effect of starch nanoparticles on the crystallization kinetics and photodegradation of high density polyethylene. *Composites Part B: Engineering* 174 (May): 106979.

41 Xu, J.T., Zhao, Y.Q., Wang, Q., and Fan, Z.Q. (2005). Isothermal crystallization of intercalated and exfoliated polyethylene/montmorillonite nanocomposites prepared by in situ polymerization. *Polymer (Guildford)* 46 (25): 11978–11985.

42 Sarafpour, A., Pircheraghi, G., Rashedi, R., and Afzali, K. (2018). Correlation between isothermal crystallization and morphological/rheological properties of bimodal polyethylene/carbon black systems. *Polymer Crystallization* 1 (3): e10014.

43 Gholami, F., Pircheraghi, G., Rashedi, R., and Sepahi, A. (2019). Correlation between isothermal crystallization properties and slow crack growth resistance of polyethylene pipe materials. *Polymer Testing* 80: 106128.

44 Supaphol, P. and Spruiell, J.E. (1998). Nonisothermal bulk crystallization studies of high density polyethylene using light depolarizing microscopy. *Journal of Polymer Science Part B: Polymer Physics* 36: 681–692.

45 Mandelkern, L., Glotin, M., and Benson, R.A. (1981). Supermolecular structure and thermodynamic properties of linear and branched polyethylenes under rapid crystallization conditions. *Macromolecules* 14 (1): 22–34.

46 Krumme, A., Lehtinen, A., and Viikna, A. (2004). Crystallisation behaviour of high density polyethylene blends with bimodal molar mass distribution: 2. Non-isothermal crystallisation. *European Polymer Journal* 40 (2): 371–378.

47 Shen, H., Xie, B., Yang, W., and Yang, M. (2013). Non-isothermal crystallization of polyethylene blends with bimodal molecular weight distribution. *Polymer Testing* 32 (8): 1385–1391.

48 Tang, X., Chen, W., and Li, L. (2019). The tough journey of polymer crystallization: battling with chain flexibility and connectivity. *Macromolecules* 52 (10): 3575–3591.

49 Lorenzo, A.T. and Müller, A.J. (2008). Estimation of the nucleation and crystal growth contributions to the overall crystallization energy barrier. *Journal of Polymer Science Part B: Polymer Physics* 46 (14): 1478–1487.

50 Reiter, G. and Strobl, G.R. (2007). *Progress in Understanding of Polymer Crystallization*. Berlin Heidelberg: Springer.

51 Schmelzer, J.W.P., Abyzov, A.S., Fokin, V.M. et al. (2015). Crystallization of glass-forming liquids: maxima of nucleation, growth, and overall crystallization rates. *Journal of Non-Crystalline Solids* 429: 24–32.

52 Canevarolo, S.V. (2019). Polymer crystallization kinetics. In: *Polymer Science, A Textbook for Engineers and Technologists*, 219–236. München: Carl Hanser Verlag GmbH & Co. KG.

53 Schulz, M., Schäfer, M., Saalwächter, K., and Thurn-Albrecht, T. (2022). Competition between crystal growth and intracrystalline chain diffusion determines the lamellar thickness in semicrystalline polymers. *Nature Communications* 13 (1): 1–10.

54 Li, M., Leenaers, P.J., Wienk, M.M., and Janssen, R.A.J. (2020). The effect of alkyl side chain length on the formation of two semi-crystalline phases in low band gap conjugated polymers. *Journal of Materials Chemistry C* 8 (17): 5856–5867.

55 Canevarolo, S.V. (2019). *Polymer Science*. Carl Hanser Verlag GmbH & Co. KG.

56 Verho, T., Paajanen, A., Vaari, J., and Laukkanen, A. (2018). Crystal growth in polyethylene by molecular dynamics: the crystal edge and lamellar thickness. *Macromolecules* 51 (13): 4865–4873.

57 Utracki, L.A. and Wilkie, C.A. (2014). *Polymer Blends Handbook*. Netherlands: Springer.

58 Yoshie, N., Asaka, A., and Inoue, Y. (2004). Cocrystallization and phase segregation in crystalline/crystalline polymer blends of bacterial copolyesters. *Macromolecules* 37 (10): 3770–3779.

59 Runt, J. and Huang, J. (2002). Polymer blends and copolymers. In: *Handbook of Thermal Analysis and Calorimetry*, vol. 3 (ed. S.Z.D. Cheng), 273–294.

60 Cherukuvada, S., Kaur, R., and Guru Row, T.N. (2016). Co-crystallization and small molecule crystal form diversity: from pharmaceutical to materials applications. *CrystEngComm* 18 (44): 8528–8555.

61 Tashiro, K., Imanish, K., Izuchi, M. et al. (1995). Cocrystallization and phase segregation of polyethylene blends between the D and H species. 8. Small-angle neutron scattering study of the molten state and the structural relationship of chains between the melt and the crystalline state. *Macromolecules* 28 (25): 8484–8490.

62 Runt, J., Jin, L., Talibuddin, S., and Davis, C.R. (1995). Crystalline homopolymer–copolymer blends: poly(tetrafluoroethylene)–poly(tetrafluoroethylene-*co*-perfluoroalkylvinyl ether). *Macromolecules* 28 (8): 2781–2786.

63 Datta, J. and Nandi, A.K. (1998). Cocrystallisation in blends of poly(vinylidene fluoride) samples: 4. Kinetics study. *Polymer (Guildford)* 39 (10): 1921–1927.

64 Pal, S. and Nandi, A.K. (2003). Cocrystallization behavior of poly(3-alkylthiophenes): influence of alkyl chain length and head to tail regioregularity. *Macromolecules* 36 (22): 8426–8432.

65 Harris, J.E. and Robeson, L.M. (1987). Isomorphic behavior of poly(aryl ether ketone) blends. *Journal of Polymer Science Part B: Polymer Physics* 25 (2): 311–323.

66 Sham, C.K., Guerra, G., Karasz, F.E., and MacKnight, W.J. (1988). Blends of two poly(aryl ether ketones). *Polymer (Guildford)* 29 (6): 1016–1020.

67 Cedeño, A. and Puig, C.C. (2006). Isothermal lamellar thickening in blends of linear and low-density polyethylene. *Journal of Macromolecular Science, Part B Physics* https://doi.org/10.1080/00222349908248116 38 (5): 505–514.

68 Arnal, M.L., Sánchez, J.J., and Müller, A.J. (2001). Miscibility of linear and branched polyethylene blends by thermal fractionation: use of the successive self-nucleation and annealing (SSA) technique. *Polymer (Guildford)* 42 (16): 6877–6890.

69 Anantawaraskul, S., Soares, J.B.P., and Wood-Adams, P.M. (2004). Cocrystallization of blends of ethylene/1-olefin copolymers: an investigation with crystallization analysis fractionation (CRYSTAF). *Macromolecular Chemistry and Physics* 205 (6): 771–777.

70 Shanks, R.A. and Amarasinghe, G. (2000). Crystallisation of blends of LLDPE with branched VLDPE. *Polymer (Guildford)* 41 (12): 4579–4587.

71 Gupta, A.K., Rana, S.K., and Deopura, B.L. (1994). Crystallization kinetics of high-density polyethylene/linear low-density polyethylene blend. *Journal of Applied Polymer Science* 51 (2): 231–239.

72 Rana, S.K. (1998). Crystallization kinetics of high-density polyethylene/linear low-density polyethylene blend. *Journal of Applied Polymer Science* 51 (2): 231–239.

73 Zhao, Y., Liu, S., and Yang, D. (1997). Crystallization behavior of blends of high-density polyethylene with novel linear low-density polyethylene. *Macromolecular Chemistry and Physics* 198 (5): 1427–1436.

74 Santonja-Blasco, L. and Monrabal, B. (2020). Assessment of the cocrystallization of ethylene/1-octene copolymer blends by CRYSTAF and TREF. *Macromolecular Symposia* 390 (1): 1900123.

75 Sweed, M. and Mallon, P.E. (2006). Co-crystallization in polyolefin blends studied by various crystallization analysis techniques. MSc thesis. University of Stellenbosch.

76 Bruaseth, I., Soares, J.B.P., and Rytter, E. (2004). Crystallization analysis fractionation of ethene/1-hexene copolymers made with the MAO-activated dual-site (1,2,4-Me$_3$Cp)$_2$ZrCl$_2$ and (Me$_5$Cp)$_2$ZrCl$_2$ system. *Polymer (Guildford)* 45 (23): 7853–7861.

77 Sun, X., Shen, G., Shen, H. et al. (2013). Co-crystallization of blends of high-density polyethylene with linear low-density polyethylene: an investigation with successive self-nucleation and annealing (SSA) technique. *Journal of*

Macromolecular Science, Part B Physics, https://doi.org/10.1080/00222348.2013 .768504 52 (10): 1372–1387.

78 Gholami, F., Pircheraghi, G., and Sarafpour, A. (2020). Long-term mechanical performance of polyethylene pipe materials in presence of carbon black masterbatch with different carriers. *Polymer Testing* 91: 106857.

79 Müller, A.J., Hernández, Z.H., Arnal, M.L., and Sánchez, J.J. (1997). Successive self-nucleation/annealing (SSA): a novel technique to study molecular segregation during crystallization. *Polymer Bulletin* 39 (4): 465–472.

80 Monrabal, B. (2006). Microstructure characterization of polyolefins. TREF and CRYSTAF. *Studies in Surface Science and Catalysis* 161: 35–42.

81 Müller, A.J., Michell, R.M., Pérez, R.A., and Lorenzo, A.T. (2015). Successive self-nucleation and annealing (SSA): correct design of thermal protocol and applications. *European Polymer Journal* 65: 132–154.

82 Lorenzo, A.T., Arnal, M.L., Müller, A.J. et al. (2006). High speed SSA thermal fractionation and limitations to the determination of lamellar sizes and their distributions. *Macromolecular Chemistry and Physics* 207 (1): 39–49.

83 Tashiro, K., Satkowski, M.M., Stein, R.S. et al. (1992). Cocrystallization and phase segregation of polyethylene blends. 2. Synchrotron-sourced X-ray scattering and small-angle light scattering study of the blends between the D and H species. *Macromolecules* 25 (6): 1809–1815.

84 Tashiro, K., Izuchi, M., Kobayashi, M., and Stein, R.S. (1994). Cocrystallization and phase segregation of polyethylene blends between the D and H species. 3. Blend content dependence of the crystallization behavior. *Macromolecules* 27 (5): 1221–1227.

85 Tashiro, K., Izuchi, M., Kaneuchi, F. et al. (1994). Cocrystallization and phase segregation of polyethylene blends between the D and H species. 6. Time-resolved FTIR measurements for studying the crystallization kinetics of the blends under isothermal conditions. *Macromolecules* 27 (5): 1240–1244.

86 Cho, K., Lee, B.H., Hwang, K.M. et al. (1998). Rheological and mechanical properties in polyethylene blends. *Polymer Engineering and Science* 38 (12): 1969–1975.

87 Lee, H., Cho, K., Ahn, T.-K. et al. (1997). Solid-state relaxations in linear low-density (1-octene comonomer), low-density, and high-density polyethylene blends. *Journal of Polymer Science Part B: Polymer Physics* 35: 1633–1642.

88 Krishnaswamy, R.K., Yang, Q., And, L.F.-B., and Kornfield, J.A. (2008). Effect of the distribution of short-chain branches on crystallization kinetics and mechanical properties of high-density polyethylene. *Macromolecules* 41 (5): 1693–1704.

89 Sun, X., Shen, H., Xie, B. et al. (2011). Fracture behavior of bimodal polyethylene: effect of molecular weight distribution characteristics. *Polymer (Guildford)* 52 (2): 564–570.

90 Vervoort, S., den Doelder, J., Tocha, E. et al. (2018). Compatibilization of polypropylene–polyethylene blends. *Polymer Engineering and Science* 58 (4): 460–465.

91 Graziano, A., Jaffer, S., and Sain, M. (2018). Review on modification strategies of polyethylene/polypropylene immiscible thermoplastic polymer blends for

enhancing their mechanical behavior. *Journal of Elastomers & Plastics*: https://doi.org/10.1177/0095244318783806 51 (4): 291–336.

92 Carmeli, E., Kandioller, G., Gahleitner, M. et al. (2021). Continuous cooling curve diagrams of isotactic-polypropylene/polyethylene blends: mutual nucleating effects under fast cooling conditions. *Macromolecules* 54 (10): 4834–4846.

93 Vega, J.F., Martínez-Salazar, J., Trujillo, M. et al. (2009). Rheology, processing, tensile properties, and crystallization of polyethylene/carbon nanotube nanocomposites. *Macromolecules* 42 (13): 4719–4727.

94 Bornani, K., Rahman, M.A., Benicewicz, B. et al. (2021). Using nanofiller assemblies to control the crystallization kinetics of high-density polyethylene. *Macromolecules* 54 (12): 5673–5682.

95 Xia, X., Xie, C., and Cai, S. (2005). Non-isothermal crystallization behavior of low-density polyethylene/copper nanocomposites. *Thermochimica Acta* 427 (1–2): 129–135.

96 Cao, X., Wu, M., Zhou, A. et al. (2017). Non-isothermal crystallization and thermal degradation kinetics of MXene/linear low-density polyethylene nanocomposites. *e-Polymers* 17 (5): 373–381.

97 Altorbaq, A.S., Krauskopf, A.A., Wen, X. et al. (2022). Crystallization kinetics and nanoparticle ordering in semicrystalline polymer nanocomposites. *Progress in Polymer Science* 128: 101527.

98 Yuan, Q., Awate, S., and Misra, R.D.K. (2006). Nonisothermal crystallization behavior of melt-intercalated polyethylene-clay nanocomposites. *Journal of Applied Polymer Science* 102 (4): 3809–3818.

99 Tarani, E., Papageorgiou, D.G., Valles, C. et al. (2018). Insights into crystallization and melting of high density polyethylene/graphene nanocomposites studied by fast scanning calorimetry. *Polymer Testing* 67: 349–358.

100 de Menezes, B.R.C., Campos, T.M.B., Montanheiro, T.L.D.A. et al. (2019). Non-isothermal crystallization kinetic of polyethylene/carbon nanotubes nanocomposites using an isoconversional method. *Journal of Composites Science* 3 (1): 21.

101 Amoroso, L., Heeley, E.L., Ramadas, S.N., and McNally, T. (2020). Crystallisation behaviour of composites of HDPE and MWCNTs: the effect of nanotube dispersion, orientation and polymer deformation. *Polymer (Guildford)* 201 (May): 122587.

102 Depan, D., Khattab, A., Simoneaux, A., and Chirdon, W. (2019). Crystallization kinetics of high-density and low-density polyethylene on carbon nanotubes. *Polymer Crystallization* 2 (4): 1–9.

5

Crystallization of PLA and Its Blends and Composites

Jesús M. Quiroz-Castillo[1], Ana D. Cabrera-González[1], Luis A. Val-Félix[2], and Tomás J. Madera-Santana[2]

[1] Universidad de Sonora, Departamento de Polímeros e Investigación en Materiales, Blv. Luis Encinas J. S/N, 83000 Hermosillo, Sonora, México
[2] Centro de Investigación en Alimentación y Desarrollo, A.C. CTAOV, Laboratorio de Envases, Carretera Gustavo Enrique Astiazarán Rosas, No. 46, Col. La Victoria, 83304 Hermosillo, Sonora, México

5.1 Introduction

Lactic acid, the basic building component of poly(lactic acid) (PLA), can be made through chemical synthesis or carbohydrate fermentation. The fermentation route currently accounts for the majority of lactic acid generation. In the literature, several purification procedures for lactic acid and lactide have been reported recently [1, 2]. Although the fermentation process now generates the majority of the lactic acid, one of the main factors driving PLA's rising utilization is the expense synthesis of large molecular weight PLA polymers (over 100 000 Da). While the generation of the lactic acid prepolymer is required for both direct polycondensation and ring-opening polymerization, polymerization by lactide creation can be performed without the use of coupling agents, as shown in Figure 5.1. The coupling agents' function is to raise the PLA's molecular weight. The PLA with a low molecular weight is used as the lactic acid prepolymer (M_w 1000–5000). This PLA's low molecular weight makes it inappropriate since it is fragile, glassy, and weak. Garlotta [4] states that the primary causes of the development of low-molecular-weight PLA and the direct reaction of prepolymers are the lack of reactivity of the groups in the structure, an abundance of water, and high viscosity of the polymer melt after polymerization. High-molecular-weight PLA polymers can be made using azeotropic dehydrative condensation, direct condensation polymerization, and/or polymerization by lactide synthesis. In this regard, the majority of available commercial high-molecular-weight PLA resins are produced using the lactide ring-opening polymerization technique [3, 4].

Ring-opening lactide polymerization was originally accomplished by Carothers in the middle of the twentieth century, and DuPont later patented this process, which enabled the industrial production of PLA. Lactide molecules undergo

Polymer Crystallization: Methods, Characterization, and Applications, First Edition.
Edited by Jyotishkumar Parameswaranpillai, Jenny Jacob, Senthilkumar Krishnasamy, Aswathy Jayakumar, and Nishar Hameed.
© 2023 WILEY-VCH GmbH. Published 2023 by WILEY-VCH GmbH.

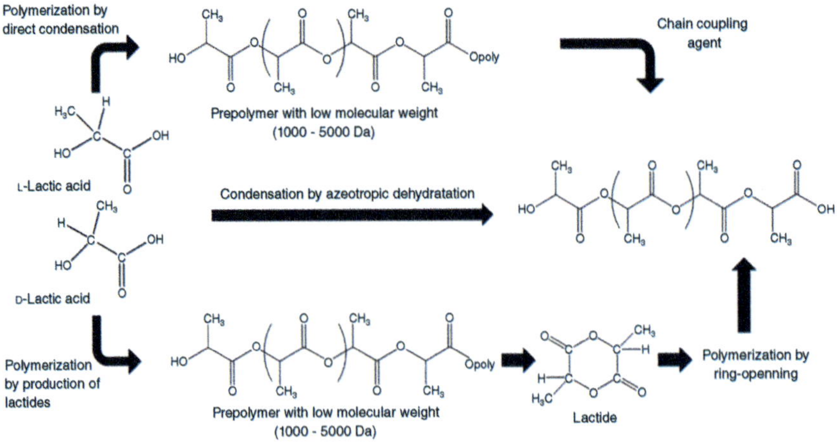

Figure 5.1 Poly(lactic acid) production pathways from lactic acid. Source: Adapted image from Lim et al. [3].

anionic or cationic ring polymerization, depending on the type of initiator used. Free radicals are produced when functional groups act as initiators to accelerate up-chain reactions, which lead to the formation of high-molecular-weight polymers. PLA polymers are poly(L-lactic acid) (PLLA) and poly(D,L-lactic acid) (PDLLA) copolymers made from L- and D,L-lactides, respectively [3]. Because the bulk of lactic acid obtained from biological sources is in this form, the L-isomer accounts for the majority of PLA derived from renewable sources.

PLA is an aliphatic-polyester synthesized from α-hydroxy acids that belongs to the aliphatic-polyester family. Lactic acid (2-hydroxy propionic acid), a PLA constituent, can occur as either optically active D- or L-enantiomers (Figure 5.2). Therefore, PLA is one of the few polymers whose stereochemical structure may be easily changed by polymerizing regulated combinations. L- or D-isomers are used to fabricate high-molecular-weight amorphous or crystalline polymers; by varying the number of enantiomers, the material characteristics can be made and be used as a

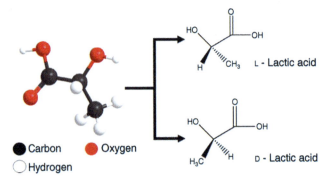

Figure 5.2 Structure and isomers of lactic acid. Source: Adapted image from Sin et al. [5].

food contact bioplastic that is regarded as GRAS (generally recognized as safe) for food applications [3, 4].

PLA crystallizes in three distinctly different forms according to the ratio of optically active L- and D,L-enantiomers' compositions (α, β, and γ). Compared to the β-structure, which has a melting temperature (T_m) of 175 °C, the α-structure is more stable and has a T_m of 185 °C [3]. The structural, barrier, mechanical, and thermal properties of the polymer are significantly influenced by the optical purity of PLA [6, 7]. Higher L-content in PLA polymers often has crystalline structures, whereas PLA polymers with reduced optical purity are more likely to have amorphous structures. When the concentration of the L-isomer diminishes, T_m, T_g (glass transition temperature), and crystallinity (χ) decrease [8]. The water vapor transmission rate (WVTR) of PLLA films was not affected by optical impurity in the range of 0–50%, according to Tsuji et al. [6]. However, WVTR values decreased as the films' crystallinity rose between 0% and 20%. As a result, a careful selection of a suitable PLA resin quality to meet the conversion process parameters is critical. Typically, PLA resins containing less than 1% D-isomer can be used to injection mold PLA items that need to be heat resistant. On the other hand, nucleating compounds or agents can be introduced to increase crystallinity growth during short molding cycles. PLA resins with a higher D-isomer concentration (from 4% to 8%) would be better suitable for thermoforming, extrusion, and blow-molding (e.g. injection molding of preforms and then blow-molding) items because these are easier to process and the crystalline structure is minimal [9].

PLA films have been widely utilized to produce biodegradable films because of their availability, excellent thermal–mechanical characteristics, naturally soft feel, easy processing, distinctive stain and filth resistance, and great film-forming capacity, among other biopolymers. The biodegradability of PLA can contribute to solving the environmental issue of increasing plastic waste. PLA has also been used as a food packaging material [10], because its biodegradability may aid in resolving the environmental problem of accumulating plastic trash. High-molecular-weight polyester chains hydrolyze into lower-molecular-weight oligomers in the first stage of PLA decomposition, which is a two-step process. In the following stage, microorganisms transform these low-molecular-weight components into carbon dioxide, water, and humus. Typical PLA-based containers and packaging include bottled water, juice, yogurt, disposable cups, cutlery, or plates, and these are now widely used in many countries [11].

5.2 Crystallization of Macromolecules

The process of crystallization has a considerable influence on the physical properties of polymers. Crystallization occurs through the steps of nucleation and growth, which can be further refined in the case of polymers to (i) primary crystal nucleation, (ii) secondary crystal nucleation/crystal development (primary crystallization), and (iii) crystal perfection and secondary crystallization [12]. Recent decades have seen research into the crystallization behavior of biodegradable polyesters. The morphology and crystalline structure of thermoplastic polymers, which can be controlled by modifying the crystallization conditions, have a significant impact on their physical properties (thermal, mechanical, and biodegradation), in particular for biodegradable plastics. As a result, investigations into the links between structure, morphology, and characteristics are critical for influencing polymeric materials' ultimate qualities.

PLA is one of the most promising bioplastics, and it is part of the aliphatic polyester group, which has a diverse range of stereochemical properties. Depending on its stereochemical structure and thermal history, PLA can be semicrystalline or completely amorphous [13]. Lactic acid or lactide is indeed a chiral molecule with two optically active stereoisomers (L-lactic and D-lactic acids), as we mentioned before may be manufactured efficiently by fermentation process from renewable resources such as starches and sugars. Lactide or lactic acid can be polymerized through ring-opening or condensation reactions to create PLA. Optically active lactide types can be polymerized to create a "family" of different polymers with different molecular weight distributions, D-lactide concentrations, and sequences in the polymer backbone. High-L-lactide-content polymers can be used to produce semicrystalline polymers, whereas those with greater D-lactide levels seem to be more amorphous.

As mentioned previously, the physicochemical properties of PLA are all affected by its crystallinity. The thermal history has an impact on the crystallization rate and the morphology in the crystalline zone. Starting with nucleation and progression through growth, the process of crystallization is connected with the partial orientation of the polymer backbone. Stereochemistry, optical purity, thermal history (annealing), as well as the amount and kind of additives, all influence the crystallization behavior of polylactides [14]. Crystal nuclei may spontaneously develop in the bulk liquid phase under the influence of random variation in the case of homogeneous crystal nucleation or on pre-existing surfaces or heterogeneities in the case of heterogeneous crystal nucleation. Following the formation of nuclei, chain segments emerge from the amorphous structure, fold into one another, and progressively adhere to the growth front to produce lamellae, an ordered structure. This ensuing lamellar crystal eventually forms spherulites, which are bigger spheroidal structures [12]. When the heterogeneities are crystals or the remnants are crystals from the same crystallizing species, the process of heterogeneous nucleation is known as self-nucleation or self-seeding [15]. At the same temperatures, heterogeneously created nuclei have a lower total surface free enthalpy than homogeneously produced nuclei. Due to the nucleus's threshold size and the labor-intensive nature of surface creation, both of which are temperature-dependent, the dominant nucleation process can change with the temperature. In order to evaluate nucleation rates and densities, it is frequently employed to count the number of crystals or spherulites per unit of volume (after they grow into larger entities). Likewise, calorimetry can provide information about nucleation [16], such as when using the Avrami equation to fit conversion-time curves derived from heat flow rate data generated by crystallization [17, 18], analyzing the crystallization onset time, or by examining the slope or form of the crystallization point during the early phases of a transition [19].

The relationship between temperature and crystal growth rate is commonly examined by polarized-light optical microscope analysis to determine the spherulite growth rate; it also reveals a qualitative dependence between the temperature and the nucleation rate. Usually, the growth rate increases as the temperature decreases, since the thermodynamic driving force increases to perform the phase transition. With the decreasing temperature, the rate of growth reaches a maximum, and after that, it drops, resulting in a reduction in the movement of molecule chains in the liquid phase close to the growth faces. Macromolecule crystallization is typically incomplete because of non-crystallizable structures close to the crystal growth front, entangled increase of the melt, and kinetic constraints or impeded diffusion of molecular chains. This suggests that amorphous and crystallized regions coincide, which is in violation of Gibbs' phase rule [19]. After the first crystallization, the amorphous phase is still present, and this leaves a thermodynamic driving force for further crystallization. This secondary crystallization, however, might endure a long period depending on the temperature and mobility of chain segments. It usually happens after mutual spherulite impingement, although it may also happen throughout their development. It can also entail the insertion of lamellae into existing stacks. Polymer crystals rapidly rearrange into more steady crystal structures with reduced free energy because lamellae and fringed micellar domains are not thermodynamically preferred crystal morphologies due to their surface-to-volume ratios and defective surface structures. Lamellar thickening is one of the rearrangement processes that can occur isothermally, overlapping basic crystal formation [20].

In the past two decades, PLA has garnered considerable attention from both academics and practitioners [13, 21]. PLA crystallizes simply during the heating process but not during the cooling process (cold crystallization). Addition of nucleating agents, such as nanoclays, nanoparticles (NPs), and inorganic chemicals, is the simplest way to increase crystallinity in the PLA matrix [22]. In the literature, there is a quantifiable thermal analysis for several PLA materials [23, 24], which includes a relationship between calorimetric information like heat capacity and molecular structure. Moreover, when PLLA is used in a variety of industrial applications, increasing the overall crystallization rate is an issue. Several studies have been conducted to determine if mixing PLLA with nucleating chemicals like talc can improve the crystallization rate. It has also been reported that the addition of PDLA improved the crystallization rate of PLLA, as well as an investigation on the use of an inorganic nucleating agent to raise the crystallization rate of PLLA. Calorimetry has been used to conduct extensive investigations, primarily on crystallization kinetics [25–27].

5.2.1 Improvement of PLA Crystallization Kinetics

PLA is a semi-crystalline polyester that exhibits multiphase transitions during heating, highlighting the T_g around 60 °C that limits several applications of this material. T_g is related to the degree of crystallinity of the PLA since a higher degree of crystallinity increases T_g. A higher crystallinity improves the mechanical, thermal, and chemical properties of PLA since the crystalline domains confer greater stability than the amorphous domains. However, the crystal formation process in PLA is slow when it is melted in extrusion or injection due to a short cooling cycle and limited orientation. The low crystallinity of PLA is attributed to its minimal crystallization time or high cooling rate compared to conventional thermoplastics. To enhance the crystallinity of PLA, crystallization kinetics have been investigated. The addition of plasticizers and/or nucleating agents to the PLA matrix, as well as the heat process known as annealing, have all been the subject of several experiments to enhance the crystallinity of PLA.

A nucleating agent lowers the free energy barrier of the surface, increasing the stable nucleation sites and the number of crystallites [28, 29]. A nucleating agent influences the crystallization kinetics of PLA, shortening the nucleation time and modifying the degree of crystallinity, the number and size of the crystallites, as well as the integrity of the spherulites [30, 31]. Table 5.1 shows the classification and main nucleating agents used to improve the crystallinity of PLA. Depending on the nucleating agent, these accelerate the crystallization kinetics through heterogeneous nucleation; they do not alter the crystallization mechanism, form transcrystalline structures and stereocomplex crystals, and do not modify the crystal structure. The effect of nucleating agents on crystallinity depends on the content of the PLA matrix. According to Sun et al. [46], the compatibility and dispersion of the type of nucleating agent in the PLA matrix can be another parameter that can influence the crystallization rate since good dispersion and compatibility increase crystallinity.

The plasticizers (low-molecular-weight compounds) allow the mobility of the PLA chains and improve the crystallization rate due to a reduction of energy for the chain rearrangement during the crystallization process [28, 52]. Table 5.2 shows the main plasticizers and their mixtures with nucleating agents to improve the crystallinity of PLA. Usually, plasticizers increase the growth rate of spherulites and the degree of crystallinity of PLA; however, when mixed with a nucleating agent, they show a synergistic effect on the crystallization rate. In addition to adding nucleating agents or plasticizers, annealing is used to improve the crystallinity of PLA during or after processing. This thermal method consists of varying the time and temperature conditions to which the PLA is subjected, wherein as the temperature increases, the crystallinity increases by recrystallization. Nevertheless, conventional annealing can cause damage to the morphology of the finished PLA product [59]; that is why modifications have been made to this method. Currently, the annealing method has been modified using solvents, supercritical fluid, or a combination of nucleating agents as shown in Table 5.3. The modification of conventional annealing can lead to an increase in crystallinity to a greater extent [59] without negative effects.

Table 5.1 Nucleating agents for PLA.

Classification	Nucleating agent	Effect on crystallization kinetics	References
Inorganic particles	Talc	Promotes the crystallization ability of the PLA matrix and the spherulitic growth from instant nuclei of heterogeneous nucleation	[32, 33]
	Al_2O_3	Enhances the crystallinity of PLA and speeds up its crystallization through heterogeneous nucleation	[34]
Nanofillers	Chitin nanocrystals	Increases nucleation capacity producing small crystallite sizes	[35]
	Nanoclay	Slightly increases crystallinity	[36]
	Graphene nanoscrolls	Accelerates the crystallization kinetics but with minimal effect on the overall crystallinity	[37]
	GNP (graphene nanoplatelets) and MWCNT (multiwalled carbon nanotubes)	MWCNT increases the degree of crystallinity slightly more than GNP	[38]
	CSNs (silk nano-discs)	The higher surface area of CSN causes an increase in crystallization rate without modifying the crystallization mechanism	[39]
Polymer blends	Unsaturated liquid crystalline polyester and dicumyl peroxide	Mixing increases the rate of crystallization through heterogeneous nucleation	[40]
	Poly D-lactic acid	Increases the crystallization due to the formation of stereocomplex crystals	[41]
	Ethylene-vinyl acetate	Generates a maximum increase in crystallinity at 5% mass	[42]

(Continued)

Table 5.1 (Continued)

Classification	Nucleating agent	Effect on crystallization kinetics	References
Organic derivatives	LAK (aromatic sulfonate derivative)	Increases the crystallinity to a greater extent than talc and calcium carbonate. The rate of crystallization increases with the increase in content, without modifying the crystalline structure	[43, 44]
	NAs (oxalamide derivatives)	They form fibrils that accelerate the crystallization process	[45]
	POCFA (furan-phosphamide derivative)	Improves crystallization rate and crystallinity by increasing the amount of POCFA	[46]
	BTCA (1,3,5-benzenetricarboxylamide derivative)	The crystallization of stereo-complexes increases with the addition of BTCA and reduces the half-time of crystallization	[47]
	Zinc oxide/zinc phenylphosphonate	The degraded PLA chains form new crystal structures	[48]
	Zinc phenylphosphonate	Improves the crystallization rate but does not improve the crystallization morphology	[49]
Natural fibers	Kenaf fiber and MWCNTs	Mixing can accelerate crystal growth and increase crystal content as well as generate transcrystalline structures	[50]
	Flax fiber	The crystallinity improves with increasing the flax fiber content, reduces the size of the spherocrystal, and generates transcrystalline structures	[31]
	Ijuk fibers	The enhancement of the interfacial area between the PLA matrix leads to a rise in crystallinity	[51]

Table 5.2 Plasticizers and their mixtures with nucleating agents for PLA.

Plasticizer	Effect on crystallization kinetics	References	Plasticizer/nucleating agent	Effect on crystallization kinetics	References
LLDPE	The growth rate of the spherulite increases but eventually decreases	[27, 53]	Polyethylene glycol (PEG)/CNFs (cellulose nanofibrils) and CNCs (cellulose nanocrystals)	Gives significantly higher crystallinity than talc	[29]
(USE) epoxidized soybean oil	The PLA/USE blend increases the degree of crystallinity due to higher chain mobility	[28, 54]	Glycerol triacetate and banana nanofibers	Increases crystallinity twofold through a dilution mechanism	[55]
Triphenyl phosphate	Increases spherulitic growth and the degree of crystallinity without modifying the crystalline structure	[30, 56]	PEG/calcium carbonate, halloysite nanotubes, talc, and LAK	Talc and LAK further increase the crystallization of PLA	[52]
Triacetine and tributyl citrate	Plasticizers increase crystallinity (tributyl citrate to a greater extent) by improving molecular mobility	[31, 57]	PEG and acetyl triethyl citrate/talc	Mixing increases crystallinity achieved even at high cooling rates	[58]

Table 5.3 Conventional annealing and modifications applied to PLA.

Annealing	Effect on crystallization kinetics	References
Conventional	The oriented segment relaxation increases the crystallinity	[60]
Using ethanol as solvent	More efficient than conventional annealing	[59]
Using supercritical carbon dioxide	Increases in pressure or annealing temperature result in a decrease in crystallinity, with higher pressure producing defective crystals	[61]
Using nucleating agents	Only with annealing, wood flour, and kaolin increase crystallization	[62]

5.3 Polylactic Acid Nucleation

Petroleum-based polymers, such as polyvinyl chloride, polyethylene, and polypropylene, have been widely used in various daily life applications, generating severe environmental pollution since many of these plastics persist in the environment as waste materials that do not degrade quickly [63, 64]. Developing biodegradable biopolymers constitutes an efficient alternative to replace petroleum-derived polymers. In comparison to other biodegradable polymers, PLA has been regarded as one of the most promising bio-based materials for the future [65]. Because of its great mechanical strength, high modulus, low toxicity, outstanding transparency, biocompatibility, and biodegradability, it is quickly processed in standard plastic equipment for the manufacture of molded parts, films, or fibers [66]. It has been widely used in various applications, such as in the field of industrial packaging, medical device market, tissue engineering, and drug-controlled release [67–69]. However, its large-scale use is somewhat restricted due to its slow crystallization rate, which at the same time means that this polymer has poor crystallinity, long injection molding time, and low heat resistance. Therefore, overcoming the drawback of its slow crystallization rate is a challenge that must be overcome to increase the performance and uses of PLA [32, 65]. Adding a nucleating chemical will decrease the surface's free energy barrier against nucleation and promote crystallization at higher temperatures, which is a simple and effective approach to accelerate the crystallization of PLA [64]. The efficiency of nucleating agents can be evidenced using differential scanning calorimetry (DSC). It is stated that a nucleating agent is more effective the higher the crystallization temperature (T_c) turns out to be and the higher and narrower the peak corresponding to the exothermic crystallization reaction [70]. Numerous nucleating agents for PLA have been developed and studied, some of which are listed below.

5.3.1 Inorganic Nucleating Agents

For PLA as well as other thermoplastics, talc acts as a physical nucleating agent quite effectively [71]. Compared to pure PLA, under resting conditions, the presence of talc increases the crystallinity rate of PLA because of the presence of heterogeneous nuclei. In addition, it is frequently chosen due to its low cost and additional reinforcing effect [72]. It is frequently utilized as a reference to compare the nucleation potential of different additives. Recently, in research carried out by Li et al. [73], they prepared PLLA/talc mixtures with different talc loads and two types of particle size and indicated that nucleated PLLA had a faster crystallization rate than pure PLLA, proving the nucleation effect of talc. They also suggested that the talc particle's size influenced the crystallization rate and the degree of crystallinity of PLLA. The crystallization rate was faster for the PLLA nucleated with the smaller particle size talc due to the greater surface area presented; however, this was the one that presented a minor improvement in the degree of crystallinity. The maximum nucleation was reached when the talc concentration of small and large particles was 0.25% and 0.5% by mass, respectively [73]. In other studies, Petchwattana and Narupai [71] proposed to increase the crystallization rate of PLA by combining talc and titanium dioxide (Ti_2O), observing that the modification of PLA with talc reduced the average crystallization time from 27 minutes to 36 seconds and that the degree of crystallinity increased from 0.92% to 20%. Similarly, the combination of talc with Ti_2O led to a drastic decrease in the average crystallization time ($t_{1/2}$ = 18 seconds); moreover, a roughly 26% increase in crystallinity. Biodegradable nanocomposites (NCs) of PLA and poly(butylene adipate-co-butylene terephthalate) (PBAT) (70/30, w/w) with diatomite or talc (1–7%) were prepared by melt processing of these bioplastics. The particles were located at the interface of two phases acting as an interface modifier to increase the interfacial binding between PLA and PBAT. As nucleating agents, talc and diatomite let the PBAT crystallize more effectively [74]. Aliotta et al. [43] examined and compared the influence of adding three different nucleating agents (talc, calcium carbonate, and a derivative of aromatic sulfonate [LAK]) in the direct crystallization process during the process of injection molding. The results demonstrated that the incorporation of the nucleating agents enhanced the crystallization of PLA in the following order: LAK > talc > calcium carbonate, observing a sharpening of the crystallization peak and a decrease in the average crystallization time. They suggest that although LAK has presented the best results in terms of improvements in the crystallinity of the material, decreases in the flexural modulus and Charpy impact resistance of PLA were recorded, but this was not the case for talc, which increased the properties. Higher values were recorded in PLA mechanics and, if a cycle time of 60 seconds was adopted, a high value of heat deflection temperature was also obtained.

Clay has also been used as a nucleating agent, increasing crystallinity and improving the barrier, mechanical, and thermal properties of many polymers [75]. For example, Taleb et al. [76] described PLA NCs reinforced with organoclays [OMt (12-4-12), OMt (10-4-10), and OMt (8-4-8)]. They formulated between 1 and 3 wt% of organoclay and reported that the addition of nanoclays considerably improved the crystallization process, demonstrating a notable reduction in the cold crystallization temperature (T_{cc}) and an increase in the degree of crystallinity (X_c), demonstrating the reinforcing effect of nanoclays as well as their function as nucleating agents [76]. In studies by Shinzawa and Mizukado [77], they demonstrated the effect of clay-induced nucleation on a PLA sample using near-infrared (NIR) image analysis based on mismatch mapping. During the study, the deregulation peaks were revealed at 5785 and 5825 cm^{-1} demonstrating the different distributions of amorphous and crystalline content in the analyzed region. The authors suggested that after heating the polymeric system, the intensity of the decoupling between the two peaks became more intense around the boundary between clay and PLA, suggesting the increase in PLA crystals and the abundance of the crystal structure in the clay region [77]. Another example is the research carried out by Coppola et al. [78], where two grades of PLA (4032D and 2003D, with a D-isomer content of 1.5 and 4, respectively) were combined with layered silicate (Cloisite-30B) at 4% by weight. It was observed that for the two peaks of PLAs, the presence of nanoclays initiated crystallization with a decrease in the cold crystallization temperature (T_{cc}) by approximately 10 °C, showing a narrower and sharper cold crystallization peak. This behavior was more marked for the PLA 4032D sample, which had a lower content of isomer D in its composition [78].

According to the literature, clay is a less effective nucleating agent for PLA than talc because it experiences a smaller reduction in the mean crystallization half-time in isothermal mode and is not suitable when high concentrations are applied to cooling rates in non-isothermal crystallization [79].

Another quite effective and investigated physical nucleation agent is sepiolite. Some properties that make it ideal for use as a nanofiller for polymers are its distinctive acicular morphology and large surface area. In the studies reported by Wu et al. [80], they evaluated the effect of the incorporation of nanosepiolite as a nucleating agent in the crystallization of PLA. The authors observed that with the addition of 1 wt% of sepiolite, the nucleation density and crystallinity improved substantially and the crystallization half-time decreased from 17.7 to 2.1 minutes at an optimal crystallization temperature of 110 °C. The presence of two crystallization peaks under non-isothermal conditions was also observed. In the isothermal crystallization studies, the value of the Avrami exponent (n) was 3, indicating that the crystal nucleation and growth mechanisms were quite similar.

Another example of promising nanofillers is carbon nanotubes (CNTs), which have recently received considerable attention from researchers due to their excellent electrical properties, good tensile strength, and high flexibility [81–83]. In studies conducted by Zhang and Zhang [84], they used sodium dodecyl benzene sulfonate (SDBS) to non-covalently modify CNTs, further investigated the modification of multiwalled carbon nanotubes (MWCNTs), and prepared surfactant-modified carbon nanotubes (SMCNT) with PLA NCs (PLA/SMCNT). The authors state that the SDBS, together with the MWCNTs, acted as nucleating agents and accelerated the crystallization process of the compounds. In addition, they state that the cold crystallization temperature of PLA/SMCNT was at higher when the SMCNT content was augmented. In studies by Shen et al. [85], CNTs were modified with the amino hyperbranched polymer (HBP N103) from an amidation reaction and prepared PLLA and PLLA/CNT compounds. The authors found that by adding CNTs, the crystallization properties of PLLA could be significantly improved. They also suggest that the modified CNTs had a better ability to regulate PLLA crystallization than the unmodified CNTs at the same concentration. The addition of modified and unmodified CNTs caused an increase in nucleation points and a significant decrease in grain size at 11 and 14 µm, respectively. The crystallization time decreased considerably in both cases, with a more marked behavior when modified CNT was incorporated due to their excellent dispersion in the material.

Despite the efforts made by researchers in the study of inorganic nucleating agents, some defects are inherent in the PLA matrix, such as its poor compatibility and dispersion, resulting in phase separation and decreased capacity of nucleation. In this sense, the nucleating agents for organic molecules or stereocomplexes based on PDLA and PLLA show better compatibility with PLA [86].

5.3.2 Organic Nucleating Agents

Other examples of physical nucleant agents for PLA are organic materials. This effect is mainly achieved by adding a low-molecular-weight substance that crystallizes faster and at a higher temperature than the polymer, incorporating organic nucleation sites [79]. An example of this is the research carried out by Jongpanya-ngam et al. [87], where they analyzed the crystallization behavior of PLA (PLA L105 and PLA 3251D) using dimethyl 5-sulfoisophthalate sodium salt (SSIPA) as a nucleating agent compared to the commercial sulfonate salt (LAK-301). The authors hypothesized that the presence of SSIPA provided the absence of the cold crystallization peak typical of pure PLA and induced a new crystallization peak. The overall crystallization rate from the isothermal crystallization studies was more marked for the PLA with a nucleating agent than for the neat polymer. Avrami exponent (n) values of PLA with a nucleating agent improved from approximately 3–4, showing that the nucleating agent increased the growth of three-dimensional crystals with homogeneous nucleation. Through images obtained by polarized optical microscopy, a higher density of the nuclei and smaller size of the spherulites were observed, concluding that the sulfonate derivative, SSIPA, could potentially be used as an effective nucleating agent for PLA [87].

Conversely, Zhao and Cai [65] used N,N'-oxalyl bis(piperonylic acid)dihydrazide (PAOD) as a novel nucleating agent to enhance PLLA crystallization. The melt crystallization process has revealed that PAOD could significantly enhance the crystallization of PLLA on cooling. An increase in PAOD concentration caused the molten state crystallization peak to undergo a shift to a higher temperature. Increasing the cooling rate negatively affected the crystallization capacity of the PLLA/PAOD samples since the crystallization peak broadened and moved toward a lower temperature value. In contrast, the cold crystallization peak moved to a lower temperature value with an increasing PAOD concentration. The effect of crystallization temperature on the melting peak was evident, and double melting peaks appeared after isothermal crystallization in the lower temperature region due to melt recrystallization [86].

Fan et al. synthesized another nucleating agent, in this case, N,N'-bis(benzoyl)-hexanedioic acid dihydrazide (BHAD), and found by non-isothermal crystallization that nucleated samples exhibited a higher maximum crystallization temperature (T_c), indicating that BHAD showed excellent heterogeneous nucleation ability.

With the help of a polarized light microscope, three different morphologies of the nucleated samples could be observed: crystal morphology (BHAD concentration of 0.1–0.2% by weight); branch and leaf morphology (BHAD concentration of 0.3–0.5% by weight); and needle crystal morphology (BHAD concentration of 0.75–2% by weight) [88].

Huang et al. [89] found in their studies that non-isothermal crystallization showed that N,N'-bis(stearic acid)-1,4-dicarboxybenzene dihydrazide (PASH) played an essential role in increasing crystallization capacity by providing sites for heterogeneous nucleation in the PLLA. They further observed that by increasing the concentration of PASH, both the melt crystallization peak and the cold crystallization peak shifted toward lower temperature levels. In isothermal melt crystallization, PLLA/PASH exhibited a shorter $t_{1/2}$ than pure PLLA, and PLLA/PASH (3%) exhibited a minimum $t_{1/2}$ of 67.3 seconds at 105 °C compared to $t_{1/2}$ for the pure PLLA it was 310.1 seconds. Finally, they concluded that the different PLLA/PASH fusion behaviors under different conditions are due to the nucleation effect of PASH within the PLLA [89].

Biobased nucleators are a subset belonging to organic nucleating agents, which have been of great interest to PLA. Some studies have reported that applying chitosan-based nucleating agents significantly improves the biodegradable characteristics of PLA compounds [90]. In this sense, Xu et al. [91] synthesized a new chitosan methyl phosphonate (CMP) and found it could act as a multifunctional biological nucleation agent. This behavior was evidenced by the Avrami equation, where the value of the Avrami exponent (n) for pure PLA was approximately 4. It demonstrates that this material formed crystals through homogeneous nucleation, and with the addition of CMP, the n values of the PLA/CMP composites decreased to approximately 3, indicating that the CMP produced a heterogeneous nucleation effect and increased the formation of spherulites in the material. Furthermore, the crystallinity of the PLA/CMP composites increased almost twofold with the introduction of only 5% by weight of CMP [91].

Wood is a natural fiber that researchers have extensively studied for use as a filler in PLA-based composites [92]. Such is the case of the research carried out by Wu et al., where they verified that the incorporation of 4% by weight of wood flour (WF) as a nucleating agent in the PLA/WF compounds improved the crystallization process, increasing the degree of crystallinity (X_c) of pure PLA from 11.5% to 15.7% of PLA/WF. In addition, they mention that the WF provided a higher density of heterogeneous nuclei; however, it decreased the size of the spherulites. Using the Avrami equation, they analyzed the isothermal crystallization kinetics of the materials. In the cold isothermal crystallization and melt isothermal crystallization processes, the resulting Avrami exponents (n) of PLA crystallization were 1.8–2.1 and 2.0–2.8, respectively, so the nucleation mechanism of the crystallization process isothermal in both processes was different; demonstrating all of the above that WF constitutes an effective nucleating agent for PLA [93].

Cellulose nanofibers (CNF) are also materials that have been extensively investigated as nucleating agents due to their properties of high flexibility, high crystallinity, and good mechanical strength [94, 95]. Considering the above, Ariffin et al.

demonstrated the promising role of CNF as a nucleating agent for PLA, proposing that the incorporation of 3% by weight of CNF to the PLA matrix increased the rate of crystallization of PLA composite material at 0.716 min^{-1} compared to that of pure PLA, which was 0.011 min^{-1}. In addition, the highest degree of crystallinity for NCs was 44.2%, which, compared to pure PLA, presented an increase of almost 95%, indicating the potential of CNF as a nanofiller in biopolymers [96].

Pectin is another material that has recently been used to modify the structure and increase PLA's biodegradability. In studies conducted by Satsum et al. [97], they prepared PLA biocomposites with different pectin contents (2%, 4%, 6%, and 8% by weight). The results showed that the presence of pectin improved the compounds' crystallinity and the crystallization time of the PLA. The pure polymer presented a gradual crystallization, while the biocomposites crystallized rapidly, reaching saturation in 6–8 minutes. The biocomposites with a pectin content of 4%, 6%, and 8% presented a mean crystallization time of 5.83, 5.82, and 5.48 minutes, respectively. A maximum crystallization rate and crystallinity were reached when the pectin content was 8 wt%.

It has been discovered that the mixture of PLLA and PDLA isomers can crystallize in the form of a stereocomplex, presenting a melting point of 50 °C higher than the homocrystals of PLLA and PDLA. In addition, it was formed at a higher temperature upon cooling than the homocrystals themselves, so small amounts of the stereocomplex constitute PLA nucleation agents [79]. Bouapao and Tsuji [98] investigated the crystallization and growth behavior of spherulites of the PLLA/PDLA stereocomplex synthesized from the mixture of low-molecular-weight PLLA and PDLA, obtaining significant results in this regard since the values of cold crystallization temperature (T_{cc}) for the mixtures obtained were lower than those of pure PLLA and PDLA, showing that the presence of both homopolymers improved crystallization during heating by DSC. In addition to the fact that the nuclei of stereocomplex crystallites played the role of nucleating agents, the nucleation constant (K_g) of the PLLA/PDLA stereocomplex obtained from the Hoffman nucleation theory was higher than that of the pure PLLA and PDLA homocrystals. Conversely, the authors stated that the increase in spherulite of stereocomplex crystallites was observed in the widest temperature range (<180 °C) compared to that of PLLA or PDLA homocrystallites (<120 or 130 °C). In other studies, Shi et al. [99] prepared asymmetric PLLA/PDLA mixtures by adding minimal amounts of PDLA to the PLLA matrix, where the formed crystallites increased in size as the PDLA concentration increased. The authors suggest that the formed stereocomplexes acted as nucleation sites and assisted in the crystallization of PLLA when the mixtures were processed at temperatures below their melting point, while at higher temperatures the mixtures with lower concentrations of PDLA presented a crystallization capacity like that of pure PLLA. However, when the PDLA concentration was higher than 8%, the crystallization ability decreased. The authors also state that the PDLA concentration and the heat treatment temperature influenced the spherulite density. In further studies, Ji et al. [100] prepared stereocomplexes of linear and four-armed PLA (scPLA). They effectively promoted PLA nucleation. The authors stated that a 10% stereocomplex increased the crystallization rates of PDLA and PLLA by more than 55% and 70%, respectively.

The four-armed PLA significantly eliminated crystallization at high temperatures due to its higher degree of branching and steric hindrance, forming defective crystallites that improved the adsorption of the matrix on the surface of the crystal, favoring nucleation and growth at higher temperatures.

5.4 Polylactic Acid Blends

Scientific research oriented to the study of biodegradable and biobased polymers and their blends has gained special interest in recent years, especially in the packaging industry [101]. Poly(lactic acid) or polylactide (PLA) is a biobased, commercially available, biodegradable, and compostable polymer with excellent mechanical properties that is also biocompatible and transparent. Nevertheless, brittleness and high elastic modulus of PLA limit its usage [102].

An inexpensive approach to coping with these limitations is preparing blends of PLA with other polymers, this is also a way to tailor the final properties of the blends [63, 103]. The aim of these blends is to prepare materials that overcome the weaknesses of pure PLA. The desirable resulting properties could be improved mechanical properties, modification of initial barrier properties, enhanced thermal resistance, or biodegradability [104].

5.4.1 Polylactic Acid Binary Blends with Biopolymers–Starch and PHAs

PLA is a biodegradable biopolymer prepared from biological materials. Other non-biodegradable biopolymers could be prepared using biological materials and other polymers could be synthetic but biodegradable [63]. In this section different binary blends of PLA with biopolymers will be reviewed.

Starch is a biocompatible and biodegradable polysaccharide prepared using renewable resources. It can be used as a filler, to obtain cheaper final products. Researchers have been preparing blends of starch with PLA in two categories, mainly PLA/starch and PLA/thermoplastic starch (TPS). PLA/starch composites were first prepared and characterized by Jacobsen and Fritz [105]. In these blends, starch acts as a filler, increasing the strength of the final product; the maximum addition of starch was 10 wt%; beyond this point, the mechanical properties of the material were seriously affected. Ke et al. [106] prepared blends using different starches with various amylose contents. It was observed that starch increased the degree of crystallinity and the crystallization rate of PLA, additionally, decreasing its melting point. A year later, Zhang and Sun [107] confirmed that starch acted as a nucleating agent, increasing the crystallization rate of PLA; they also observed an enhancement in strain at break. Poly(ethylene glycol) has been used as a plasticizer for PLA/starch blends, producing an improvement in processability and a reduction in T_g [105].

An improvement in the miscibility of PLA/starch blends can be achieved using compatibilizers; their presence leads to an improvement in mechanical properties and thermal stability. Some compatibilizers are methylene diphenyl diisocyanate

[108]. PLA grafted maleic anhydride and acetyl triethyl citrate [109], PLA grafted acrylic acid [110], PLA grafted amylose [111], PLA grafted glycidyl methacrylate [112], and hexamethylene diisocyanate grafted starch [113].

TPS is prepared by mixing different plasticizers (glycerol is the most used) with starch, usually at high temperatures, with the objective of restricting the presence of hydrogen bonds in starch. Palai et al. [114] prepared blends using PLA and TPS and observed an improvement in mechanical properties such as strain at break compared to blends with starch. They concluded that starch is a nucleating agent for PLA. In a study by Serra-Parareda et al. [115], it was reported that the addition of TPS affected the tensile strength and the elastic modulus of the blends, while PLA retained its brittleness. In this case, the addition of 30% TPS into a PLA matrix was meant to increase the compostability of pure PLA; nevertheless, this incorporation severely affected the tensile strength (decreasing 37.3%), Young's modulus (decreasing 34.9%), and strain at break (decreasing 12.6%) of the blends.

Compatibilization of the blends is used to improve miscibility in PLA/TPS blends. This effect leads to an improvement in thermal degradation resistance and mechanical behavior, especially strain at break. Some compatibilizers are: PLA grafted maleic anhydride [116]; methylene diphenyl diisocyanate [117], formamide [118], maleic anhydride grafted starch [119], PLA grafted starch copolymer [120], and maleic anhydride grafted poly(ethylene glycol) [121].

Polyhydroxyalkanoates (PHAs) are biodegradable, biobased polymers. They have been used in food packaging and many biomedical applications, including surgical sutures, tissue engineering, and drug delivery systems. Considering the above-mentioned disadvantages of PLA, the aim of the blends with PHA is enhance the biocompatibility and biodegradability of PLA. Due to the high thermal stability of PLA, PLA/PHA blends have greater thermal stability compared with pure PHAs [90]. Blending PLA with PHA improves the performance of the composite material while also preserving its biodegradable characteristics.

Poly(3-hydroxybutyrate) (PHB) is the most studied PHA. Its T_g is around 5 °C, and its T_m is 180 °C; during heating, it presents cold crystallization [122]. Several authors have proposed blending PLA with PHB to improve the properties of each pure polymer [122, 123]. PLA/PHB mixtures show an improvement in barrier properties compared to pure PLA [123], nevertheless, mechanical properties are severely affected [124]. Zhang et al. [125] reported the immiscibility of PLA/PHB in solution-cast blends, unlike melt-blended mixtures, which were miscible. Later, Kikkawa et al. [126] observed that blends of PLA and PHB presented different phases, including miscible, partially miscible, and immiscible zones. This was highly dependent on the composition of the blends and the molecular weight of the polymers. Kervran et al. [127] recently reported also the immiscibility of PLA/PHB blends and that the addition of PHB to the blends increases the biodegradability of PLA but negatively affects the mechanical properties. Bartczak et al. [128] also reported an improvement in biodegradability by adding PHB to PLA. Recently, Zheng et al. [102] reported miscible blends of PLA with 20 wt% of PHA, it was observed to have a strain at break of 23.36% and a tensile strength of 39.33 MPa.

D-Limonene, acetyltributylcitrate [124], tributyrin [101], and oregano essential oil [102] have been used as compatibilizers for the PLA/PHB blends, leading to a reduction in T_g, an improvement in processability, elongation at break, and degradability in compost.

5.4.2 Polylactic Acid Binary Blends with Biodegradable Polymers – PCL, PBAT, and PBS

Binary blends of PLA and biodegradable polymers will be studied in this section. Polycaprolactone (PCL), a synthetic semi-crystalline polyester, is a highly preferred polymer in biomedical applications due to its good thermal behavior, biocompatibility, nano-size fiber formation, bioactivity, and non-toxicity [129]. PCL has high strain at break, but low tensile strength. Preparing PLA/PCL blends can result in an improvement of the properties of the neat polymers, for example, higher toughness and ductility than PLA and higher tensile strength and processability than PCL [90]. Besides, the long decay time of PCL provides an advantage for some applications [129]; composition in the blends could be used to control the degradation time of the material.

PLA/PCL blends have been prepared extensively over the last 25 years. Tsuji and Ikada [130] studied the degradation of PLA/PCL films in phosphate buffer solution. They concluded that terminal carboxyl groups present in PCL, accelerated the degradation of the blend films, this effect was restricted to a 25 wt% PCL in the blends. Gaona et al. [131] also studied the degradation rate of PLA/PCL membranes in PBS. They observed a faster crystallization rate in degraded PLA molecules, due to the molecular weight reduction. Degradation of PLA/PCL blends was also studied by Fukushima et al. [132], they concluded that several factors affected the biotic and abiotic degradation of the materials, including: chemical structure, miscibility, glass transition temperature, temperature of the degradation medium, and presence of enzymes.

Sivalingham [133] observed that PCL was more thermally stable than PLA. They also concluded that the incorporation of PLA did not affect the degradation rate of PCL in the blends. Vieira et al. [134] degraded PLA/PCL fibers and measured the mechanical properties along the degradation process. They observed a decrease in tensile strength of the fibers depending on the decrease in molecular weight.

Triphenyl phosphite [135], PCL/PEG block copolymer [136], dicumyl peroxide [137], lysine triisocyanate [138], glycidyl methacrylate [139], epoxidized palm oil [140], and PLA grafted maleic anhydride [141] have been used as compatibilizers for the PLA/PCL blends, leading to an improvement in strain at break and impact strength.

PBAT is an aromatic/aliphatic copolymer prepared using 1,4-butanediol, adipic acid, and terephthalic acid. It is a biodegradable co-polyester with high toughness, high strain at break, and water resistance [142]. Blending PBAT with PLA complements some disadvantages of PLA, such as low impact resistance, maintaining the biodegradability in the material [143].

Jiang et al. [144] studied immiscible extruded blends of PLA/PBAT. They concluded that strain at break and toughness increased with PBAT content in the blends. Nevertheless, they observed a decrease in tensile strength and elastic modulus. PBAT and PLA are biodegradable polymers with contrasting properties, this can be used to prepare synergistic materials. Compatibilizers can be used to improve the miscibility of these blends [143].

Nofar et al. [63] prepared PLA/PBAT blends using a twin-screw extruder, they observed an increase of 265% in ductility after incorporating 25 wt% PBAT in the blends. Deng et al. [145] also observed that as PBAT content increases from 10 to around 20 wt%, ductility in the blend system dramatically increases up to 300%. In recent studies, PLA was blended at different compositions (from 10 to 25wt%) with high molecular weight PBAT, they observed an improvement of 200% in elongation at break for the mixture with 20% and 25% of PBAT. Regarding Charpy impact characteristics, they observed a value of $9\,kJ\,m^{-2}$ for PLA with 25 wt% and a value of $3\,kJ\,m^{-2}$ of pure PLA [146].

Four main strategies are reported in the literature to improve compatibility in PBAT and PLA blends: (i) the use of substances with epoxy functionalities, (ii) the use of anhydrides and peroxides, (iii) the use of ionic liquids, and (iv) the use of both nano- and micro-particles [143].

Poly(butylene succinate) (PBS) is an aliphatic polyester with properties similar to polypropylene but is fully biodegradable. It is a highly flexible polymer with good processability, thermal and chemical resistance, and excellent impact strength. Park and Im [147] prepared PLA/PBS extruded blends and studied the crystallization behavior and compatibility of this system. PBS increased the crystallization rate of PLA, and PLA/PBS blends showed good compatibility, evidenced by the presence of a single T_g over the entire composition range. While Bhatia et al. [148] prepared PLA/PBS blends using a twin-screw extruder, they also studied the mechanical, morphological, and rheological properties of the blends. Flow curves for blends containing 50 wt% PLA and 50 wt% PBS showed strong shear thinning behavior at low frequencies, but all other compositions showed Newtonian behavior. Viscosity analysis at 200 °C and $0.1\,s^{-1}$ showed that blends with low PBS content (below 20 wt%) displayed viscosities between those of the pure polymers; these findings suggest a high degree of compatibility in blends with low PBS content. Yokohara and Yamaguchi [149] also confirmed the compatibility of the blends using rheological measurements.

Dicumyl peroxide [150] and lysine triisocyanate [151] have been used as compatibilizers in PLA/PBS blends; this improvement in miscibility leads to higher strain at break and impact strength values.

5.5 Polylactic Acid Composites

5.5.1 Polylactic Acid – Natural Fiber Composites

Natural fibers are used as reinforcements in PLA to improve mechanical, functional, and biodegradability properties while reducing production costs and providing an environmentally friendly material [152]. Natural fibers are mainly composed of cellulose, hemicelluloses, and lignin, which vary in composition according to the fiber source. Cellulose is the main component responsible for the rigidity and stability of the fibers; hemicelluloses are short-branched chains that embed cellulose in a matrix; and lignin gives rigidity to the fibers since it works as a chemical adhesive and generates a higher mechanical resistance [153]. The most used natural fibers for producing PLA-natural fiber composites are flax, jute, kenaf, hemp, sisal, bamboo, and coir. These composites are mainly obtained by compression molding, direct or extrusion injection molding, and hot pressing. Table 5.4 presents a list of single and hybrid PLA-natural fiber composites that have been developed. The addition of natural fibers to the PLA matrix affects the physicochemical properties, such as the mechanical, thermal, structural, and degradation properties, mainly. Usually, the mechanical properties improve as the content of fiber increases to a limit, whereas the property tends to decrease [155, 159]. The interfacial adhesion between the fiber and PLA is affected when the maximum fiber load is reached since the interaction between the fibers is greater than the fiber–matrix interaction, so the mechanical load does not transfer from the matrix to the fiber. Different treatments have been applied to modify natural fibers and improve fiber compatibilities, such as the physical (e.g. plasma treatment and the electric discharge) and chemical methods (e.g. alkali and silane treatments) [168]. The mechanical properties of the PLA-natural fiber composites are also influenced by crystallinity since the fibers can act as nucleating agents (e.g. sisal), increasing the crystallinity and improving some properties, but decreasing others. In addition, the increase in crystallinity improves the thermal stability of the composites. The effect of natural fibers on the crystallinity of PLA is affected by the length of the fibers [160], among other parameters. Regarding the degradation of the composites, the water present in fibers promotes the hydrolysis of the PLA, which can accelerate the biodegradability of the composites [157]. Due to the benefits of these composites in terms of costs, processing, weight, performance, and renewability, the PLA-natural fiber composites are very useful in aerospace, automotive, building construction, 3D printing, packaging, and biomedical industries. There is still a wide area of study to improve the performance of these materials and allow them to be used in new applications, replacing conventional synthetic polymers that cause significant environmental damage.

5.5.2 Polylactic Acid – Nanocomposites

Biodegradable polymeric NCs are commonly used in food packaging and agricultural films, but they are also used in artificial joints, wound therapy, medicine administration, and orthopedic devices. PLA has several advantages that make it a popular polymer, but it also has certain drawbacks that limit its application. High brittleness, low glass transition temperature, and susceptibility to hydrolytic breakdown during the melting process are just a few of them.

The use of inorganic, organic, and hybrid nanofillers is seen as a viable solution to these issues. The use of inorganic fillers in the production of PLA-based

Table 5.4 Improvements in the properties of PLA-natural fiber composites.

Fibers	Content	Processing method	Property improvements	References
Single fiber composites				
Sugarcane bagasse	Up to 30%	Injection molding	Increase the stiffness and impact resistance, while the tensile strength is constant	[154]
Flax	7.9–17.6%	Compression molding	The tensile strength increases as the fiber content increases, while the flexural strength is constant	[155]
Hemp	≤3%	Melt compounding	Traces of residual water bound to the fibers accelerate the hydrolysis of PLA	[156]
Jute	≤50%	Compression molding	Water absorption and mechanical properties increase as fiber content increases	[157]
Sisal	5–15%	Extrusion injection molding	As the fiber content increases, the dynamic and mechanical properties improve, and sisal causes an increase in crystallinity	[158]
Elephant grass	5–25%	Injection molding	The tensile and flexural strengths of the individual composites increased up to 20%, then decreased	[159]
Bamboo	Up to 40%	Extrusion injection molding	Longer fibers increase flexural strength, while shorter fibers promote crystallinity	[160]

(*Continued*)

Table 5.4 (Continued)

Fibers	Content	Processing method	Property improvements	References
Kenaf	30–70%	Hot pressing	Composites are thermally stable and biodegradable, and the mechanical properties increase up to 50% fiber	[161]
Ramie	30%	Hot pressing	The mechanical and thermo-mechanical properties are higher than those of pure PLA	[162]
Hybrid composites				
Flax/jute	Up to 50%	Compression molding	Improve the impact strength	[163]
	Up to 30%		Higher flexural strength, impact strength, and modulus	[164]
Pineapple leaf/coir	30%	Hot pressing	The hybrid composite in a 1 : 1 ratio improves the mechanical properties and thermal stability of neat PLA	[165]
Kenaf/bamboo/coir	20%	Hot pressing	The hybrid composite improves the mechanical properties in comparison with the individual materials	[166]
Sisal/hemp	30%	Extrusion injection molding	The hybrid composite presents thermal resistance and low water absorption, increases impact resistance, and decreases crystallinity	[167]

Table 5.5 Organic and inorganic nanoparticles in PLA matrix and blends.

Organic NPs	Inorganic NPs
Peptides, protein, lipids, polysaccharides (starch, chitosan, etc.) [63]	Metals, metal oxides, and minerals. Including zinc oxide, magnesium oxide, titanium dioxide, zirconium oxide, silver, silver–copper, halloysite nanotubes, hydroxyl apatite, silica, alumina, magnetite, and calcium carbonate [169]
	Nanoclay [63]: kaolinite [170], montmorillonite-MMT [171], organo-modified-MMT and cloisite 30B [172, 173], cloisite 20A [174], cloisite 15A and 30B [175], halloysite nanotube (HNT) [176], mixtures cloisite 10A, 20A, and 30B [177]
	Carbon nanotubes (CNT) functionalized and un-functionalized [178–181], chemical modification [178, 182]
	Graphite NPs [183, 184], graphene oxide [185, 186]
	Cellulose nanocrystals [187], surface modification [188, 189], coating [190]

composites has opened several possibilities for modification, resulting in the polymer's widespread application.

Nanofillers are capable of altering the crystalline structure, increasing thermo-mechanical characteristics, thermal conductivity, permeability, and flammability. The principle behind the reinforcement is that NPs have a high surface area (SA) and SA/volume (Vs) ratio, which offers a larger interfacial area and a longer synergy between the polymer backbone and NPs; this longer connectivity enhances NCs' characteristics [169]. Table 5.5 describes the organic and inorganic NPs incorporated in PLA matrix and blends.

It has been demonstrated that these inorganic fillers enhance the properties of PLA, enabling NCs to replace fossil-based polymers in a broad range of food and nonfood applications. The authors endeavored to study the rules and literature based on these findings to evaluate, appraise, and record the scope of the research and its applicability. Additionally, inorganic NPs (nanoclay, CNTs, graphite, graphene, etc.) have been employed to enhance PLA's low melt strength, slow crystallization kinetics, and foaming behavior. Through heterogeneous cell-nucleation, the introduction of NPs increased cell density while controlling the expansion ratio by modifying the storage modulus and melt strength of PLA.

5.6 Conclusions

A biodegradable polymer known as PLA can be produced in a variety of resin grades and processed into a wide range of goods. The infrastructure of conventional polymer processing can be used to process PLA. From an environmental point of view, when recycling, reuse, and product recovery are not practical, PLA is compostable and well suited for the purpose. A rise in PLA resin demand will have a favorable impact on the world's agriculture economy because the PLA feedstock is based on agricultural raw resources. The applications of PLA as a replacement for current thermoplastics, however, are one area that has to be addressed. High-barrier protection, for instance, is crucial in packaged foods. Due to its fragility, PLA cannot be used in applications where toughness and impact resistance are essential.

The understanding of the kinetics of melt PLA crystallization is summarized in this review because crystallization is a multi-step process that involves the nucleation of primary crystals and crystal growth. Having PDLA and PLLA acting as homopolymers, PLA is best understood from the perspective of crystallization as a copolymer of monomer units of D- and L-lactic acid. Numerous studies have examined heterogeneous nucleation and plasticization as methods for improving the PLA crystallization process' kinetics. Minerals like talc, organics like hydrazide compounds, and organic-mineral hybrids could be able to boost the nucleation rate. However, adding plasticizers and nucleates to PLA led to the highest rates of crystallization. The biodegradability and biocompatibility of the PLA-based mix systems, which may be tailored for particular applications and requirements, have revealed different behaviors with and without the introduction of various additives. In addition, the potential addition of functional qualities to the mix is accelerating the creation of ternary and hybrid blends. Due to the synergy created by the interfacial contacts of the various phases in the mix, these functional qualities may have been achieved; thus, different attributes may have been influenced.

References

1 El-Sheshtawy, H.S., Fahim, I., Hosny, M., and El-Badry, M.A. (2022). Optimization of lactic acid production from agro-industrial wastes produced by *Kosakonia cowanii*. *Current Research in Green and Sustainable Chemistry* 5: 100228. https://doi.org/10.1016/j.crgsc.2021.100228.

2 More, N., Avhad, M., Utekar, S., and More, A. (2022). Polylactic acid (PLA) membrane – significance, synthesis, and applications: a review. *Polymer Bulletin* https://doi.org/10.1007/s00289-022-04135-z.

3 Lim, L.-T., Auras, R., and Rubino, M. (2008). Processing technologies for poly(lactic acid). *Progress in Polymer Science* 33: 820–852. https://doi.org/10.1016/j.progpolymsci.2008.05.004.

4 Garlotta, D. (2001). A literature review of poly(lactic acid). *Journal of Polymers and the Environment* 9: 63–84. https://doi.org/10.1023/A:1020200822435.

5 Sin, L.T., Rahmat, A.R., and Rahman, W.A.W.A. (2012). *Polylactic Acid-PLA Biopolymer Technology and Applications*. Waltham, MA: Elsevier.

6 Tsuji, H., Okino, R., Daimon, H., and Fujie, K. (2005). Water vapor permeability of poly(lactide)s: effects of molecular characteristics and crystallinity. *Journal of Applied Polymer Science* 99: 2245–2252. https://doi.org/10.1002/app.22698.

7 Sarasua, J.R., Arraiza, A.L., Balerdi, P., and Maiza, I. (2005). Crystallinity and mechanical properties of optically pure polylactides and their blends. *Polymer Engineering and Science* 45: 745–753. https://doi.org/10.1002/pen.20331.

8 Urayama, H., Moon, S.I., and Kimura, Y. (2003). Microstructure and thermal properties of polylactides with different L- and D-unit sequences: importance of the helical nature of the L-sequenced segments. *Macromolecular Materials and Engineering* 288: 137–143. https://doi.org/10.1002/mame.200390006.

9 Drumright, R.E., Gruber, P.R., and Henton, D.E. (2000). Polylactic acid technology. *Advanced Materials* 12: 1841–1846. https://doi.org/10.1002/1521-4095(200012)12:23<1841::AID-ADMA.

10 Frederiksen, C.S., Haugaard, V.K., Poll, L., and Becker, E.M. (2003). Light-induced quality changes in plain yoghurt packed in polylactate and polystyrene. *European Food Research and Technology* 217: 61–69. https://doi.org/10.1007/s00217-003-0722-3.

11 Jamshidian, M., Tehrany, E.A., and Desobry, S. (2012). Release of synthetic phenolic antioxidants from extruded poly lactic acid (PLA) film. *Food Control* 28: 445–455. https://doi.org/10.1016/j.foodcont.2012.05.005.

12 Mandelkern, L. (2004). *Crystallization of Polymers*. Cambridge, UK: Cambridge University Press.

13 Magoń, A. and Pyda, M. (2009). Study of crystalline and amorphous phases of biodegradable poly(lactic acid) by advanced thermal analysis. *Polymer* 50: 3967–3973. https://doi.org/10.1016/j.polymer.2009.06.052.

14 Tsuji, H. and Ikada, Y. (1995). Properties and morphologies of poly(L-lactide): 1. Annealing condition effects on properties and morphologies of poly(L-lactide). *Polymer* 36: 2709–2716. https://doi.org/10.1016/0032-3861(95)93647-5.

15 Blundell, D.J., Keller, A., and Kovacs, A.J. (1966). A new self-nucleation phenomenon and its application to the growing of polymers crystals from solution. *Journal of Polymer Science Part B: Polymer Letters* 4: 481–486. https://doi.org/10.1002/pol.1966.110040709.

16 Lorenzo, A.T., Arnal, M.L., Albuerne, J., and Müller, A.J. (2007). DSC isothermal polymer crystallization kinetics measurements and the use of the Avrami equation to fit the data: guidelines to avoid common problems. *Polymer Testing* 26: 222–231. https://doi.org/10.1016/j.polymertesting.2006.10.005.

17 Avrami, M. (1939). Kinetics of phase change I. General theory. *Journal of Chemical Physics* 7: 1103–1112. https://doi.org/10.1063/1.1750380.

18 Avrami, M. (1940). Kinetics of phase change. II. Transformation-time relations for random distribution of nuclei. *Journal of Chemical Physics* 8: 212–224. https://doi.org/10.1063/1.1750631.

19 Androsch, R., Schick, C., and Di Lorenzo, M.L. (2018). Kinetics of nucleation and growth of crystals of poly(L-lactic acid). *Advances in Polymer Science* 279: 235–272. https://doi.org/10.1007/12_2016_13.

20 Hikosaka, M., Amano, K., Rastogi, S., and Keller, A. (2000). Lamellar thickening growth of an extended chain single crystal of polyethylene (II): ΔT dependence of lamellar thickening growth rate and comparison with lamellar

thickening. *Journal of Materials Science* 35: 5157–5168. https://doi.org/10.1023/A:1004804420369.
21 Tsuji, H. (2005). Poly(lactide) stereocomplexes: formation, structure, properties, degradation, and applications. *Macromolecular Bioscience* 5: 569–597. https://doi.org/10.1002/mabi.200500062.
22 Ebadi-Dehaghani, H., Barikani, M., Khonakdar, H.A., and Jafari, S.H. (2015). Microstructure and non-isothermal crystallization behavior of PP/PLA/clay hybrid nanocomposites. *Journal of Thermal Analysis and Calorimetry* 121: 1321–1332. https://doi.org/10.1007/s10973-015-4554-8.
23 Pyda, M., Bopp, R.C., and Wunderlich, B. (2004). Heat capacity of poly(lactic acid). *Journal of Chemical Thermodynamics* 36: 731–742. https://doi.org/10.1016/j.jct.2004.05.003.
24 De Santis, F., Pantani, R., and Titomanlio, G. (2011). Nucleation and crystallization kinetics of poly(lactic acid). *Thermochimica Acta* 522: 128–134. https://doi.org/10.1016/j.tca.2011.05.034.
25 Refaa, Z., Boutaous, M., Xin, S., and Siginer, D.A. (2017). Thermophysical analysis and modeling of the crystallization and melting behavior of PLA with talc: kinetics and crystalline structures. *Journal of Thermal Analysis and Calorimetry* 128: 687–698. https://doi.org/10.1007/s10973-016-5961-1.
26 Kalish, J.P., Aou, K., Yang, X., and Hsu, S.L. (2011). Spectroscopic and thermal analyses of α′ and α crystalline forms of poly(L-lactic acid). *Polymer* 52: 814–821. https://doi.org/10.1016/j.polymer.2010.12.042.
27 Díaz-Díaz, A.M., López-Beceiro, J., Li, Y. et al. (2021). Crystallization kinetics of a commercial poly(lactic acid) based on characteristic crystallization time and optimal crystallization temperature. *Journal of Thermal Analysis and Calorimetry* 145: 3125–3132. https://doi.org/10.1007/s10973-020-10081-7.
28 Battegazzore, D., Bocchini, S., and Frache, A. (2011). Crystallization kinetics of poly(lactic acid)-talc composites. *Express Polymer Letters* 5: 849–858. https://doi.org/10.3144/expresspolymlett.2011.84.
29 Clarkson, C.M., el Awad Azrak, S.M., Schueneman, G.T. et al. (2020). Crystallization kinetics and morphology of small concentrations of cellulose nanofibrils (CNFs) and cellulose nanocrystals (CNCs) melt-compounded into poly(lactic acid) (PLA) with plasticizer. *Polymer* 187: 122101. https://doi.org/10.1016/j.polymer.2019.122101.
30 Bai, H., Zhang, W., Deng, H. et al. (2011). Control of crystal morphology in poly(L-lactide) by adding nucleating agent. *Macromolecules* 44: 1233–1237. https://doi.org/10.1021/ma102439t.
31 Xia, X., Shi, X., Liu, W. et al. (2017). Effect of flax fiber content on polylactic acid (PLA) crystallization in PLA/flax fiber composites. *Iranian Polymer Journal* 26: 693–702. https://doi.org/10.1007/s13726-017-0554-9.
32 Somsunan, R. and Mainoiy, N. (2020). Isothermal and non-isothermal crystallization kinetics of PLA/PBS blends with talc as nucleating agent. *Journal of Thermal Analysis and Calorimetry* 139: 1941–1948. https://doi.org/10.1007/s10973-019-08631-9.

33 Yu, W., Wang, X., Ferraris, E., and Zhang, J. (2019). Melt crystallization of PLA/talc in fused filament fabrication. *Materials and Design* 182: 108013. https://doi.org/10.1016/j.matdes.2019.108013.

34 Wen, B., Ma, L., Zou, W., and Zheng, X. (2020). Enhanced thermal conductivity of poly(lactic acid)/alumina composite by synergistic effect of tuning crystallization of poly(lactic acid) crystallization and filler content. *Journal of Materials Science – Materials in Electronics* 31: 6328–6338. https://doi.org/10.1007/s10854-020-03189-x.

35 Singh, S., Patel, M., Schwendemann, D. et al. (2020). Effect of chitin nanocrystals on crystallization and properties of poly(lactic acid)-based nanocomposites. *Polymers* 12: 726. https://doi.org/10.3390/polym12030726.

36 Shayan, M., Azizi, H., Ghasemi, I., and Karrabi, M. (2019). Influence of modified starch and nanoclay particles on crystallization and thermal degradation properties of cross-linked poly(lactic acid). *Journal of Polymer Research* 26: 10. https://doi.org/10.1007/s10965-019-1879-1.

37 Ajala, O., Werther, C., Nikaeen, P. et al. (2019). Influence of graphene nanoscrolls on the crystallization behavior and nano-mechanical properties of polylactic acid. *Polymers for Advanced Technologies* 30: 1825–1835. https://doi.org/10.1002/pat.4615.

38 Kotsilkova, R., Petrova-Doycheva, I., Menseidov, D. et al. (2019). Exploring thermal annealing and graphene-carbon nanotube additives to enhance crystallinity, thermal, electrical and tensile properties of aged poly(lactic) acid-based filament for 3D printing. *Composites Science and Technology* 181: 107712. https://doi.org/10.1016/j.compscitech.2019.107712.

39 Patwa, R., Kumar, A., and Katiyar, V. (2018). Crystallization kinetics, morphology, and hydrolytic degradation of novel bio-based poly(lactic acid)/crystalline silk nano-discs nanobiocomposites. *Journal of Applied Polymer Science* 135: 16590. https://doi.org/10.1002/app.46590.

40 Yang, R., Cao, H., Zhang, P. et al. (2021). Highly toughened and heat-resistant poly(lactic acid) with balanced strength using an unsaturated liquid crystalline polyester via dynamic vulcanization. *ACS Applied Polymer Materials* 3: 299–309. https://doi.org/10.1021/acsapm.0c01095.

41 Aliotta, L., Cinelli, P., Coltelli, M.B. et al. (2017). Effect of nucleating agents on crystallinity and properties of poly(lactic acid) (PLA). *European Polymer Journal* 93: 822–832. https://doi.org/10.1016/j.eurpolymj.2017.04.041.

42 Tábi, T. (2019). The application of the synergistic effect between the crystal structure of poly(lactic acid) (PLA) and the presence of ethylene vinyl acetate copolymer (EVA) to produce highly ductile PLA/EVA blends. *Journal of Thermal Analysis and Calorimetry* 138: 1287–1297. https://doi.org/10.1007/s10973-019-08184-x.

43 Aliotta, L., Sciara, L.M., Cinelli, P. et al. (2022). Improvement of the PLA crystallinity and heat distortion temperature optimizing the content of nucleating agents and the injection molding cycle time. *Polymers* 14: 1–16. https://doi.org/10.3390/polym14050977.

44 Nagarajan, V., Mohanty, A.K., and Misra, M. (2016). Crystallization behavior and morphology of polylactic acid (PLA) with aromatic sulfonate derivative. *Journal of Applied Polymer Science* 133: 43673. https://doi.org/10.1002/app.43673.

45 Ma, P., Xu, Y., Wang, D. et al. (2014). Rapid crystallization of poly(lactic acid) by using tailor-made oxalamide derivatives as novel soluble-type nucleating agents. *Industrial and Engineering Chemistry Research* 53: 12888–12892. https://doi.org/10.1021/ie502211j.

46 Sun, J., Li, L., and Li, J. (2019). Effects of furan-phosphamide derivative on flame retardancy and crystallization behaviors of poly(lactic acid). *Chemical Engineering Journal* 369: 150–160. https://doi.org/10.1016/j.cej.2019.03.036.

47 Xie, Q., Han, L., Shan, G. et al. (2016). Polymorphic crystalline structure and crystal morphology of enantiomeric poly(lactic acid) blends tailored by a self-assemblable aryl amide nucleator. *ACS Sustainable Chemistry & Engineering* 4: 2680–2688. https://doi.org/10.1021/acssuschemeng.6b00191.

48 Chen, P., Yu, K., Wang, Y. et al. (2018). The effect of composite nucleating agent on the crystallization behavior of branched poly(lactic acid). *Journal of Polymers and the Environment* 26: 3718–3730. https://doi.org/10.1007/s10924-018-1251-2.

49 Liu, Y., Jiang, S., Yan, W. et al. (2020). Crystallization morphology regulation on enhancing heat resistance of polylactic acid. *Polymers (Basel)* 12: 1–11. https://doi.org/10.3390/polym12071563.

50 Chen, P.Y., Lian, H.Y., Shih, Y.F. et al. (2017). Preparation, characterization and crystallization kinetics of Kenaf fiber/multi-walled carbon nanotube/polylactic acid (PLA) green composites. *Materials Chemistry and Physics* 196: 249–255. https://doi.org/10.1016/j.matchemphys.2017.05.006.

51 Chalid, M., Yuanita, E., and Pratama, J. (2015). Study of alkalization to the crystallinity and the thermal behavior of arenga pinnata "ijuk" fibers-based polylactic acid (PLA) biocomposite. *Materials Science Forum* 827: 326–331. https://doi.org/10.4028/www.scientific.net/MSF.827.326.

52 Shi, X., Zhang, G., Phuong, T.V., and Lazzeri, A. (2015). Synergistic effects of nucleating agents and plasticizers on the crystallization behavior of poly(lactic acid). *Molecules* 20: 1579–1593. https://doi.org/10.3390/molecules20011579.

53 Bhasney, S.M., Bhagabati, P., Kumar, A., and Katiyar, V. (2019). Morphology and crystalline characteristics of polylactic acid [PLA]/linear low density polyethylene [LLDPE]/microcrystalline cellulose [MCC] fiber composite. *Composites Science and Technology* 171: 54–61. https://doi.org/10.1016/j.compscitech.2018.11.028.

54 Darie-Niţa, R.N., Vasile, C., Irimia, A. et al. (2016). Evaluation of some eco-friendly plasticizers for PLA films processing. *Journal of Applied Polymer Science* 133: 43223. https://doi.org/10.1002/app.43223.

55 Farid, T., Herrera, V.N., and Kristiina, O. (2018). Investigation of crystalline structure of plasticized poly(lactic acid)/Banana nanofibers composites. *IOP Conference Series: Materials Science and Engineering* 369: 012031. https://doi.org/10.1088/1757-899X/369/1/012031.

56 Xiao, H., Lu, W., and Yeh, J.T. (2009). Effect of plasticizer on the crystallization behavior of poly(lactic acid). *Journal of Applied Polymer Science* 113: 112–121. https://doi.org/10.1002/app.29955.

57 Ljungberg, N. and Wesslén, B. (2002). The effects of plasticizers on the dynamic mechanical and thermal properties of poly(lactic acid). *Journal of Applied Polymer Science* 86: 1227–1234. https://doi.org/10.1002/app.11077.

58 Li, H. and Huneault, M.A. (2007). Effect of nucleation and plasticization on the crystallization of poly(lactic acid). *Polymer* 48: 6855–6866. https://doi.org/10.1016/j.polymer.2007.09.020.

59 Vadas, D., Nagy, Z.K., Csontos, I. et al. (2020). Effects of thermal annealing and solvent-induced crystallization on the structure and properties of poly(lactic acid) microfibres produced by high-speed electrospinning. *Journal of Thermal Analysis and Calorimetry* 142: 581–594. https://doi.org/10.1007/s10973-018-7528-9.

60 Ali, A.M. (2022). The impact of the thermal annealing conditions on the structural properties of polylactic acid fibers. *Microscopy Research and Technique* 85: 875–881. https://doi.org/10.1002/jemt.23956.

61 Yang, Y., Li, X., Zhang, Q. et al. (2019). Foaming of poly(lactic acid) with supercritical CO_2: the combined effect of crystallinity and crystalline morphology on cellular structure. *Journal of Supercritical Fluids* 145: 122–132. https://doi.org/10.1016/j.supflu.2018.12.006.

62 Pastorek, M. and Kovalcik, A. (2018). Effects of thermal annealing as polymer processing step on poly(lactic acid). *Materials and Manufacturing Processes* 33: 1674–1680. https://doi.org/10.1080/10426914.2018.1453153.

63 Nofar, M., Sacligil, D., Carreau, P.J. et al. (2019). Poly(lactic acid) blends: processing, properties and applications. *International Journal of Biological Macromolecules* 125: 307–360. https://doi.org/10.1016/j.ijbiomac.2018.12.002.

64 Feng, Y., Ma, P., Xu, P. et al. (2018). The crystallization behavior of poly(lactic acid) with different types of nucleating agents. *International Journal of Biological Macromolecules* 106: 955–962. https://doi.org/10.1016/j.ijbiomac.2017.08.095.

65 Zhao, L. and Cai, Y. (2021). Thermal performances and fluidity of biodegradable poly(L-lactic acid) filled with N,N'-oxalyl bis(piperonylic acid) dihydrazide as a nucleating agent. *Materials Science (Medžiagotyra)* 27: 458–465. https://doi.org/10.5755/j02.ms.25139.

66 Farah, S., Anderson, D.G., and Langer, R. (2016). Physical and mechanical properties of PLA, and their functions in widespread applications – a comprehensive review. *Advanced Drug Delivery Reviews* 107: 367–392. https://doi.org/10.1016/j.addr.2016.06.012.

67 Varga, N., Árpád, T., Hornok, V., and Csapó, E. (2019). Vitamin E-loaded PLA- and PLGA-based core-shell nanoparticles: synthesis, structure optimization and controlled drug release. *Pharmaceutics* 11: 1–14. https://doi.org/10.3390/pharmaceutics11070357.

68 Luo, W., Cheng, L., Yuan, C. et al. (2019). Preparation, characterization and evaluation of cellulose nanocrystal/poly(lactic acid) in situ nanocomposite scaffolds for tissue engineering. *International Journal of Biological Macromolecules* 134: 469–479. https://doi.org/10.1016/j.ijbiomac.2019.05.052.

69 Swaroop, C. and Shukla, M. (2019). Development of blown polylactic acid–MgO nanocomposite films for food packaging. *Composites Part A: Applied Science and Manufacturing* 124: 1–9. https://doi.org/10.1016/j.compositesa.2019.105482.

70 Ageyeva, T., Kovács, J.G., and Tábi, T. (2021). Comparison of the efficiency of the most effective heterogeneous nucleating agents for poly(lactic acid). *Journal of Thermal Analysis and Calorimetry* 1: 1–13. https://doi.org/10.1007/s10973-021-11145-y.

71 Petchwattana, N. and Narupai, B. (2019). Synergistic effect of talc and titanium dioxide on poly(lactic acid) crystallization: an investigation on the injection molding cycle time reduction. *Journal of Polymers and the Environment* 27: 837–846. https://doi.org/10.1007/s10924-019-01396-0.

72 Refaa, Z., Boutaous, M., Xin, S., and Fulchiron, R. (2017). Synergistic effects of shear flow and nucleating agents on the crystallization mechanisms of poly(lactic acid). *Journal of Polymer Research* 24: 1–13. https://doi.org/10.1007/s10965-016-1179-y.

73 Li, Y., Han, C., Yu, Y. et al. (2019). Effect of content and particle size of talc on nonisothermal melt crystallization behavior of poly(L-lactide). *Journal of Thermal Analysis and Calorimetry* 135: 2049–2058. https://doi.org/10.1007/s10973-018-7365-x.

74 Ding, Y., Zhang, C., Luo, C. et al. (2021). Effect of talc and diatomite on compatible, morphological, and mechanical behavior of PLA/PBAT blends. *e-Polymers* 21: 234–243. https://doi.org/10.1515/epoly-2021-0022.

75 Mayekar, P.C., Castro-Aguirre, E., Auras, R. et al. (2020). Effect of nano-clay and surfactant on the biodegradation of poly(lactic acid) films. *Polymers* 12: 1–16. https://doi.org/10.3390/polym12020311.

76 Taleb, K., Pillin, I., Grohens, Y., and Saidi-Besbes, S. (2021). Polylactic acid/Gemini surfactant modified clay bio-nanocomposites: morphological, thermal, mechanical and barrier properties. *International Journal of Biological Macromolecules* 177: 505–516. https://doi.org/10.1016/j.ijbiomac.2021.02.135.

77 Shinzawa, H. and Mizukado, J. (2020). Near-infrared (NIR) spectroscopic imaging analysis of poly(lactic acid)-nanocomposite using disrelation mapping. *Journal of Molecular Structure* 1217: 1–7. https://doi.org/10.1016/j.molstruc.2020.128332.

78 Coppola, B., Cappetti, N., Di Maio, L. et al. (2018). 3D printing of PLA/clay nanocomposites: influence of printing temperature on printed samples properties. *Materials* 11: 1–17. https://doi.org/10.3390/ma11101947.

79 Saeidlou, S., Huneault, M.A., Li, H., and Park, C.B. (2012). Poly(lactic acid) crystallization. *Progress in Polymer Science* 37: 1657–1677. https://doi.org/10.1016/j.progpolymsci.2012.07.005.

80 Wu, J., Zou, X., Jing, B., and Dai, W. (2015). Effect on sepiolite on the crystallization behavior of biodegradable poly(lactic acid) as an efficient nucleating agent. *Polymer Engineering and Science* 55: 1104–1112. https://doi.org/10.1002/pen.23981.

81 Binh, T., Thi, N., Ata, S. et al. (2020). Tailoring the electrically conductive network of injection-molded polymer–carbon nanotube composite at low filler content. *Materials Today Proceedings* 40: 5–8. https://doi.org/10.1016/j.matpr.2020.02.137.

82 Khan, T., Irfan, M.S., Ali, M. et al. (2021). Insights to low electrical percolation thresholds of carbon-based polypropylene nanocomposites. *Carbon* 176: 602–631. https://doi.org/10.1016/j.carbon.2021.01.158.

83 Rathinavel, S., Priyadharshini, K., and Panda, D. (2021). A review on carbon nanotube: an overview of synthesis, properties, functionalization, characterization, and the application. *Materials Science and Engineering B* 268: 1–28. https://doi.org/10.1016/j.mseb.2021.115095.

84 Zhang, Q. and Zhang, S. (2022). Electrical conductivity and crystallization of polylactic acid nanocomposites containing surfactant modified carbon nanotubes. *Journal of Materials Science and Chemical Engineering* 10: 30–43. https://doi.org/10.4236/msce.2022.104003.

85 Shen, B., Lu, S., Sun, C. et al. (2022). Effects of amino hyperbranched polymer-modified carbon nanotubes on the crystallization behavior of poly(L-lactic acid) (PLLA). *Polymers* 14: 1–13. https://doi.org/10.3390/polym14112188.

86 Zhao, L.-S., Cai, Y.-H., and Lio, H.-L. (2019). N,N′-Sebacic bis(hydrocinnamic acid) dihydrazide: a crystallization accelerator for poly(L-lactic acid). *e-Polymers* 19: 141–153. https://doi.org/10.1515/epoly-2019-0016.

87 Jongpanya-ngam, P., Khankrua, R., and Seadan, M. (2022). Effect of synthesized sulfonate derivatives as nucleating agents on crystallization behavior of poly(lactic acid). *Designed Monomers and Polymers* 25: 115–127. https://doi.org/10.1080/15685551.2022.2072697.

88 Fan, Y., Zhu, J., Yan, S. et al. (2015). Nucleating effect and crystal morphology controlling based on binary phase behavior between organic nucleating agent and poly(L-lactic acid). *Polymer* 67: 63–71. https://doi.org/10.1016/j.polymer.2015.04.062.

89 Huang, H.A.O., Liu, S., Li, C. et al. (2021). Poly(L-lactic acid) modified by N,N′-bis(stearic acid)-1,4-dicarboxybenzene dihydrazide: studies of crystallization, melting behavior and thermal decomposition. *Materiale Plastice* 58: 73–83. https://doi.org/10.37358/Mat.Plast.1964.

90 Hamad, K., Kaseem, M., Ayyoob, M. et al. (2018). Polylactic acid blends: the future of green, light and tough. *Progress in Polymer Science* 85: 83–127. https://doi.org/10.1016/j.progpolymsci.2018.07.001.

91 Xu, Y., Zhang, W., Qiu, Y. et al. (2022). Preparation and mechanism study of a high efficiency bio-based flame retardant for simultaneously enhancing flame retardancy, toughness and crystallization rate of poly(lactic acid). *Composites Part B: Engineering* 238: 1–15. https://doi.org/10.1016/j.compositesb.2022.109913.

92 Csizmadia, R., Faludi, G., Renner, K. et al. (2013). PLA/wood biocomposites: improving composite strength by chemical treatment of the fibers. *Composites Part A: Applied Science and Manufacturing* 53: 46–53. https://doi.org/10.1016/j.compositesa.2013.06.003.

93 Wu, W., Wu, G., and Zhang, H. (2017). Effect of wood flour as nucleating agent on the isothermal crystallization of poly(lactic acid). *Polymers for Advanced Technologies* 28: 252–260. https://doi.org/10.1002/pat.3881.

94 Ariffin, H., Yasim-Anuar, T.A.T., Amadi, N.I., and Nadia, F. (2019). Characterization of cellulose nanofiber from various tropical plant resources. In: *Lignocellulose for Future Bioeconomy*, vol. 1 (ed. H. Ariffin, S.M. Sapuan, and M.A. Hassan), 71–89. Elsevier. https://doi.org/10.1016/B978-0-12-816354-2.00005-0.

95 Agwuncha, S., Anusionwu, C., Owonubi, S.J. et al. (2019). Extraction of cellulose nanofibers and their eco-friendly polymer composites. In: *Sustainable Polymer Composites and Nanocomposites*, vol. 1, 653–691. https://doi.org/10.1007/978-3-030-05399-4_23.

96 Ariffin, H., Shazleen, S.S., Yasim-Anuar, T.A.T. et al. (2021). Functionality of cellulose nanofiber as bio-based nucleating agent and nano-reinforcement material to enhance crystallization and mechanical properties of polylactic acid nanocomposite. *Polymers* 13: 1–19. https://doi.org/10.3390/polym13030389.

97 Satsum, A., Busayaporn, W., Rungswang, W. et al. (2022). Structural and mechanical properties of biodegradable poly(lactic acid) and pectin composites: using bionucleating agent to improve crystallization behavior. *Polymer Journal* 54: 921–930. https://doi.org/10.1038/s41428-022-00637-9.

98 Bouapao, L. and Tsuji, H. (2009). Stereocomplex crystallization and spherulite growth of low molecular weight poly(L-lactide) and poly(D-lactide) from the melt. *Macromolecular Chemistry and Physics* 210: 993–1002. https://doi.org/10.1002/macp.200900017.

99 Shi, X., Jing, Z., and Zhang, G. (2018). Influence of PLA stereocomplex crystals and thermal treatment temperature on the rheology and crystallization behavior of asymmetric poly(L-lactide)/poly(D-lactide) blends. *Journal of Polymer Research* 25: 1–16. https://doi.org/10.1007/s10965-018-1467-9.

100 Ji, N., Hu, G., Li, J., and Ren, J. (2019). Influence of poly(lactide) stereocomplexes as nucleating agents on the crystallization behavior of poly(lactide)s. *RSC Advances* 9: 6221–6227. https://doi.org/10.1039/c8ra09856e.

101 Iglesias Montes, M.L., Cyras, V.P., Manfredi, L.B. et al. (2020). Fracture evaluation of plasticized polylactic acid/poly(3-hydroxybutyrate) blends for commodities replacement in packaging applications. *Polymer Testing* 84: 106375. https://doi.org/10.1016/j.polymertesting.2020.106375.

102 Zheng, H., Tang, H., Yang, C. et al. (2022). Evaluation of the slow-release polylactic acid/polyhydroxyalkanoates active film containing oregano essential oil on the quality and flavor of chilled pufferfish (*Takifugu obscurus*) fillets. *Food Chemistry* 385: 132693. https://doi.org/10.1016/j.foodchem.2022.132693.

103 Martinez Villadiego, K., Arias Tapia, M.J., Useche, J., and Escobar Macias, D. (2022). Thermoplastic starch (TPS)/polylactic acid (PLA) blending methodologies: a review. *Journal of Polymers and the Environment* 30: 75–91. https://doi.org/10.1007/s10924-021-02207-1.

104 Macosko, C.W. (2000). Morphology development and control in immiscible polymer blends. *Macromolecular Symposia* 149: 171–184. https://doi.org/10.1002/15213900(200001)149:1b171::aid-masy171N3.0.co;2-8.

105 Jacobsen, S. and Fritz, H.G. (1996). Filling of poly(lactic acid) with native starch. *Polymer Engineering and Science* 36: 2799–2804. https://doi.org/10.1002/pen.10680.

106 Ke, T., Sun, S.X., and Seib, P. (2003). Blending of poly(lactic acid) and starches containing varying amylose content. *Journal of Applied Polymer Science* 89: 3639–3646. https://doi.org/10.1002/app.12617.

107 Zhang, J.F. and Sun, X. (2004). Mechanical properties and crystallization behavior of poly(lactic acid) blended with dendritic hyperbranched polymer. *Polymer International* 53: 716–722. https://doi.org/10.1002/pi.1457.

108 Wang, H., Sun, X., and Seib, P. (2001). Strengthening blends of poly(lactic acid) and starch with methylenediphenyl diisocyanate. *Journal of Applied Polymer Science* 82: 1761–1767. https://doi.org/10.1002/app.2018.abs.

109 Zhang, J.F. and Sun, X. (2004). Mechanical properties of poly(lactic acid)/starch composites compatibilized by maleic anhydride. *Biomacromolecules* 5: 1446–1451. https://doi.org/10.1021/bm0400022.

110 Wu, C.S. (2005). Improving polylactide/starch biocomposites by grafting polylactide with acrylic acid – characterization and biodegradability assessment. *Macromolecular Bioscience* 5: 352–361. https://doi.org/10.1002/mabi.200400159.

111 Schwach, E., Six, J.L., and Avrous, L. (2008). Biodegradable blends based on starch and poly(lactic acid): comparison of different strategies and estimate of compatibilization. *Journal of Polymers and the Environment* 16: 286–297. https://doi.org/10.1007/s10924-008-0107-6.

112 Liu, J., Jiang, H., and Chen, L. (2012). Grafting of glycidyl methacrylate onto poly(lactide) and properties of PLA/starch blends compatibilized by the grafted copolymer. *Journal of Polymers and the Environment* 20: 810–816. https://doi.org/10.1007/s10924-012-0438-1.

113 Xiong, Z., Sonqi, L., Ma, S. et al. (2013). Effect of castor oil enrichment layer produced by reaction on the properties of PLA/HDI-g-starch blends. *Carbohydrate Polymers* 94: 235–243. https://doi.org/10.1016/j.carbpol.2013.01.038.

114 Palai, B., Biswal, M., Mohanty, S., and Nayak, S.K. (2019). In situ reactive compatibilization of polylactic acid (PLA) and thermoplastic starch (TPS) blends; synthesis and evaluation of extrusion blown films thereof. *Industrial Crops and Products* 141: 111748. https://doi.org/10.1016/j.indcrop.2019.111748.

115 Serra-Parareda, F., Delgado-Aguilar, M., Espinach, F.X. et al. (2022). Sustainable plastic composites by polylactic acid-starch blends and bleached kraft hardwood fibers. *Composites Part B: Engineering* 238: 109901. https://doi.org/10.1016/j.compositesb.2022.109901.

116 Huneault, M.A. and Li, H. (2007). Morphology and properties of compatibilized polylactide/thermoplastic starch blends. *Polymer* 48: 270–280. https://doi.org/10.1016/j.polymer.2006.11.023.

117 Yu, L., Dean, K., Yuan, Q. et al. (2007). Effect of compatibilizer distribution on the blends of starch/biodegradable polyesters. *Journal of Applied Polymer Science* 103: 812–818. https://doi.org/10.1002/app.25184.

118 Ning, W., Jiugao, Y., and Xiaofei, M. (2008). Preparation and characterization of compatible thermoplastic dry starch/poly(lactic acid). *Polymer Composites* 29: 551–559. https://doi.org/10.1002/pc.20399.

119 Gao, H., Hu, S., Su, F. et al. (2011). Mechanical, thermal, and biodegradability properties of PLA/modified starch blends. *Polymer Composites* 32: 2093–2100. https://doi.org/10.1002/pc.21241.

120 Shin, B., Jang, S., and Kim, B. (2011). Thermal, morphological, and mechanical properties of biobased and biodegradable blends of poly(lactic acid) and chemically modified thermoplastic starch. *Polymer Engineering and Science* 51: 826–834. https://doi.org/10.1002/pen.21896.

121 Akrami, M., Ghasemi, I., Azizi, H. et al. (2016). A new approach in compatibilization of the poly(lactic acid)/thermoplastic starch (PLA/TPS) blends. *Carbohydrate Polymers* 144: 254–262. https://doi.org/10.1016/j.carbpol.2016.02.035.

122 Abdelwahab, M., Flynn, A., Chiou, B. et al. (2012). Thermal, mechanical and morphological characterization of plasticized PLA/PHB blends. *Polymer Degradation and Stability* 97: 1822–1828. https://doi.org/10.1016/j.polymdegradstab.2012.05.036.

123 Armentano, I., Fortunati, E., Burgos, N. et al. (2015). Processing and characterization of plasticized PLA/PHB blends for biodegradable multiphase systems. *Express Polymer Letters* 9: 583–596. https://doi.org/10.3144/expresspolymlett.2015.55.

124 Arrieta, M., Lopez, J., Hernandez, A., and Rayon, E. (2014). Ternary PLA-PHB-limonene blends intended for biodegradable food packaging applications. *European Polymer Journal* 50: 255–270. https://doi.org/10.1016/j.eurpolymj.2013.11.009.

125 Zhang, L., Xiong, C., and Deng, X. (1996). Miscibility, crystallization, and morphology of poly(β-hydroxybutyrate)/poly(D,L-lactide) blends. *Polymer* 37: 235–241. https://doi.org/10.1016/0032-3861(96)81093-7.

126 Kikkawa, Y., Suzuki, T., Tsuge, T. et al. (2009). Effect of phase structure on enzymatic degradation in poly(L-lactide)/atactic poly(3-hydroxybutyrate) blends with different miscibility. *Biomacromolecules* 10: 1013–1018. https://doi.org/10.1021/bm900117J.

127 Kervran, M., Vagner, C., Cochez, M. et al. (2022). Thermal degradation of polylactic acid (PLA)/polyhydroxybutyrate (PHB) blends: a systematic review. *Polymer Degradation and Stability* 201: 109995. https://doi.org/10.1016/j.polymdegradstab.2022.109995.

128 Bartczak, Z., Galeski, A., Kowalczuk, M. et al. (2013). Tough blends of poly(lactide) and amorphous poly([R,S]-3-hydroxy butyrate)–morphology and

properties. *European Polymer Journal* 49: 3630–3641. https://doi.org/10.1016/j.eurpolymj.2013.07.033.

129 Oztemur, J. and Yalcin-Enis, I. (2021). Morphological analysis of fibrous webs electrospun from polycaprolactone, polylactic acid and their blends in chloroform based solvent systems. *Materials Today Proceedings* 46: 2161–2166. https://doi.org/10.1016/j.matpr.2021.02.638.

130 Tsuji, H. and Ikada, Y. (1998). Blends of aliphatic polyesters. II. Hydrolysis of solution-cast blends from poly(L-lactide) and poly(ε-caprolactone) in phosphate-buffered solution. *Journal of Applied Polymer Science* 67: 405–415. https://doi.org/10.1002/(SICI)1097-4628(19980118)67:3%3C405::AID-APP3%3E3.0.CO;2-Q.

131 Gaona, L., Ribelles, J., Perilla, J., and Lebourg, M. (2012). Hydrolytic degradation of PLLA/PCL microporous membranes prepared by freeze extraction. *Polymer Degradation and Stability* 97: 1621–1632. https://doi.org/10.1016/j.polymdegradstab.2012.06.031.

132 Fukushima, K., Feijoo, J., and Yang, M. (2013). Comparison of abiotic and biotic degradation of PDLLA, PCL and partially miscible PDLLA/PCL blend. *European Polymer Journal* 49: 706–717. https://doi.org/10.1016/j.eurpolymj.2012.12.011.

133 Sivalingham, G. (2004). Thermal degradation of binary physical mixtures and copolymers of poly(3-caprolactone), poly(D,L-lactide), poly(glycolide). *Polymer Degradation and Stability* 84: 393–398. https://doi.org/10.1016/s0141-3910(03)00411-7.

134 Vieira, A., Vieira, J., Ferra, J. et al. (2011). Mechanical study of PLA–PCL fibers during in vitro degradation. *Journal of the Mechanical Behavior of Biomedical Materials* 4: 451–460. https://doi.org/10.1016/j.jmbbm.2010.12.006.

135 Wang, L., Ma, W., Gross, R., and Mccarthy, S. (1998). Reactive compatibilization of biodegradable blends of poly(lactic acid) and poly(ε-caprolactone). *Polymer Degradation and Stability* 59: 161–168. https://doi.org/10.1016/s0141-3910(97)00196-1.

136 Na, Y., He, Y., Shuai, X. et al. (2002). Compatibilization effect of poly(ε-caprolactone)-b-poly(ethylene glycol) block copolymers and phase morphology analysis in immiscible poly(lactide)/poly(ε-caprolactone) blends. *Biomacromolecules* 3: 1179–1186. https://doi.org/10.1021/bm020050r.

137 Semba, T., Kitagawa, K., Ishiaku, U., and Hamada, H. (2006). The effect of crosslinking on the mechanical properties of polylactic acid/polycaprolactone blends. *Journal of Applied Polymer Science* 101: 1816–1825. https://doi.org/10.1002/app.23589.

138 Tuba, F., Olh, L., and Nagy, P. (2011). Characterization of reactively compatibilized poly(D,L-lactide)/poly(ε-caprolactone) biodegradable blends by essential work of fracture method. *Engineering Fracture Mechanics* 78: 3123–3133. https://doi.org/10.1016/j.engfracmech.2011.09.010.

139 Shin, B. and Han, D. (2013). Compatibilization of immiscible poly(lactic acid)/poly(ε-caprolactone) blend through electron-beam irradiation with

the addition of a compatibilizing agent. *Radiation Physics and Chemistry* 83: 98–104. https://doi.org/10.1016/j.radphyschem.2012.10.001.

140 Al-Mulla, E., Ibrahim, N., Shameli, K. et al. (2013). Effect of epoxidized palm oil on the mechanical and morphological properties of a PLA–PCL blend. *Research on Chemical Intermediates* 40: 689–698. https://doi.org/10.1007/s11164-012-0994-y.

141 Gardella, L., Calabrese, M., and Monticelli, O. (2014). PLA maleation: an easy and effective method to modify the properties of PLA/PCL immiscible blends. *Colloid and Polymer Science* 292: 2391–2398. https://doi.org/10.1007/s00396-014-3328-3.

142 Teamsinsungvon, A., Ruksakulpiwat, Y., and Jarukumjorn, K. (2013). Preparation and characterization of poly(lactic acid)/poly(butylene adipate-co-terepthalate) blends and their composite. *Polymer-Plastics Technology and Engineering* 52: 1362–1367. https://doi.org/10.1080/03602559.2013.820746.

143 Aversa, C., Barletta, M., Cappiello, G., and Gisario, A. (2022). Compatibilization strategies and analysis of morphological features of poly(butylene adipate-co-terephthalate) (PBAT)/poly(lactic acid) PLA blends: a state-of-art review. *European Polymer Journal* 173: 1–27. https://doi.org/10.1016/j.eurpolymj.2022.111304.

144 Jiang, L., Wolcott, M.P., and Zhang, J. (2006). Study of biodegradable polylactide/poly(butylene adipate-co-terephthalate) blends. *Biomacromolecules* 7: 199–207. https://doi.org/10.1021/bm050581q.

145 Deng, Y., Yu, C., Wongwiwattana, P., and Thomas, N. (2018). Optimising ductility of poly(lactic acid)/poly(butylene adipate-co-terephthalate) blends through co-continuous phase morphology. *Journal of Polymers and the Environment* 26: 3802–3816. https://doi.org/10.1007/s10924-018-1256-x.

146 Gigante, V., Canesi, I., Cinelli, P. et al. (2019). Rubber toughening of polylactic acid (PLA) with poly(butylene adipate-co-terephthalate) (PBAT): mechanical properties, fracture mechanics and analysis of ductile-to-brittle behavior while varying temperature and test speed. *European Polymer Journal* 115: 125–137. https://doi.org/10.1016/j.eurpolymj.2019.03.015.

147 Park, J. and Im, S. (2002). Phase behavior and morphology in blends of poly(L-lactic acid) and poly(butylene succinate). *Journal of Applied Polymer Science* 86: 647–655. https://doi.org/10.1002/app.10923.

148 Bhatia, A., Gupta, R., Bhattacharya, S., and Choi, H. (2007). Compatibility of biodegradable poly(lactic acid) (PLA) and poly(butylene succinate) (PBS) blends for packaging application. *Korea-Australia Rheology Journal* 19: 125–131.

149 Yokohara, T. and Yamaguchi, T. (2008). Structure and properties for biomass-based polyester blends of PLA and PBS. *European Polymer Journal* 44: 677–685. https://doi.org/10.1016/j.eurpolymj.2008.01.008.

150 Wang, R., Wang, S., Zhang, Y. et al. (2009). Toughening modification of PLLA/PBS blends via in situ compatibilization. *Polymer Engineering and Science* 49: 26–33. https://doi.org/10.1002/pen.21210.

151 Harada, M., Ohya, T., Iida, K. et al. (2007). Increased impact strength of biodegradable poly(lactic acid)/poly(butylene succinate) blend composites by using isocyanate as a reactive processing agent. *Journal of Applied Polymer Science* 106: 1813–1820. https://doi.org/10.1002/app.26717.

152 Siakeng, R., Jawaid, M., Ariffin, H. et al. (2019). Natural fiber reinforced polylactic acid composites: a review. *Polymer Composites* 40: 446–463. https://doi.org/10.1002/pc.24747.

153 Ilyas, R., Zuhri, M., Aisyah, H. et al. (2022). Natural fiber-reinforced polylactic acid, polylactic acid blends and their composites for advanced applications. *Polymers (Basel)* 14: 202. https://doi.org/10.3390/polym14010202.

154 Bartos, A., Nagy, K., Anggono, J. et al. (2021). Biobased PLA/sugarcane bagasse fiber composites: effect of fiber characteristics and interfacial adhesion on properties. *Composites Part A: Applied Science and Manufacturing* 143: 106273. https://doi.org/10.1016/j.compositesa.2021.106273.

155 Motru, S., Adithyakrishna, V., Bharath, J., and Guruprasad, R. (2020). Development and evaluation of mechanical properties of biodegradable PLA/flax fiber green composite laminates. *Materials Today Proceedings* 24: 641–649. https://doi.org/10.1016/j.matpr.2020.04.318.

156 Mazzanti, V., Salzano de Luna, M., Pariante, R. et al. (2020). Natural fiber-induced degradation in PLA-hemp biocomposites in the molten state. *Composites Part A: Applied Science and Manufacturing* 137: 105990. https://doi.org/10.1016/j.compositesa.2020.105990.

157 Singh, J., Singh, S., and Dhawan, V. (2020). Influence of fiber volume fraction and curing temperature on mechanical properties of jute/PLA green composites. *Polymers and Polymer Composites* 28: 273–284. https://doi.org/10.1177/0967391119872875.

158 Samouh, Z., Molnar, K., Boussu, F. et al. (2019). Mechanical and thermal characterization of sisal fiber reinforced polylactic acid composites. *Polymers for Advanced Technologies* 30: 529–537. https://doi.org/10.1002/pat.4488.

159 Gunti, R., Ratna, A., and Gupta, A. (2018). Mechanical and degradation properties of natural fiber-reinforced PLA composites: jute, sisal, and elephant grass. *Polymer Composites* 39: 1125–1136. https://doi.org/10.1002/pc.24041.

160 Gamon, G., Evon, P., and Rigal, L. (2013). Twin-screw extrusion impact on natural fibre morphology and material properties in poly(lactic acid) based biocomposites. *Industrial Crops and Products* 46: 173–185. https://doi.org/10.1016/j.indcrop.2013.01.026.

161 Ochi, S. (2008). Mechanical properties of kenaf fibers and kenaf/PLA composites. *Mechanics of Materials* 40: 446–452. https://doi.org/10.1016/j.mechmat.2007.10.006.

162 Yu, T., Li, Y., and Ren, J. (2009). Preparation and properties of short natural fiber reinforced poly(lactic acid) composites. *Transactions of Nonferrous Metals Society of China* 19: s651–s655. https://doi.org/10.1016/S1003-6326(10)60126-4.

163 Ejaz, M., Azad, M., Shah, A. et al. (2020). Mechanical and biodegradable properties of jute/flax reinforced PLA composites. *Fibers and Polymers* 21: 2635–2641. https://doi.org/10.1007/s12221-020-1370-y.

164 Manral, A., Ahmad, F., and Chaudhary, V. (2019). Static and dynamic mechanical properties of PLA bio-composite with hybrid reinforcement of flax and jute. *Materials Today Proceedings* 25: 577–580. https://doi.org/10.1016/j.matpr.2019.07.240.

165 Siakeng, R., Jawaid, M., Ariffin, H., and Sapuan, S. (2019). Mechanical, dynamic, and thermomechanical properties of coir/pineapple leaf fiber reinforced polylactic acid hybrid biocomposites. *Polymer Composites* 40: 2000–2011. https://doi.org/10.1002/pc.24978.

166 Yusoff, R., Takagi, H., and Nakagaito, A. (2016). Tensile and flexural properties of polylactic acid-based hybrid green composites reinforced by kenaf, bamboo and coir fibers. *Industrial Crops and Products* 94: 562–573. https://doi.org/10.1016/j.indcrop.2016.09.017.

167 Pappu, A., Pickering, K., and Thakur, V. (2019). Manufacturing and characterization of sustainable hybrid composites using sisal and hemp fibres as reinforcement of poly(lactic acid) via injection moulding. *Industrial Crops and Products* 137: 260–269. https://doi.org/10.1016/j.indcrop.2019.05.040.

168 Sanivada, U., Mármol, G., Brito, F., and Fangueiro, R. (2020). PLA composites reinforced with flax and jute fibers – a review of recent trends, processing parameters and mechanical properties. *Polymers (Basel)* 12: 1–29. https://doi.org/10.3390/polym12102373.

169 Mulla, M., Rahman, M., Marcos, B. et al. (2021). Polylactic acid (PLA) nanocomposites: effect of inorganic nanoparticles reinforcement on its performance and food packaging applications. *Molecules* 26: 1967. https://doi.org/10.3390/molecules26071967.

170 Cabedo, L., Feijoo, J., Villanueva, M. et al. (2006). Optimization of biodegradable nanocomposites based on aPLA/PCL blends for food packaging applications. *Macromolecular Symposia* 233: 191–197. https://doi.org/10.1002/masy.200690017.

171 Salehiyan, R. and Hyun, K. (2013). Effect of organoclay on non-linear rheological properties of poly(lactic acid)/poly(caprolactone) blends. *Korean Journal of Chemical Engineering* 30: 1013–1022. https://doi.org/10.1007/s11814-013-0035-6.

172 Urquijo, J., Dagréou, S., Guerrica-Echevarría, G., and Eguiazábal, J. (2016). Structure and properties of poly(lactic acid)/poly(ε-caprolactone) nanocomposites with kinetically induced nanoclay location. *Journal of Applied Polymer Science* 133: 43815. https://doi.org/10.1002/app.43815.

173 Wokadala, O., Ray, S., Bandyopadhyay, J. et al. (2015). Morphology, thermal properties and crystallization kinetics of ternary blends of the polylactide and starch biopolymers and nanoclay: the role of nanoclay hydrophobicity. *Polymer* 71: 82–92. https://doi.org/10.1016/j.polymer.2015.06.058.

174 Kumar, M., Mohanty, S., Nayak, S., and Parvaiz, M. (2010). Effect of glycidyl methacrylate (GMA) on the thermal, mechanical and morphological property of biodegradable PLA/PBAT blend and its nanocomposites. *Bioresource Technology* 101: 8406–8415. https://doi.org/10.1016/j.biortech.2010.05.075.

175 Ebadi-Dehaghani, H., Khonakdar, H., Barikani, M., and Jafari, S. (2015). Experimental and theoretical analyses of mechanical properties of PP/PLA/clay nanocomposites. *Composites Part B: Engineering* 69: 133–144. https://doi.org/10.1016/j.compositesb.2014.09.006.

176 Rashmi, B., Prashantha, K., Lacrampe, M., and Krawczak, P. (2016). Toughening of poly(lactic acid) without sacrificing stiffness and strength by melt-blending with polyamide 11 and selective localization of halloysite nanotubes. *AIP Conf. Proc.* 1713: 060001. https://doi.org/10.1063/1.4942284.

177 Ock, H., Ahn, K., Lee, S., and Hyun, K. (2016). Characterization of compatibilizing effect of organoclay in poly(lactic acid) and natural rubber blends by FT-rheology. *Macromolecules* 49: 2832–2842. https://doi.org/10.1021/acs.macromol.5b02157.

178 Raja, M., Ryu, S.H., and Shanmugharaj, A. (2013). Thermal, mechanical and electroactive shape memory properties of polyurethane (PU)/poly(lactic acid) (PLA)/CNT nanocomposites. *European Polymer Journal* 49: 3492–3500. https://doi.org/10.1016/j.eurpolymj.2013.08.009.

179 Shi, Y., Li, Y., Xiang, F. et al. (2011). Carbon nanotubes induced microstructure and mechanical properties changes in cocontinuous poly(L-lactide)/ethylene-*co*-vinyl acetate blends. *Polymers for Advanced Technologies* 23: 783–790. https://doi.org/10.1002/pat.1959.

180 Wu, D., Zhang, Y., Zhang, M., and Yu, W. (2009). Selective localization of multiwalled carbon nanotubes in poly(ε-caprolactone)/polylactide blend. *Biomacromolecules* 10: 417–424. https://doi.org/10.1021/bm801183f.

181 Wu, D., Lin, D., Zhang, J. et al. (2011). Selective localization of nanofillers: effect on morphology and crystallization of PLA/PCL blends. *Macromolecular Chemistry and Physics* 212: 613–626. https://doi.org/10.1002/macp.201000579.

182 Jang, M.G., Lee, Y.K., and Kim, W.N. (2015). Influence of lactic acid-grafted multi-walled carbon nanotube (LA-g-MWCNT) on the electrical and rheological properties of polycarbonate/poly(lactic acid)/LA-g-MWCNT composites. *Macromolecular Research* 23: 916–923. https://doi.org/10.1007/s13233-015-3129-7.

183 Kelnar, I., Kratochvíl, J., Kaprálková, L. et al. (2017). Graphite nanoplatelets-modified PLA/PCL: effect of blend ratio and nanofiller localization on structure and properties. *Journal of the Mechanical Behavior of Biomedical Materials* 71: 271–278. https://doi.org/10.1016/j.jmbbm.2017.03.028.

184 Forouharshad, M., Gardella, L., Furfaro, D. et al. (2015). A low environmental-impact approach for novel bio-composites based on PLLA/PCL blends and high surface area graphite. *European Polymer Journal* 70: 28–36. https://doi.org/10.1016/j.eurpolymj.2015.06.016.

185 Wu, W., Wu, C., Peng, H. et al. (2017). Effect of nitrogen-doped graphene on morphology and properties of immiscible poly(butylene succinate)/polylactide blends. *Composites Part B: Engineering* 113: 300–307. https://doi.org/10.1016/j.compositesb.2017.01.037.

186 Botlhoko, O.J., Ramontja, J., and Ray, S.S. (2017). Thermal, mechanical, and rheological properties of graphite- and graphene oxide-filled biodegradable polylactide/poly(ε-caprolactone) blend composites. *Journal of Applied Polymer Science* 134: 45373. https://doi.org/10.1002/app.45373.

187 Arrieta, M., Fortunati, E., Dominici, F. et al. (2014). Multifunctional PLA–PHB/cellulose nanocrystal films: processing, structural and thermal properties. *Carbohydrate Polymers* 107: 16–24. https://doi.org/10.1016/j.carbpol.2014.02.044.

188 Luzi, F., Fortunati, E., Jiménez, A. et al. (2016). Production and characterization of PLA-PBS biodegradable blends reinforced with cellulose nanocrystals extracted from hemp fibres. *Industrial Crops and Products* 93: 276–289. https://doi.org/10.1016/j.indcrop.2016.01.045.

189 Bitinis, N., Fortunati, E., Verdejo, R. et al. (2013). Poly(lactic acid)/natural rubber/cellulose nanocrystal bionanocomposites. Part II: properties evaluation. *Carbohydrate Polymers* 96: 621–627. https://doi.org/10.1016/j.carbpol.2013.03.091.

190 Heshmati, V., Kamal, M.R., and Favis, B.D. (2017). Tuning the localization of finely dispersed cellulose nanocrystal in poly(lactic acid)/bio-polyamide 11 blends. *Journal of Polymer Science Part B: Polymer Physics* 56: 576–587. https://doi.org/10.1002/polb.24563.

6

Crystallization in PLLA-Based Blends, and Composites

Pratick Samanta[1] and Bhanu Nandan[2]

[1]KTH Royal Institute of Technology, Department of Fiber and Polymer Technology, KTH Campus Teknikringen 56–58, Stockholm, 10044, Sweden
[2]Indian Institute of Technology Delhi, Department of Textile and Fibre Engineering, IIT Delhi Main Rd, IIT Campus, Hauz Khas, New Delhi, Delhi 110016, India

6.1 Introduction

The applications of fossil-based polymeric materials are developing stress on environment day by day, and this has become a serious concern now [1–3]. Therefore, bio-based polymeric materials are getting significant attention in recent past. Poly(lactic acid) (PLA) derived from bio-resources like, corn starch or sugar cane is one such interesting polymer [4, 5]. It has comparable mechanical strength, optical transparency, and modulus as poly(ethylene terephthalate) (PET). But slow crystallization kinetics, poor stability against heat, low elongation at break, and process complexity limit its practical applications. PLA shows three types of optical isomers, which are poly(L-lactic acid) (PLLA), poly(D-lactic acid) (PDLA), and poly(D, L-lactic acid) (PDLLA). The PLLA and PDLA are both semi-crystalline in nature but the blend of 50% L and 50% D forms amorphous phase. Commercially available PLA contains major fraction of L-form and minor fraction of D-form. Presence of "D" units in PLLA influences the thermal and mechanical behavior [6, 7]. Moreover, the hydrolytic degradation of PLLA is controlled by the crystallinity percentage [8, 9]. It is proposed that the crystal structure and lamellar arrangement have critical role in controlling the rate of degradation. The crystallized PLA film showed the reduced permeability coefficient (water and oxygen) than amorphous PLA [10]. Therefore, knowledge of crystallization of PLLA is important from fundamental point and material design aspect. In this chapter, crystallization behavior of pure PLLA, and its different forms (in blends, in composites, and after adding nucleating agents, etc.) are described here.

Polymer Crystallization: Methods, Characterization, and Applications, First Edition.
Edited by Jyotishkumar Parameswaranpillai, Jenny Jacob, Senthilkumar Krishnasamy, Aswathy Jayakumar, and Nishar Hameed.
© 2023 WILEY-VCH GmbH. Published 2023 by WILEY-VCH GmbH.

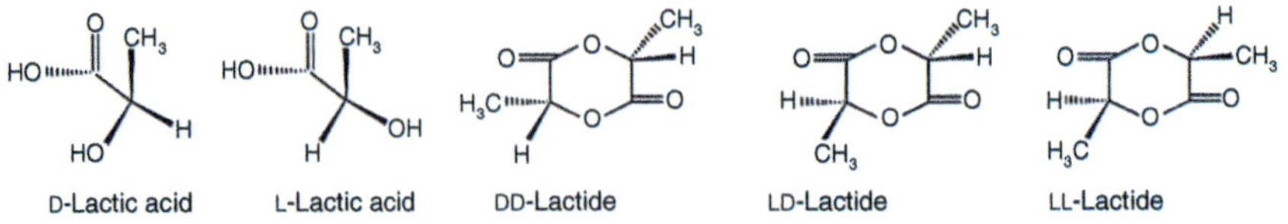

6 Crystallization in PLLA-Based Blends, and Composites

Figure 6.1 Chemical structure of the lactic acid and lactide.

6.2 Chemical and Crystal Structure of PLLA

The knowledge of chemical structure of PLLA is essential to understand its crystallization behavior. PLA is synthesized mainly through polycondensation of lactic acid or ring-opening polymerization of lactide [11]. PLA consists of asymmetric carbon atoms in the chemical structure. Hence, it shows optical isomers, which are L-lactic acid and D-lactic acid (Figure 6.1). Once, PLA is prepared from lactide, it shows three forms, such as LL-lactide is prepared from two L-lactates, DD-lactide is produced from two DD-lactates, and LD-lactide or meso-lactide is prepared from one L-lactates and one D-lactates. The schematic of lactic acid and lactide structures is shown in Figure 6.1.

The PLLA and PDLA are prepared from pure L and D forms of PLA, respectively. Industrially prepared PLA chips contain PLLA-rich fraction because bacteria follow the fermentation route to produce lactic acid. PLA may contain 1–2% D-form due to complexity of the purification process.

PLLA shows mainly three (α, β, and γ) types of crystal structures. The α-form of crystals is formed during crystallization from solution or melt [12–17]. Here, two polymer chains form 10_3 helical conformation and are arranged in orthorhombic unit crystal with the dimensions $a = 10.7$ Å, $b = 6.45$ Å, and $c = 27.8$ Å (fiber axis) [15]. Slightly modified form of α crystal is reported as $\acute{\alpha}$ crystal form of PLA [18]. This crystal structure is similar to α phase but has less densely packed chains. Such types of crystals are developed during crystallization below 120 °C. However, other reports show that in the crystallization temperature range 100–120 °C both crystal phases (α and $\acute{\alpha}$) can be developed [19, 20]. The β phase is formed during crystallization at high temperature or during high-speed stretching process [17, 21]. In β phase, three polymer chains are packed in a trigonal cell $a = b = 610.52$ Å and $c = 8.8$ Å [22]. The γ crystal form is obtained through epitaxial crystallization on hexamethylbenzene [21]. In γ phase, two anti-parallel 3_1 helical polymer chains are arranged in orthorhombic unit cell with dimensions $a = 9.95$ Å, $b = 6.25$ Å, and $c = 8.8$ Å.

6.3 PLLA Properties: Glass Transition and Melting Temperature

6.3.1 Glass Transition Temperature

The glass transition temperature (T_g) of PLLA decides the crystallization window since the chains mobility is related to melting temperature (T_m) – T_g. The $T_{g\,PLLA}$

($M_n \sim 17 \times 10^3$) is $\sim 59\,°C$, and it increases with increasing molecular weight [23] and reaches to equilibrium level after certain higher molecular weight [24]. Also, the glass transition temperature is controlled by the polymer chains architecture; hence, branched polymer shows lower transition temperature than linear chains due to more free volume present [25, 26]. The $T_{g\,PLLA}$ reduces upon blending with copolymer such as D-lactate or PDLLA [23, 24, 27, 28].

6.3.2 Melting Temperature

The melting temperature of PLLA depends on its molecular weight. Sample with $M_w < 3000$ showed weak endothermic peak due to unstable crystallized region and this transformed into sharp single melting peak with increasing molecular weight [24]. Melting temperature increased with increasing molecular weight and approached toward constant value of $\sim 184\,°C$. The below Flory equation (Eq. (6.1)) informs about the melting temperature of polymers with different molecular weights.

$$1/T_m^\infty - 1/T_m = 2RM_0/\Delta H_m M_n \tag{6.1}$$

Here, T_m^∞ is the melting temperature with infinite M_w, M_0 denotes molecular mass of repeating unit, "R" is the universal gas constant, and ΔH_m is the heat of fusion. The plot $1/M_n$ against $1/T_m$ will provide straight line. Slope of the plot signifies $2RM_0/\Delta H_m$ and intercept represents T_m^∞. The Flory equation is suitable for high M_w polymer. The measured melting temperature and heat of fusion from the plot were $184\,°C$ and $3.5\,kcal\,mol^{-1}$ [24].

6.4 PLLA Crystallization

6.4.1 PLLA Crystallization Study Through Spherulite Growth

The polarized optical microscope is used to study isothermal crystallization through the determination of nucleation density and spherulite growth rate [29, 30]. Here, polymer is heated above the melting temperature and held for some time to erase previous thermal history. The same is then quenched to pre-decided temperature to evaluate the spherulite growth rate and nucleation rate with time. Several articles in the past have studied the spherulite growth with crystallization temperature for PLLA. It was observed that spherulite density increased with decreasing temperature or crystallization temperature (Figure 6.2a,b) [31, 33].

With increase in spherulite density, induction time for primary nucleation reduces. The nucleation process becomes slow with increasing crystallization temperature and this becomes more difficult near melting temperature because of lower degree of undercooling [34]. The crystal growth process is evaluated through the measurement of spherulite radius (r) and crystallization time (t). Typically, diameter of spherulite increases linearly with increase of time at particular crystallization temperature. The slope of "r" vs "t" reveals about growth rate constant (G)

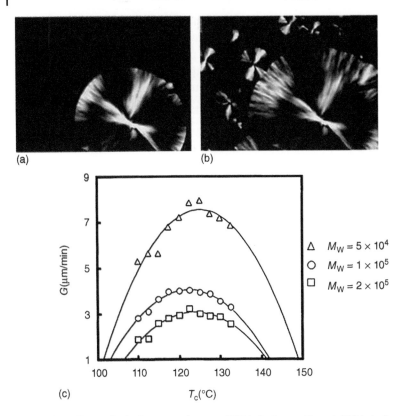

Figure 6.2 The optical microscope image of PLLA during cooling at 5 °C/min from melt phase (a) at 120 °C and (b) at 110 °C. Source: Di Lorenzo [31]/Reproduced with permission from Elsevier and (c) the growth rate with crystallization temperature for different molecular weights PLLA. Source: Miyata and Masuko [32]/with permission of Elsevier.

for spherulite. This means "G" is constant at particular crystallization temperature [35, 36]. Figure 6.2c represents the spherulite growth rate as function of crystallization temperature for various molecular weights of PLLA. The spherulite growth rate (G) increases with increasing crystallization temperature for certain temperature range and then starts decreasing. Once degree of supercooling increases, the driving force for secondary nucleation enhances the "G." Moreover, the diffusion of polymer chains toward growth front of spherulite is challenging due to high melt viscosity at higher supercooling. This led to decrease in "G" value. Also, increase in molecular weight of polymer chains reduces the spherulite growth rate because of limited polymer chains' mobility at particular temperature [37].

6.4.2 Lauritzen and Hoffman Theory in PLLA Crystallization

The classical Lauritzen and Hoffman (LH) theory is often used to discuss the crystallization kinetics of liner polymer chains through nucleation and growth rate. It consists of three regimes according to the ratio of nucleation rate and chain deposition rate on the crystal surface. Low undercooling is regime I where nucleation rate is much lower than chain growth rate. As temperature reduces, the nucleation rate increases and polymer chains growth rate decreases. Both can be comparable after certain cooling. Such temperature range represents regime II. Whereas, large degree of supercooling denotes regime III where nucleation rate is higher than polymer chains growth rate. Based on LH theory, crystal growth rate (G) is shown in Eq. (6.2).

$$G = G_0 \exp\left(-\frac{U^*}{R(T_c - T_\infty)}\right) \exp\left(-\frac{K_g}{T_c \Delta T f}\right) \quad (6.2)$$

The first part of Eq. (6.2) describes the chains diffusion behavior, and second part says about secondary nucleation barrier. The "G_0" is pre-exponential factor, "R" is universal gas constant, "T_c" is isothermal crystallization temperature, "$\Delta T = (T_m^0 - T_c)$" (T_m^0 symbolizes equilibrium melting temperature) is degree of supercooling, "U^*" signifies activation energy for chain diffusion, $T_\infty = T_g - 30$ K is temperature where flow ceases, K_g is nucleation constant, and "f" ($f = \frac{2T_c}{T_m^0 + T_c}$) is the correction factor for temperature (Eq. (6.3)).

$$K_g = \frac{jb\sigma\sigma_e T_m^0}{K_B \Delta H_m} \quad (6.3)$$

where, "j" is constant and it depends on the regime ("j" is equal to 2 for regime II and 4 for regime I and III), "K_B" is Boltzmann constant, "ΔH_m" is the heat of fusion of 100% crystalline, T_m^0 denotes equilibrium melting temperature, "b" is the thickness of surface nucleus, here, "σ" indicates surface free energy, and "σ_e" stands for fold surface free energy. The parameter values of LH measurement for PLLA is shown in Table 6.1.

Table 6.1 Parameters values of Lauritzen and Hoffman measurement.

Parameters in Lauritzen and Hoffman equation	Description	Value	References
K_B	Boltzmann constant	1.38×10^{-23} J K^{-1}	
ΔH_m	Heat of fusion of 100% crystalline	82 J g^{-1}	[38]
T_m^0	Equilibrium melting temperature	200 °C, Hoffman-Weeks	[38]
		205 °C, Hoffman-Weeks	[39]
		215 °C, Hoffman-Weeks	[40]
σ_e	Fold surface free energy	60.89×10^{-3} J m^{-2}	• [41]
σ	Surface free energy	12.03×10^{-3} J m^{-2}	• [41]
b	Thickness of surface nucleus	5.17×10^{-10} m	[40]
U^*	Activation chain diffusion	6.27×10^3 J mol^{-1}	[41]
R	Gas constant	8.314 J K^{-1} mol^{-1}	

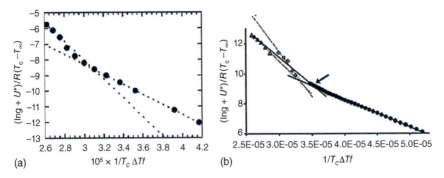

Figure 6.3 Lauritzen and Hoffman (LH) graphs of different PLLA samples. (a) $M_w = 206$ kg mol^{-1}. Source: Kawai et al. [19], American Chemical Society and (b) $M_w = 101$ kg mol^{-1}. Source: Di Lorenzo [31]/with permission of Elsevier.

The plot "$(\ln g + U^*)/R(T_c - T_\infty)$" against "$1/T_c \Delta Tf$" gives straight line. The slope and intercept of the plot symbolizes "$-K_g$" and "$\ln(G_0)$," respectively. PLLA crystallization kinetics study using LH theory and LH graphs for different PLLA samples are shown in Figure 6.3.

Kawai et al. [19] showed transition of PLLA ($M_w = 206$ kg mol^{-1}) occurred from regime III to regime II at 120 °C and regime III to regime II at 147 °C. The calculated "K_g" values were 6.02×10^5 and 3.22×10^5 K^2 for the temperatures below and above 120 °C, respectively. The alteration in crystallization regimes of PLLA depends on tacticity, co-monomer content, and molecular weight [23, 42]. It was noticed in previous study that PLLA did not change crystal morphology at 120 °C during transition from regime III to regime II. But transition in crystal morphology occurred during changing from regime II to regime I at 147 °C [23]. Moreover, hexagonal-shaped crystal morphology was found above 147 °C whereas, spherulite crystal morphology was noticed below 147 °C for PLLA. Di Lorenzo [31] classified the three temperature ranges with "K_g" values for PLLA. In the range of 75 ° C $\leq T_c \leq$ 100 ° C : $K_g = 4.38 \times 10^5$ K^2, for 108 ° C $\leq T_c \leq$ 120 ° C : $K_g = 5.97 \times 10^5$ K^2 and for, $T_c \geq$ 128 ° C : $K_g = 1.85 \times 10^5$ K^2.

6.4.3 Crystallization Kinetics Through Calorimetry Study

The calorimetry method is an alternative approach to study the crystallization kinetics through isothermal or non-isothermal techniques. In isothermal crystallization process, polymer is quenched to predefined crystallization temperature (T_c) from melt stage. The heat flow is measured with time until crystallization process reaches saturation level. The heat flow during crystallization is converted into crystallinity fraction and plotted against time (or temperature). Typical crystallinity development with temperature for PLLA during isothermal crystallization study is shown in Figure 6.4a. This plot can be fitted using classical Avrami equation (Eq. (6.4)) to describe isothermal crystallization kinetics for PLLA (Figure 6.4b).

$$X_t = 1 - \exp[(-kt)^n] \tag{6.4}$$

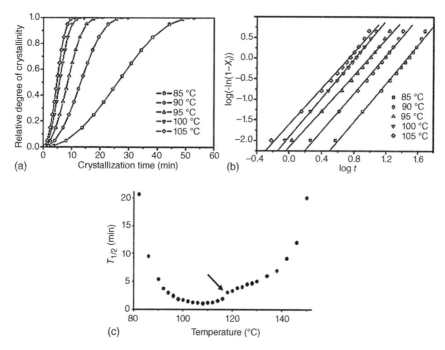

Figure 6.4 Isothermal crystallization behavior of pure PLLA. (a) Development of crystallinity with time at different temperatures. Source: Yu and Qiu [43]/American Chemical Society, (b) Avrami plots at different crystallization temperatures [43], and (c) half crystallization time at different crystallization temperatures. Source: Di Lorenzo [31]/with permission of Elsevier.

where, "X_t" is the fraction of crystallinity, "n" and "k" are Avrami exponent and kinetic rate constant, respectively. The plot $\lg([(-\ln(1-X_t)]$ against $\lg(t)$ gives straight line (Figure 6.4b). The slope and intercept are used to determine the Avrami exponent (n) value and growth rate (k). The Avrami exponent describes the nucleation mechanism and growth dimension of crystals. The exponent "$n \leq 1$" indicates homogeneous nucleated crystallization with one-dimensional growth, Avrami exponent (n) for PLLA was measured from 2.53 to 3.01 in the temperature range 105–125 °C [44]. Another study showed the Avarmi exponent value for PLLA ($M_n = 1.23 \times 10^5$ g mol^{-1}) varied from 2.5 to 2.8 with crystallization temperature range of 90–130 °C [45]. These Avrami exponent values revealed heterogeneous nucleated crystallization with two- to three-dimensional growth (i.e. disc). The exponent values for PLLA were also found to be in the 3.5–4.5 range. This suggests heterogeneous nucleated crystallization with three-dimensional growth (i.e. spherulite) [32, 46, 47]. The crystallization rate can be compared among different PLLA samples based on the time required to reach 50% crystallinity. The crystallization half-time of polymer could be measured using the Eq. (6.5) below.

$$t_{1/2} = [\ln2/k]^{1/n} \tag{6.5}$$

The crystallization half-time ($t_{1/2}$) against crystallization temperature for PLLA is shown in Figure 6.4c [31]. The graph shows the crystallization rate is slow near melting and glass transition temperatures as the process depends on polymer chains mobility and nucleation rate. The faster crystallization rate is observed near 108 °C [31]. In non-isothermal process, sample is cooled from melt phase at different cooling rates. The crystallization curves shifted to lower temperature range with increasing cooling rate. Transition from liquid to solid happened for PLLA ($M_w = 101$ kg mol^{-1}) at 140 °C during cooling at 5 °C/min, and the crystallization rate became faster near 118 °C. However, the phase transition occurred at 130 °C during cooling at 10 °C/min. Further, faster rate of crystallization was noticed at 118 °C [31]. Same Avrami equation and extent of Avarmi equations are applied to measure the growth rate constant (k) and exponent value (n) to describe the crystallization kinetics for PLLA under non-isothermal conditions [48–51].

6.5 Crystallization of PLLA in Blends

The crystallization behavior of PLLA after blending with different polymers [52–58] is shown in Table 6.2. Shibata et al. [60] prepared PLLA blends with poly(butylene succinate) (PBS) and poly(butylene succinate-co-L-lactate) (PBSL) separately. The crystallization of PLLA was enhanced (isothermal and non-isothermal crystallization) after addition of small fraction of PBSL, whereas PBS did not significantly affect the crystallization of PLLA. Ikada and Tsuji [62] studied the effect of molecular weight of atactic PDLLA on crystallization behavior of PLLA ($M_v = 3.9 \times 10^5$).

Table 6.2 Crystallization behavior of PLLA blended with different polymers.

Blend	Composition	T_g	T_c (°C)	H_c (Jg^{-1})	T_m (°C)	H_m (Jg^{-1})	n	k	References
PLLA	100		115	45	179	47	3.0	3.5×10^{-4}	[59]
PLLA:PCL	90 : 10		100	35	178	55	2.7	2.1×10^{-3}	[59]
PLLA:PCL	80 : 20		100	35	179	56	3.0	8.1×10^{-4}	[59]
PLLA:PCL	70 : 30		100	34	177	56	3.0	7.8×10^{-4}	[59]
PLLA	100	60.5	136.3	6.38	164.5	7.02	2.59	2.47×10^{-8}	[60]
PLLA:PBS	99 : 01	60.6	133.9	2.53	164.0	4.56	2.66	2.07×10^{-8}	[60]
PLLA:PBS	95 : 05	59.9	135.6	14.9	164.8	16.6	2.45	1.20×10^{-7}	[60]
PLLA:PBS	90 : 10	58.2	128.9	30.6	162.8	33.4	2.36	1.77×10^{-7}	[60]
PLLA:PBSL	99 : 01	59.4	132.0	23.1	162.8	25.3	2.67	5.38×10^{-8}	[60]
PLLA:PBSL	95 : 05	58.2	124.2	33.2	162.4	38.1	2.69	2.48×10^{-8}	[60]
PLLA:PBSL	90 : 10	58.1	127.3	28.4	163.8	31.6	2.81	2.86×10^{-8}	[60]
PLLA	100		98.9	23.2	162.4	33.3	2.46	1.9×10^{-4}	[61]
PLLA:PDLA	99 : 01		98.6	25.9	162.7	32.2	2.89	5.69×10^{-5}	[61]
PLLA:PDLA	97 : 03		97.9	28.3	162.4	28.5	2.81	2.45×10^{-4}	[61]
PLLA:PDLA	95 : 05		109.3	10.4	160.8	21.9	2.52	3.78×10^{-3}	[61]
PLLA:PDLA	92 : 08		132.4		164.7	29.1	2.82	6.40×10^{-3}	[61]
PLLA:PDLA	90 : 10		128.6		165.3	26.9	2.62	1.65×10^{-3}	[61]

PLLA crystallized and formed spherulites in presence of PDLLA. The size of spherulites was larger in blended system as compared to non-blended system, irrespective of molecular weight. With increase in PDLLA molecular weight, the spherulites became more disordered. Also, the crystallization process of PLLA became slower in blended system and initial increase in overall crystallinity turned into less significant with increase in molecular weight of PDLLA. Moreover, reduction in equilibrium melting temperature was noticed after blending with PDLLA. Crystallinity increased for PLLA blend with the lowest molecular weight PDLLA as compared to other blends and non-blended systems. Saturation time for PLLA crystallization became shorter for blend-annealed sample after quenching from melt as compared to directly annealed from melt without quenching. Sun et al. [63] investigated crystallization behavior of PLLA and PDLA blends. The cooling and heating cycles of PLLA/PDLA blend with different compositions are shown in Figure 6.5a,b, respectively. In the heating cycle, PLLA showed single melting peak at 179 °C and double melting peaks (broad 178 and 225 °C) were observed for blends. During cooling cycle, single crystallization peak was detected at 106 °C for pure PLLA. All blend samples showed crystallization peak at 100 °C except for PLLA/PDLA (70/30). PLLA/PDLA (70/30) showed crystallization temperature of 122 °C. Once, the blended samples were cooled at various temperatures like 250, 240, and 190 °C at the rate of 5 °C/min, the stereocomplex (sc) crystallites remained at diverse stages. The "sc" behaved as nucleating agent and significantly influenced PLLA crystallization. The temperature-dependent XRD study at 240 °C showed that all PLLA/PDLA blended samples contained stereocomplex (sc) crystallites though, this was not detected in DSC study. Wei et al. [64] studied stereocomplex crystallites (sc) in PLLA/PDLA blends. The PLLA/PDLA blend formed "sc," which showed 50 °C higher melting temperature than PLLA or PDLA crystallites; hence, it reserved in PLLA melt phase in the asymmetric blend of PLLA/PDLA and behaved as nucleating agent for PLLA. The nucleation efficiency of "sc" on PLLA with different PDLA weight concentrations is shown in Figure 6.5c. As PDLA concentration increased, the nucleation efficiency increased and then saturated. The development of relative crystallinity with time during isothermal crystallization at 137 °C for PLLA/PDLA blends is shown in Figure 6.5d. The figure clearly shows that the overall crystallization improved remarkably with the "sc" crystallites. The overall crystallization rate was independent of PDLA concentration.

Li et al. [67] studied PLLA/PDLA blend to prepare foam with oriented crystal structure. Incorporation of molecular orientation and stereocomplex crystals enhanced the crystallization of PLLA. Ikada and coworkers [68] studied poly(ethylene oxide) (PEO)/PLLA solution-cast blended films. During this study, it was observed that small fraction of PEO was trapped inside the amorphous phase between crystallites of PLLA and, hence, PLLA crystallization was hindered. Further, the miscibility and crystallization behavior of PLLA with poly(ethylene glycol) (PEG) and poly(ε-caprolactone) (PCL) were studied in blend system, where PLLA was not less than 50% [69]. The molecular weight of PEG and PCL was $M_w \sim 10\,000$. Change in crystallization rate of PLLA was noticed for PLLA/PEG blend. The spherulite growth rate increased with increasing PEG composition.

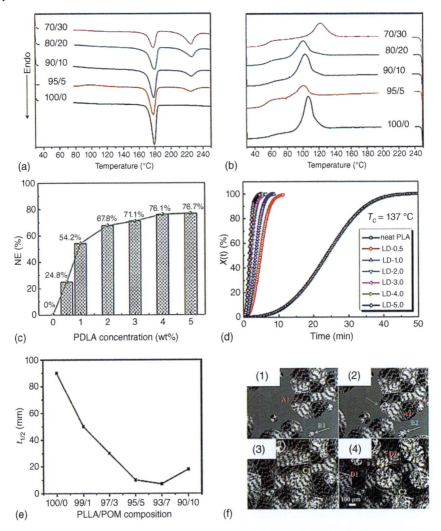

Figure 6.5 The DSC plots of pure PLLA and PLLA/PDLA blends from RT to 250 °C at the heating/cooling rate of 5 °C. (a) heating (b) cooling cycles. Source: Sun et al. [63]/American Chemical Society. (c) the efficiency of nucleation of "sc" crystallites on PLLA with different PDLA weight concentrations. [64]. (d) relative crystallinity with time during isothermal crystallization at 137 °C for PLLA/PDLA blends. Source: Wei et al. [64]/American Chemical Society. (e) Crystallization half-time of pure PLLA and PLLA/POM blends. Source: Qiu et al. [65]/American Chemical Society. (f) The PLLA/POM (90/10) sample crystallized isothermally at 145° for different times under polarized light ((1) 45, (2) 50, (3) 85, and (4) 165 minutes). Source: Ye et al. [66]/Reproduced with permission from American Chemical Society.

The crystallization ability of PEG reduced after blending with PLLA. But the crystallization ability was unaffected for PLLA after blending with PEG. The PLLA/PCL blend showed enhancement in crystallization rate for PLLA due to the partial miscibility with PCL. Dell'Erba et al. [59, 70] studied similar PLLA/PCL blends where PLLA was major fraction and PCL was present as minor fraction.

PCL weight fraction was varied from 10% to 30% in the blend. The PLLA/PCL blend was immiscible in nature but it was not incompatible. Improvement in crystallization rate of PLLA was noticed because of increase in nucleation rate. Sakurai and Banpean [71] studied crystallization behavior of PLLA/PEG (50/50) blend. Here, sample was heated to 180 °C and held for five minutes to achieve complete melt phase without any phase separation. Later, sample was first cooled to 127 °C and then to 43 °C. In such typical condition, PEG crystallized on the preformed PLLA spherulites. But crystallization was not observed in the amorphous phase of PLLA/PEG because of the suppression of freezing temperature (T_f) of PEG in the PLLA/PEG (50/50) blend. During crystallization of PLLA at 127 °C, the PEG amount increased in the amorphous region inside of PLLA spherulites due to formation of PLLA crystals. Specifically, PLLA fraction was reduced in amorphous region of PLLA/PEG blend. Later, T_f of PEG improved and PEG crystallized only inside the spherulite of PLLA. With increase in crystallization time at 127 °C, the crystallinity of PEG increased and induction time of PEG crystallization decreased.

Qiu et al. [65] studied improvement in crystallization rate of PLLA using polyoxymethylene (POM) crystals fragment in PLLA/POM blends. POM formed finely dispersed crystal fragments in the blend during cooling from melt. The POM crystal fragments enhanced the crystallization of PLLA matrix and increased the crystallinity of PLLA in the blend. PLLA crystallization depended on the POM fraction in the blend. The highest crystallization rate was found for 7 wt% POM contained sample. At low fraction, POM acted as nucleating agent. However, at POM fraction, POM formed large spherulites, which could not effectively nucleate PLLA. The crystallization half-time (isothermal study at 135 °C) of pure PLLA and PLLA/POM different compositions is shown in Figure 6.5e. The crystallization rate increased with increasing POM content. The highest rate of crystallization was noticed for 7% POM-contained sample. The PLLA/POM-blended sample with lower weight fraction of POM showed excellent optical transmission and mechanical properties. Ye et al. [66] studied the crystallization of PLLA with POM at different blend compositions. PLLA and POM both have close melting temperatures and, thus, they can crystallize simultaneously or individually based on their blend composition and crystallization temperature. This led to formation of fascinating morphology. At, 3 wt% $< \varphi_{POM} < 20$ wt%, crystallization kinetics of POM was comparable with PLLA. Hence, two types of spherulites showed side-by-side concurrent growth with diffusion of PLLA spherulites into the POM crystals. The growth rate of POM crystals was slightly higher than that of PLLA. At $\varphi_{POM} = 3$ wt% ,PLLA/POM formed novel core–shell morphology. At, 20 wt% $\leq \varphi_{POM} < 80$ wt%, PLLA/POM formed interpenetrated blended spherulite morphology. For $\varphi_{POM} \geq 80$ wt%, PLLA chains were confined by POM chains and crystallized into small crystals. The PLLA/POM (90/10) spherulites formed after isothermal crystallization at 145 °C for different times under polarized light ((1) 45, (2) 50, (3) 85, and (4)165 minutes) are shown in Figure 6.5f.

Pan and coworkers [72] studied liquid crystal (LC) and PLLA blend. LC formed metastable phase at lower crystallization temperature of PLLA and stable phase formed at higher crystallization temperature of PLLA in the mixture of PLLA/LC.

The formation of stable and metastable phases of LC gave information related to interspherulitic and interlamellar phases, respectively, of the polymer crystals.

6.6 Crystallization of PLLA in Nanocomposites

PLLA crystallization in nanocomposites was studied in detail in recent past [73–81]. For example, ternary blend of polymers was prepared with PLLA, poly(methyl methacrylate) (PMMA), and PEO. Later, the blend was used as matrix for two types of clays such as organically modified vermiculite (OVMT) and organically modified montmorillonite (Clo10A) to develop nanocomposites [82]. Addition of PMMA in the blend resulted in the reduced chains mobility and crystallization ability of PLLA/PEO mixture. The effect was less significant at higher PEO-contained blend. Enhancement in the crystallization process was noticed in presence of OVMT and Clo10A in the blend. Crystallization kinetics was described successfully using Mo method. Tsuji et al. [83] investigated the influence of type of solvent and percentage of fullerene (C_{60}) loading in crystallization of PLLA during cooling from molten phase, heating from RT, and time of solvent evaporation. The C_{60} improved the crystallization of PLLA during heating of melt-quenched film, time of cooling of as-cased film from melt stage, and at the time of solvent evaporation. For solvent evaporation, the difference in crystallinity between solvents with and without containing C_{60} became higher for lower boiling point solvents. For heating of melt-quenched film, after addition of C_{60}, less significant impact was noticed on PLLA crystallinity. Pan et al. [84] prepared nanocomposites with PLLA and organically modified layered double hydroxide (LDH) through melt mixing. The layer distance of dodecyl sulfate-modified LDH (LDH-DS) increased in PLLA/LDH nanocomposites. Pan et al. [85] also studied composites prepared with PLLA, zinc phenylphosphonate (PPZn), and layered metal phosphonate. The cooling cycle and heating cycle of PLLA nanocomposites prepared with different materials (1% PDLA, 1% PPZn, and 1% talc) are shown in Figure 6.6a,b. PLLA did not show any crystallization peak during cooling process but in subsequent heating cycle, cold crystallization peak was detected at 115 °C. Once, 1% PDLA was added in PLLA, broad crystallization peak was noticed at 97 °C. The crystallization process was enhanced after addition of 1% talc but promising result was noticed after incorporation of 1% PPZn. After addition of only 0.02% PPZn, PLLA crystallized at undercooling rate of 10 °C/min. Also, PLLA crystallization rate increased with increasing PPZn loading. The thermal behavior of PLLA/PPZn nanocomposites with different weight percentages of PPZn is shown in Figure 6.6c,d. The crystallization half-time reduced from 28 to only 0.33 minutes at 130 °C and from 60.2 to 1.4 minutes at 140 °C after incorporating 15% PPZn compared to pure PLLA. In the presence of PPZn, the number of nuclei increased and spherulite size reduced. Based on crystal structure analysis, assumed nucleation mechanism was epitaxial for PLLA/PPZn.

Li et al. [86] studied PLLA nanocomposites prepared by the melt blending technique. By incorporating twice-functionalized organoclay (TFC) in nanocomposites,

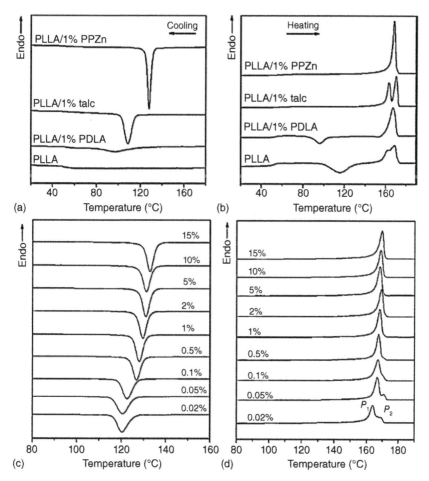

Figure 6.6 The DSC (a) cooling and (b) heating curves of pure PLLA and PLLA composites (containing PDLA, PPZn, and talc) [85]. The cooling and heating rates were 10 °C/min. The thermal behavior of PLLA/PPZn composites with different percentages PPZn in the composites. (c) cooling cycle (d) heating cycle. Both cycles were tested under 10 °C/min. Source: Pan et al. [85]/American Chemical Society.

the crystallization half-time and induction period were reduced for PLLA. On addition of more TFC, cracks were generated in PLLA crystals in the nanocomposites. The melting temperature of PLLA decreased with increasing TFC contained in the nanocomposite. Percentage crystallinity reached the highest level after incorporation of 5% TCF but the percentage crystallinity decreased immensely after loading of 10% TFC. Fu and coworkers [87] investigated the influence of matrix crystallization on mechanical and electrical conductive behaviors of PLLA/multiwalled carbon nanotubes (MWCNTs) composites. After adding small fraction of nucleating agent (0.15 wt%) and adjusting time (0.1–8 minutes) in isothermal crystallization of PLLA matrix in a hot mold at 130 °C, injection molded nanocomposites bars were prepared with different matrix crystallinity

(5–45%). The electrical conductivity improved with rise of matrix crystallinity in the nanocomposites at lower concentrations of MWCNTs (less than percolation threshold). The electrical conductivity was not increased effectively with higher loading of MMCNTs (more than percolation threshold). Huang et al. [88] tried to gain knowledge on the effect of graphene oxide nanosheet (GONS) loading on crystallization behavior of PLLA in PLLA/GONS composites. The GONS loading was varied from 0 to 4.0 wt%. The GONS acted as heterogeneous nucleating agent for PLLA and it reduced the mobility of PLLA chains. At lower weight fractions of GONS (0.25 and 0.5 wt%), the GONS acted as templates where PLLA chains were landed because of high specific surface area. This led to reduction in nucleation barrier and conformational ordering for PLLA chains. After addition of GONS, crystallization rate of PLLA increased due to the dominance of nucleation by the GONS. Once GONS weight fraction was raised to 1.0%, it started to form network structure inside PLLA. Such events limited the mobility and diffusion of PLLA chains. In such conditions, PLLA crystallization mechanism turned from promoting to restricting crystallization. Further rise in GONS loading (4.0 wt%), formed denser network, which pushed the PLLA lamella to grow two-dimensionally. Li et al. [89] studied PLLA and graphene nanosheet (GNS) composites prepared through solution blending technique. It was found that ά phase was more favorable than α-form of PLLA in PLLA/GNS composites. The GNS enhanced the crystallization rate (nucleation density and spherulites growth rate) of PLLA in PLLA/GNS composite. The PLLA/GNS composite showed lower long-range ordering compared to pure PLLA. Recently, the PLLA/Zinc Oxide (ZnO) composite was developed for UV-shielding coating for packaging applications [90]. ZnO was considered the most appropriate for UV-shielding applications among other metallic nanoparticles. Here, ZnO acted as nucleating agent for PLLA; thus, in addition to 0.05% ZnO, half-time for crystallization was reduced from 7.4 to 4.7 minutes only. Moreover, 0.45 wt% loaded PLLA/ZnO nanocomposite showed ability to restrict 61.2% UV radiation and allowed 95.9% visible light to pass through the sample. The $T_{g\,PLLA}$ raised after adding ZnO in PLLA because of the confinement effect on PLLA chains. Jin et al. [91] investigated crystallization behavior of PLLA with surface-modified carbon nanofibers. Here, carboxyl functional carbon nanofibers (f-CNFs) and those further grafted with PEG (g-CNFs) were used with PLLA. The surface modification enhanced the dispersion phase of f-CNFs or g-CNFs in PLLA and increased the interfacial adhesion in PLLA/f-CNFs or PLLA/g-CNFs composites. In PLLA, the raw carbon nanofibers (r-CNFs) and f-CNFs showed activation energies for nucleation −12.49 and −11.81 kJ mol^{-1}, respectively. The nucleation efficiency was weak, and activation energy for nucleation was found −5.45 kJ mol^{-1} for g-CNFs. In isothermal study, PLLA/g-CNFs showed higher spherulites growth rate in comparison with PLLA/f-CNFs. The spherulite morphology of pure PLLA and PLLA/CNTs-g-PLLA nanocomposites during isothermal crystallization study was observed under polarized optical microscopy [79]. This clearly showed that spherulites density decreased with rise in crystallization temperature (from 130 to 140 °C) for pure PLLA. The spherulite density was higher in PLLA/CNTs-g-PLLA nanocomposites compared to pure PLLA. Such event indicated that CNTs-g-PLLA acted as heterogeneous nucleating agent and assisted the crystallization process at higher temperatures.

Shi et al. [92] discussed PLLA/TSOS (trisilanolheptaphenyl POSS) nanocomposites prepared through casting and solution methods. Here, TSOS acted as nucleating agent for PLLA and supported recrystallization of quenched PLLA during cold crystallization. In isothermal crystallization study, the crystallization rate of PLLA was suppressed after addition of TSOS molecules in PLLA due to formation of hydrogen bonding between carbonyls groups in PLLA chain and hydroxyls groups in TSOS. In the mixture of PLLA/TSOS, the α and ά crystals were found. The ά phase was detected here because, the TSOS was responsible for hydrogen bond formation. This led to generation of distorted packing of chains.

6.7 Crystallization of PLLA in Block Copolymer

The crystallization of PLLA has also been extensively studied in block copolymers with PLLA as one of the blocks. Ho and coworkers [93] synthesized polystyrene-b-poly(L-lactide) (PS-b-PLLA) diblock copolymers with low and high molecular weights to prepare weekly and strongly separated thin film morphologies (disordered texture and ordered lamellar). Large size, well-aligned lamellar morphology was developed using crystallizable solvents (benzoic acid [BA] and hexamethylbenzene [HMB]). For strongly segregated PS-b-PLLA (high molecular weight), oriented morphology was found due to the micro-phase separation of PS-b-PLLA. For weekly segregated PS-b-PLLA (low molecular weight), oriented microstructure was detected. This group also investigated the chiral effect on self-assembly in PS-b-PLLA diblock copolymer [94]. A phase with hexagonally packed PLLA helices in PS matrix was observed through self-assembly of PS-rich diblock copolymer having volume fraction of PLLA $f^v_{PLLA} = 0.34$. However, no such morphology was observed for polystyrene-block-poly(D.L-lactide) (PS-b-PLA). By controlling the PLLA crystallization temperature, interesting nanostructures were obtained such as crystalline helices when $T_{c,PLLA} < T_{g,PS}$ and crystalline cylinder when $T_{c,PLLA} \geq T_{g,PLLA}$. PS-b-PLLA having PLLA-rich ($f^v_{PLLA} = 0.65$) phase developed core–shell cylinders packed in hexagonally with helical sense where PS formed shell and PLLA formed matrix and core. The self-assembled superstructure was decided by the competition between microphase segregation and crystallization in PS-b-PLLA [95]. Through the slow self-assembly (i.e. the slow rate of nonsolvent addition) single-crystal lozenge lamellae were prepared by PS-b-PLLA having large PS chain. The fast self-assembly process (i.e. the fast rate of nonsolvent addition) developed amorphous helical ribbon. Also, achiral diblock copolymer (poly(styrene)-b-poly(DL-lactide) [PS-PLA]) formed amorphous flat ribbon structure. This indicated that chirality plays critical role in development of helical ribbon structures. For short PS chain in PS-b-PLLA, the crystallization of PLLA dominates, irrespective of nonsolvent addition rate. This showed that the crystallization mechanism of PLLA depends on PS chain length. The formation of helical structure through self-assembly in PS-b-PLLA reflected the importance of the chirality on phase segregation. Huang et al. [96] explored the crystallization behavior of PLLA in poly(L-lactide)-*block*-methoxy poly(ethylene glycol) (PLLA-*b*-MePEG) diblock copolymers. In the block copolymer, irrespective of block ratio, no significant change in crystals was found for either polymer. The isothermal kinetics were greatly influenced by the link between PLLA and MePEG. The melting temperature of PLLA ($T_{m,PLLA}$) decreased in the block copolymer. The crystallinity of PLLA also decreased with its decreasing molecular weight. At lower crystallization temperatures, where crystallization is controlled by the growth mechanism, the crystallization ability of PLLA in block copolymer was greater than pure PLLA.

Cai et al. [97] studied the crystallization behavior of poly(L-lactide)-*b*-poly(ethylene oxide) (PLLA-*b*-PEO) block copolymers with different branching. They synthesized six arms (6sPLLA-*b*-PEO), four arms (4sPLLA-*b*-PEO), linear containing one arm (LPLLA-*b*-PEO), and two arms (2LPLLA-*b*-PEO) block copolymers. As number of branching increased, the cold crystallization temperature, melting temperature, and percentage crystallinity of PLLA in the block copolymer decreased. Also, spherulite growth rate decreased with increasing number of branching in the block copolymer ($G_{L\ PLLA-b-PEO} > G_{2LPLLA-b-PEO} > G_{4sLPLLA-b-PEO} > G_{6sLPLLA-b-PEO}$). Hamley et al. [98] studied the sequential crystallization of PLLA followed by PCL in PLLA-*b*-PCL diblock copolymers. Three different samples were investigated here. The shortest PLLA block (~32 wt%) started crystallization from homogeneous melt stage, whereas other two samples (~44 and 60 wt%), began from microphase segregated morphology. The microphase segregated morphology at melt stage varied as crystallization of PLLA proceeded at 122 °C. Once PCL was present as major fraction, crystallization of PCL followed at 42 °C after crystallization of PLLA. This led to rearrangement of lamellar morphology. The modified unit crystal of PLLA was found by the crystallization of PCL in the block copolymer having higher PCL weight fraction. The crystallization rate of PLLA decreased with increasing PCL weight fraction in the block copolymer. Wang et al. [99] studied the crystallization behavior of poly(ester urethane) copolymer based on PLLA as soft part. PLLA crystallization was restricted by phase segregation and the crystallization of urethane. In another study, crystallization behavior of PLLA was investigated in triblock copolymers poly(L-lactide)-*block*-poly(vinylidene fluoride)-*block*-poly(L-lactide) (PLLA-*b*-PVDF-*b*-PLLA) [100]. Through the ring-opening polymerization, double-crystalline PLLA-*b*-PVDF-*b*-PLLA was synthesized. The block copolymers were miscible in the melt phase, and alternative lamellae morphology was formed during crystallization from homogeneous melt stage. Asymmetric block copolymer with lower PLLA-contained sample showed reduction in crystallization temperature due to the fractionated crystallization effect. Yang et al. [101] studied the double-crystalline poly(L-lactide)-*block*-poly(ethylene glycol) copolymer (PLLA-*b*-PEG). The PEG crystallized in the interlamellar or interfibrillar region of PLLA crystals (Figure 6.7a,b). The crystallization temperature of PLLA was controlled by the microstructure formation in PLLA-*b*-PEG. Also, the nucleation and crystallization of PEG segments were controlled by the PLLA crystallization temperature. Boissé et al. [102] prepared poly(L-lactide-*b*-2-dimethylaminoethylmethacrylate) (PLLA-*b*-PDMAEMA) and

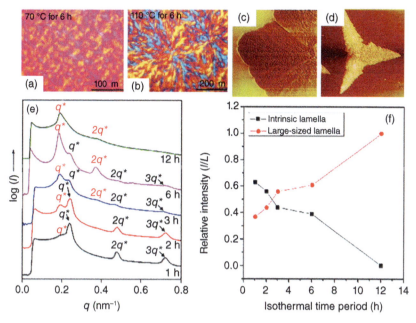

Figure 6.7 The polarized optical microscope (POM) images of PLLA-*b*-PEG diblock copolymer after isothermal crystallization for six hours. (a) 70 °C, (b) 110 °C. Source: Yang et al. [101]/Reproduced with permission from American Chemical Society. The atomic force microscopy (AFM) images of poly(L-lactide-*b*-2-dimethylaminoethyl methacrylate) (PLLA-*b*-PDMAEMA)/PDLA stereocomplexes of 30 nm thin film. The film was melted, then it was in the AFM hot stage and crystallized at 200 °C. (c) 19 K–5 K/19 K with PLLA/PDLA molar ratio of 50/50 [102], (d) 19 K–5 K/19 K with PLLA/PDLA molar ratio 25/75. Source: Boissé et al. [102]/ Reproduced with permission from American Chemical Society. The width of the image (c) is 90 μm and (d) 100 μm. (e) 1D small angle X-ray scattering (SAXS) pattern of syndiotactic poly(p-methylstyrene)-*block*-poly(L-lactide) (sPPMS-*b*-PLLA) crystallized isothermally at 140 °C for different times [103], (f) the relative intensity of dual-sized lamellae (primary peaks) changed with time periods during isothermal crystallization at 140 °C. Source: Huang et al. [103]/American Chemical Society.

(PDMAEMA-*b*-PLLA-*b*-PDMAEMA) copolymers. The molar mass of prepared PLLA varied from 13 K to 19 K, and the molar mass of PDMAEMA was 5 K to 35 K. The isothermal crystallization kinetics and morphology (Figure 6.7c,d) were investigated using above-mentioned block copolymers in bulk, thin film, thick film, and solution stages. In bulk phase, reduction in crystallization rate and distorted morphology were observed. Such events became more significant in tri-block copolymer compared to di-block copolymer. The local growth rate (G) and overall growth rate (K) varied in the similar path with respect to temperature or PDMAEMA content. Surprisingly inverse effect was noticed in quaternization. Here, "G" decreased because of reduction in chain mobility, and "K" increased due to generation of higher number of nuclei in the process. In tri-block copolymer, "G" was less and "K" was almost constant with same PDMAEMA contained. The crystallization in thin film (30 nm) and in solution led to creation of single crystals which were distorted relatively more in tri-block copolymer as compared to di-block copolymer. Incorporation of PDLA homopolymer again showed the formation of single crystal in thin film with diblock and triblock copolymers. The enantiomeric compositions gave hexagonal-shaped crystals, and nonenantiomeric compositions provided triangular-shaped crystals with diblock copolymers in blends. The triblock copolymers also showed triangular-shaped crystals for enantiomeric compositions. Steinhart and coworkers [104] investigated the crystallization behavior of PLLA in PS-b-PLLA lamella forming diblock copolymer inside cylindrical nanoporous alumina membrane (AAO) under a feeble confinement condition (nanopore diameter (D)/equilibrium PS-b-PLLA period (L_0) $\geq$ 4.8). The concentric lamellae aligned along the AAO nanopores at melt. At ($D/L_0 \approx 7.3$) intertwined helices were found for PS-b-PLLA. Once PS-b-PLLA was inside AAO pores and quenched to below glass transition temperature of PS, the PLLA crystallized under the vitrified PS domains. Above glass transition temperature of PS, the crystallization of PLLA was templated through the melt morphology of PS-b-PLLA inside nanopores. The double, single, and coincident crystallizations under confinement (hard or soft) were studied by the syndiotactic poly(p-methylstyrene)-block-poly(L-lactide) (sPPMS-b-PLLA) diblock copolymers [103]. 1D small-angle X-ray scattering (SAXS) pattern of isothermally crystallized at 140 °C for different times is shown in Figure 6.7e. The reflection in lower "q" region suggested the large size lamella (Figure 6.7e). With increasing time (one to six hours), intensity increased at the lower "q" peak but intrinsic reflection at higher "q" region reduced (Figure 6.7e). The change in relative intensity of dual-sized lamellae (primary peaks) with time is shown in Figure 6.7f. The opposite trend was noticed for both lamellae. Crystallization of sPPMS matrix led to formation of distorted PLLA cylinder under soft confinement. The lamellar morphology was not disturbed by the crystallization of PLLA, irrespective of whether it occurred under hard or soft confinement conditions. The crystallization-induced morphology transition was noticed from confined lamella to metastable dual-size lamella and breakout structure.

6.8 Crystallization of PLLA After Adding Nucleating Agents

PLLA has slower crystallization kinetics and lower crystallinity; hence, to improve the crystallinity, often thermal annealing approach is followed. Incorporation of nucleating agents is an alternative approach to increase the crystallization rate. Therefore, significant research attention has been given to escalate the crystallization rate of PLLA using various types of nucleating agents in past [105–116]. For example, Tsuji et al. [117] investigated the effect of PDLA, fullerene (C_{60}), talc, montmorillonite, and different polysaccharides on crystallization kinetics (non-isothermal) of PLLA. Once the samples were heated after melt quenching from RT, PLLA showed enhancement in overall crystallization rate after adding PDLA and the mixture of talc/montmorillonite. Crystallization of PLLA became faster after addition of these additives due to increase in density of PLLA spherulites. The cast film showed rise in PLLA crystallization on addition of talc, PDLA, montmorillonite, C_{60}, and their mixtures. The improvement in overall crystallization of PLLA was found after introducing these additives because of higher density of PLLA spherulites and nucleation at higher temperature. However, no significant effect on crystallization of PLLA was noticed after addition of polysaccharides. Based on the study, polysaccharides could be used as low-cost filler for PLLA without much compromising the crystallization of PLLA. The effect of additives on overall crystallization of PLLA during cooling from melt phase was followed in the following order: PDLA > talc > C_{60} > montmorillonite > polysaccharides. Hillmyer and coworkers [118, 119] studied the effect of stereocomplex crystallites as nucleating agent on crystallization of PLLA. The melt blending method was followed to prepare PLLA/PDLA blend, where PLLA composition was ≥95%. Also, three different molecular weights of PDLA (≤5%) were used with PLLA. The effectiveness of in-situ developed stereocomplex crystallites as nucleating agents on crystallization of PLLA was measured through the non-isothermal and self-nucleation methods. The heating cycle of PLLA with different molecular weight PDLA blends is shown in Figure 6.8a.

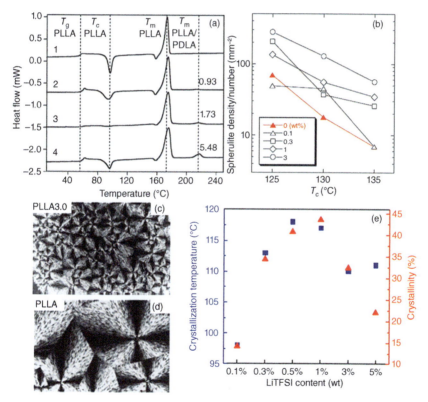

Figure 6.8 (a) The DSC heating cycle of pure PLLA and PLLA/PDLA blends. (1: pure PLLA, 2: PLLA/PDLA5.8 (99.5/0.5), 3: PLLA/PDLA5.8 (99.0/1.0), 4: PLLA/PDLA5.8 (97.0/3.0)). Source: Anderson and Hillmyer [118]/with permission of Elsevier. (b) The number of spherulites per unit area against crystallization temperatures for pure PLLA and PLLA containing different fractions of PDLA [119]. The polarized microscope images of isothermally crystallized PLLA with different nucleating agents. (c) Pure PLLA at 135 °C [119], (d) PLLA containing 3% PDLA at 135 °C. Source: Tsuji et al. [119]/Reproduced with permission from Elsevier, and (e) Change in crystallinity and crystallization temperature against nucleating agent fraction in the samples. Source: Xie et al. [120]/American Chemical Society.

Particularly 3% PDLA with molecular weight of 14 kg mol^{-1} showed efficiency close to 100%. The fast crystallization kinetics were found during isothermal crystallization kinetics at 140 °C. The stereocomplex crystallites showed higher ability to increase the crystallization rate of PLLA than talc and other common nucleating agents. The number of spherulites per unit area against crystallization temperatures for pure PLLA and PLLA containing different fractions of PDLA is shown in Figure 6.8b. The polarized microscope images of isothermally crystallized PLLA with different nucleating agents are shown in Figure 6.8c,d. Hillmyer and coworkers [121] further investigated the poly(D-lactide)-*block*-poly(menthide)-*bock*-poly(D-lactide) (PDLA-*b*-PM-*b*-PDLA) triblock copolymer as novel nucleating agent for PLLA. Through the bath melt mixer, PLLA/ PDLA-*b*-PM-*b*-PDLA blend containing ≥85 wt% PLLA was prepared. The triblock copolymer end group PDLA, having molecular weight of 15 kg mol^{-1} showed the nucleation efficiency near to 100%. In comparison to homopolymer PLLA/PDLA blend, PLLA/PDLA containing triblock copolymer showed higher nucleation efficiency. Kimura and coworkers [122] studied benzenetricarboxylamide (BTA) derivatives as a nucleator for crystallization of PLLA. Though BTA-cyclohexyl (BTA-cHe) was the promising nucleating agent but the developed material lost the transparency. Whereas, BTA-nhexyl (BTA-nHe) improved the crystallization behavior with small rise of haze. The BTA-cHe and BTA-nHe enhanced the crystallization of PLLA in α-form of crystals and ά-form of crystals having smaller dimension spherulites, respectively. BTA-nHe was dissolved entirely in PLLA at melt stage. They recrystallized during thermal annealing process. The size of BTA-nHe was in nanometer range and could nucleate effectively. The synthesized $N'N'$-Bis(benzoyl) suberic acid dihydrazide (NA) from benzoyl hydrazine and suberoyl chloride was tried as alternative nucleating agent for PLLA [47]. Samples were prepared through melt-blending and hot press forming techniques. The peak crystallization temperature became sharper and moved to higher temperature for PLLA after adding nucleating agents. After incorporation of nucleating agent (NA), the crystallization of PLLA accelerated during non-isothermal study. In case of isothermal study, nucleating agent reduced the crystallization time and increased the overall crystallization rate. The crystallization time reduced from 26.5 to 1.4 minutes after adding 0.8% NA at 115 °C. The spherulites number of PLLA increased and spherulites size decreased in presence of NA. Ning et al. [123] reported on the significance of double-walled carbon nanotube (DWNT) fibers bundle structure as nucleating agent and orientation substrate during crystallization of PLLA. After adding DWNT in PLLA, a fine nanohybrid shish kebab (NHSK) morphology with precise lateral dimension and period of kebabs was observed. Such interesting morphology was found might be due to the geometrical confinement effect of outer crinkly surface morphology in the bundle of DWNT.

Lai [124] studied the effect of 1,3:2,4-dibenzylidene-D-sorbitol (DBS) on thermal behavior and crystal structure of PLLA. By changing PLLA crystallization temperature and loading of DBS, different crystal forms (α and ά) were prepared. As DBS was added, the α-form of crystals was favored in PLLA and more α crystals were formed at lower temperature. Hence, the transition from disorder to order (ά to α) moved to lower temperature range as more DBS was added. According to the proposed mechanism, the π–π interaction directed the DBS molecules to stack together and form strand with PLLA through hydrogen bonding. However, crystallization and glass transition temperatures did not change much after adding DBS in PLLA. But the crystallization rate of PLLA was influenced by DBS due to change in crystal structure. Cai and Zhang [125] studied the effect of N,N,N'-Tris(benzoyl) trimesic acid hydrazide (TTAD) as organic nucleating agent on crystallization of PLLA. The overall crystallization rate increased after loading TTAD in PLLA. In comparing pure PLLA with PLLA/TTDA (1%), the onset crystallization temperature shifted from 101.36 to 125.26 °C, the peak melt-crystallization temperature raised from 94.49 to 117.56 °C, and the enthalpy of crystallization improved from 0.1023 to 33.44 J g^{-1} at the cooling rate of 1 °C/min. The non-isothermal crystallization study showed the crystallization peak became wider and moved to lower temperature with higher cooling rate. Gao et al. [126] prepared high melting temperature PLLA (hPLLA) fibers with high nucleation efficiency. Later, fibers were incorporated inside PLLA matrix to create biocomposites. Nucleating surface of hPLLA fibers supported the chain assembling and lamellar arrangement process. This led to generation of ordered PLLA transcrystallinity on the outer face of hPLLA fiber in quiescent environment. Such organization of transcrystallinity enhanced the crystallinity percentage and interfacial bonding. Improvement in percentage crystallinity and interfacial bonding, increased tensile strength, heat resistance, and gas barrier properties of the composites were also observed. On addition of 1 wt% hPLLA, the storage modulus increased 82 times (4–330 MPa) at 80 °C. The oxygen and water permeability coefficient decreased by 52% and 51% compared to pure PLLA. The transparency was not disturbed after adding hPLLA fibers in the composite.

Li and coworkers [120] showed interesting approach to increase the crystallization rate. They accelerated the PLLA crystallization through the conformational pre-ordering of polymer chains at melt stage. The organic salt (bistrifluoromethysulfonyl) imide lithium salt (LiTFSI) was added into PLLA and its effect on PLLA crystallization was inspected. The salt was miscible with PLLA, and the anion part of the salt interacted with polymer chains. The crystallization of PLLA improved largely on addition of small fraction of salt. For PLLA, the best nucleation efficiency was found with addition of 0.5 wt% salt. Once LiTFSI salt was added in PLLA, the pre-ordering of PLLA chains was noticed at melt condition. Due to the interaction between PLLA and salt, PLLA/LiTFSI composite formed more "gt" conformers (means 10$_3$ helical conformation of α crystal) than pure PLLA. This led to lower nucleation barrier for PLLA. The change in crystallinity and crystallization temperature against nucleating agent percentage (LiTFSI) is shown in Figure 6.8e. Cai et al. [127] studied N,N,N'-Tris(1H-benzotriazole) trimesinic acid acethydrazide (BD) as an organic nucleating agent for PLLA. The presence of BD in PLLA increased the overall crystallization rate of PLLA. With the addition of 0.5 wt% BD, the onset crystallization temperature of PLLA moved from 101.4 to 111.3 °C and the enthalpy of crystallization (non-isothermal) raised from 0.1 to 38.6 J g^{-1} at cooling rate of 1 °C/min. The isothermal study showed that the crystallization half-time decreased from 49.9 to 1.1 minutes at 105 °C after addition of 0.5 wt% BD. The equilibrium melting temperature of PLLA/BD (0.5 wt%) was lower than that of pure PLLA. Two melting peaks of PLLA/BD (0.5 wt%) proposed the melt and recrystallization behavior.

6.9 PLLA Plasticization

Inherent brittle nature restricts the potential applications of PLLA. Therefore, Kulinski et al. [128] studied about poly(propylene glycol) (PPG) as plasticizer for PLLA. PPG does not crystallize but it is miscible with PLLA. The thermal behavior of the PLLA was investigated with different PPG contents (5–12.5 wt%). The PLLA/PPG blend showed lower glass transition temperature than pure PLLA. It was also observed that PPG enhanced the crystallizibility of PLLA. PEG has been used as an alternative solution as plasticizer for PLLA. PEG ($M_w = 1000$ g mol^{-1}) with different weight fractions (5–20 wt%) was added in PLLA [129, 130]. After incorporation of PEG, crystallinity and impact resistance increased, whereas glass transition temperature and stiffness decreased. Nevertheless, other report shows the effect of molecular weight ($M_n = 0.6 - 20$ kg mol^{-1}) and weight fraction (10–20 wt%) on plasticization in PLLA/PEG blends [131]. Here, PEG acted as plasticizer in the PLLA/PEG blend. It was found that 10 wt% PEG was not sufficient to plasticize PLLA, but 20 wt% PEG plasticized PLLA efficiently. The PEG with low molecular weight ($M_n \leq 2$ kg mol^{-1}) chains was miscible with PLLA chains, whereas, high molecular weight ($M_n \leq 8$ kg mol^{-1}) PEG entangled with PLLA chains to reduce the phase separation. In intermediate molecular weight ($M_n \leq 5$ kg mol^{-1}), the crystallization of PEG and phase separation both incidents occurred simultaneously. This led to less plasticization in PLLA/PEG-15% blend. In PLLA/PEG-15% blend, the mixture of α and ά-forms of crystals were found during heating. Martino et al. [132] studied poly(L-lactide-co-D, L-lactide) (PLLA/PDLLA) blended films with different adipates to get improved thermal property. All plasticizers were compatible with matrix phase till critical composition and molar mass. On addition of the plasticizer, decrease in tensile stress and elastic modulus was noticed. Ljungberg et al. [133, 134] studied the plasticization of PLA with malonate oligomers. Plasticizer suppressed the glass transition temperature of PLLA, and the plasticizer with low molar mass showed better efficiency. The amorphous region was saturated by the plasticizer after loading certain concentration, and then phase separation event began. The saturation level is reached at lower concentration for high molar mass plasticizer in PLA with malonate oligomer blends. Diep et al. [135] studied the effect of plasticizer on improvement of PLLA crystallization. The partial enhancement in crystallizability was found after adding plasticizer (organic acid monoglyceride) in PLLA. Also, the crystallization rate increased after incorporation of dioctyl phthalate (DOP) as plasticizer in the PLLA system.

6.10 Conclusion and Future Outlook

PLLA is an environment-friendly, biocompatible, and biodegradable polymer. It is derived from bio-resources like corn starch or sugar cane. Therefore, PLLA has importance as a sustainable and environment-friendly polymer. However, inadequate mechanical performance, limited thermal stability, poor crystallization rate, and inherent brittle nature restrict its practical applications. A lot of work has been done in the past to solve these issues in PLLA. Here, the focus was on understanding crystallization behavior of PLLA in blends, composites, and block copolymer, after incorporation of nucleating agents and after addition of plasticizer. Previous studies on crystallization behavior of PLA show that it has different crystal forms like α, ά, β, and γ. The common form is α. The glass transition temperature and melting temperature of PLLA are decided by the molecular weight and copolymer contained (PDLA). The faster crystallization rate was detected in the temperature range of 100–130 °C. The PLLA and PDLA mixture have ability to form stereocomplex crystallites, which showed higher melting temperature than melting temperature of respective homopolymer. The rate of crystallization of PLLA can be regulated through blending or preparing composites or adding nucleating agents or covalently attaching with other polymers (block polymer) or incorporating plasticizers, etc. The block copolymer showed different morphology and distinct crystallization behavior because of number of variables like molecular weight, segregation strength between two blocks, transition temperature of each block, and composition that control the properties of the block copolymer. The nucleating agents like talc, metal phosphonates, and hydrazide compounds increased the nucleation rate of PLLA to enhance crystallization behavior.

Future research might be on injection molding application of PLLA. PLLA with improved processability and thermal stability needs careful attention. Moreover, the crystallization of PLLA in complex conditions like under pressure, flow condition, and during fast cooling could be studied. Furthermore, crystallization behavior of PLLA-based mixtures under confinement also demands serious attention. The knowledge of crystallization behavior of PLLA will be helpful in future to develop PLLA-based materials.

References

1 Fewrreira-Filipe, D.A., Paço, A., Duarte, A.C. et al. (2021). Are biobased plastics green alternatives?—a critical review. *International Journal of Environmental Research and Public Health* 18: 7729. https://doi.org/10.3390/ijerph18157729.
2 Spierling, S., Knüpffer, E., Behnsen, H. et al. (2018). Bio-based plastics – a review of environmental, social and economic impact assessments. *Journal of Cleaner Production* 185: 476–491. https://doi.org/10.1016/j.jclepro.2018.03.014.
3 Adekunle, K.F. (2014). Bio-based polymers for technical applications: a review — Part 2. *Open Journal Polymer Chemistry* 4: 95–101. https://doi.org/10.4236/ojpchem.2014.44011.
4 Garlotta, D. (2001). A literature review of poly(lactic acid). *Journal of Polymers and the Environment* 9: 63–84. https://doi.org/10.1023/A:1020200822435.
5 Ashothaman, A., Sudha, J., and Senthilkumar, N. (2021). A comprehensive review on biodegradable polylactic acid polymer matrix composite material reinforced with synthetic and natural fibers. *Materials Today: Proceedings* https://doi.org/10.1016/j.matpr.2021.07.047.

6 Saeidlou, S., Huneault, M.A., Li, H., and Park, C.B. (2012). Poly(lactic acid) crystallization. *Progress in Polymer Science* 37: 1657–1677. https://doi.org/10.1016/j.progpolymsci.2012.07.005.

7 Müller, A.J., Ávila, M., Saenz, G., and Salazar, J. (2014). Chapter 3. Crystallization of PLA-based materials. In: *Polymer Chemistry Series* (ed. A. Jiménez, M. Peltzer, and R. Ruseckaite), 66–98. Royal Society of Chemistry, Cambridge http://ebook.rsc.org/?DOI=10.1039/9781782624806-00066 (accessed 7 May 2022).

8 Tsuji, H., Nakahara, K., and Ikarashi, K. (2001). Poly(L-Lactide), 8. High-temperature hydrolysis of poly(L-lactide) films with different crystallinities and crystalline thicknesses in phosphate-buffered solution. *Macromolecular Materials and Engineering* 286: 398–406. https://doi.org/10.1002/1439-2054(20010701)286:7<398::AID-MAME398>3.0.CO;2-G.

9 Cam, D., Hyon, S., and Ikada, Y. (1995). Degradation of high molecular weight poly(L-lactide) in alkaline medium. *Biomaterials* 16: 833–843. https://doi.org/10.1016/0142-9612(95)94144-A.

10 Drieskens, M., Peeters, R., Mullens, J. et al. (2009). Structure versus properties relationship of poly(lactic acid). I. Effect of crystallinity on barrier properties: structure vs barrier properties of PLA. *Journal of Polymer Science Part B: Polymer Physics* 47: 2247–2258. https://doi.org/10.1002/polb.21822.

11 Mehta, R., Kumar, V., Bhunia, H., and Upadhyay, S.N. (2005). Synthesis of poly(lactic acid): a review. *Journal of Macromolecular Science Polymer Reviews* 45: 325–349. https://doi.org/10.1080/15321790500304148.

12 Miyata, T. and Masuko, T. (1997). Morphology of poly(L-lactide) solution-grown crystals. *Polymer* 38: 4003–4009. https://doi.org/10.1016/S0032-3861(96)00987-1.

13 De Santis, P. and Kovacs, A.J. (1968). Molecular conformation of poly(S-lactic acid). *Biopolymers* 6: 299–306. https://doi.org/10.1002/bip.1968.360060305.

14 Sasaki, S. and Asakura, T. (2003). Helix distortion and crystal structure of the α-form of poly(L-lactide). *Macromolecules* 36: 8385–8390. https://doi.org/10.1021/ma0348674.

15 Kobayashi, J., Asahi, T., Ichiki, M. et al. (1995). Structural and optical properties of poly lactic acids. *Journal of Applied Physics* 77: 2957–2973. https://doi.org/10.1063/1.358712.

16 Brizzolara, D., Cantow, H.-J., Diederichs, K. et al. (1996). Mechanism of the stereocomplex formation between enantiomeric poly(lactide)s. *Macromolecules* 29: 191–197. https://doi.org/10.1021/ma951144e.

17 Hoogsteen, W., Postema, A.R., Pennings, A.J. et al. (1990). Crystal structure, conformation and morphology of solution-spun poly(L-lactide) fibers. *Macromolecules* 23: 634–642. https://doi.org/10.1021/ma00204a041.

18 Zhang, J., Duan, Y., Sato, H. et al. (2005). Crystal modifications and thermal behavior of poly(L-lactic acid) revealed by infrared spectroscopy. *Macromolecules* 38: 8012–8021. https://doi.org/10.1021/ma051232r.

19 Kawai, T., Rahman, N., Matsuba, G. et al. (2007). Crystallization and melting behavior of poly (L-lactic acid). *Macromolecules* 40: 9463–9469. https://doi.org/10.1021/ma070082c.

20 Zhang, J., Tashiro, K., Tsuji, H., and Domb, A.J. (2008). Disorder-to-order phase transition and multiple melting behavior of poly(L-lactide) investigated by simultaneous measurements of WAXD and DSC. *Macromolecules* 41: 1352–1357. https://doi.org/10.1021/ma0706071.

21 Cartier, L., Okihara, T., Ikada, Y. et al. (2000). Epitaxial crystallization and crystalline polymorphism of polylactides. *Polymer* 41: 8909–8919. https://doi.org/10.1016/S0032-3861(00)00234-2.

22 Puiggali, J., Ikada, Y., Tsuji, H. et al. (2000). The frustrated structure of poly(L-lactide). *Polymer* 41: 8921–8930. https://doi.org/10.1016/S0032-3861(00)00235-4.

23 Abe, H., Kikkawa, Y., Inoue, Y., and Doi, Y. (2001). Morphological and kinetic analyses of regime transition for poly[(S)-lactide] crystal growth. *Biomacromolecules* 2: 1007–1014. https://doi.org/10.1021/bm015543v.

24 Jamshidi, K., Hyon, S.-H., and Ikada, Y. (1988). Thermal characterization of polylactides. *Polymer* 29: 2229–2234. https://doi.org/10.1016/0032-3861(88)90116-4.

25 Korhonen, H., Helminen, A., and Seppälä, J.V. (2001). Synthesis of polylactides in the presence of co-initiators with different numbers of hydroxyl groups. *Polymer* 42: 7541–7549. https://doi.org/10.1016/S0032-3861(01)00150-1.

26 Pitet, L.M., Hait, S.B., Lanyk, T.J., and Knauss, D.M. (2007). Linear and branched architectures from the polymerization of lactide with glycidol. *Macromolecules* 40: 2327–2334. https://doi.org/10.1021/ma0618068.

27 Nakafuku, C. and Takehisa, S. (2004). Glass transition and mechanical properties of PLLA and PDLLA-PGA copolymer blends. *Journal of Applied Polymer Science* 93: 2164–2173. https://doi.org/10.1002/app.20687.

28 Dorgan, J.R., Janzen, J., Clayton, M.P. et al. (2005). Melt rheology of variable L-content poly(lactic acid). *Journal of Rheology* 49: 607–619. https://doi.org/10.1122/1.1896957.

29 Gedde, U.W. (2001). *Polymer Physics, Repr*. Dordrecht: Kluwer Academic Publishers.

30 Wunderlich, B. (1976). *Macromolecular Physics*, vol. 2. New York: Academic Press http://www.123library.org/book_details/?id=100252 (accessed 9 May 2022).

31 Di Lorenzo, M.L. (2005). Crystallization behavior of poly(L-lactic acid). *European Polymer Journal* 41: 569–575. https://doi.org/10.1016/j.eurpolymj.2004.10.020.

32 Miyata, T. and Masuko, T. (1998). Crystallization behaviour of poly(L-lactide). *Polymer* 39: 5515–5521. https://doi.org/10.1016/S0032-3861(97)10203-8.

33 Pan, P., Zhu, B., Kai, W. et al. (2008). Effect of crystallization temperature on crystal modifications and crystallization kinetics of poly(L-lactide). *Journal of Applied Polymer Science* 107: 54–62. https://doi.org/10.1002/app.27102.

34 Li, X., Li, Z., Zhong, G., and Li, L. (2008). Steady–shear-induced isothermal crystallization of poly(L-lactide) (PLLA). *Journal of Macromolecular Science, Part B Physics* 47: 511–522. https://doi.org/10.1080/00222340801955313.

35 Yasuniwa, M., Tsubakihara, S., Iura, K. et al. (2006). Crystallization behavior of poly(L-lactic acid). *Polymer* 47: 7554–7563. https://doi.org/10.1016/j.polymer.2006.08.054.

36 Di Lorenzo, M.L. (2001). Determination of spherulite growth rates of poly(L-lactic acid) using combined isothermal and non-isothermal procedures. *Polymer* 42: 9441–9446. https://doi.org/10.1016/S0032-3861(01)00499-2.

37 Pan, P., Kai, W., Zhu, B. et al. (2007). Polymorphous crystallization and multiple melting behavior of poly(L-lactide): molecular weight dependence. *Macromolecules* 40: 6898–6905. https://doi.org/10.1021/ma071258d.

38 Tsuji, H. and Ikada, Y. (1996). Crystallization from the melt of poly(lactide)s with different optical purities and their blends. *Macromolecular Chemistry and Physics* 197: 3483–3499. https://doi.org/10.1002/macp.1996.021971033.

39 Tsuji, H. and Ikada, Y. (1995). Blends of isotactic and atactic poly(lactide). I. Effects of mixing ratio of isomers on crystallization of blends from melt. *Journal of Applied Polymer Science* 58: 1793–1802. https://doi.org/10.1002/app.1995.070581018.

40 Kalb, B. and Pennings, A.J. (1980). General crystallization behaviour of poly(L-lactic acid). *Polymer* 21: 607–612. https://doi.org/10.1016/0032-3861(80)90315-8.

41 Vasanthakumari, R. and Pennings, A.J. (1983). Crystallization kinetics of poly(L-lactic acid). *Polymer* 24: 175–178. https://doi.org/10.1016/0032-3861(83)90129-5.

42 Tsuji, H., Tezuka, Y., Saha, S.K. et al. (2005). Spherulite growth of L-lactide copolymers: effects of tacticity and comonomers. *Polymer* 46: 4917–4927. https://doi.org/10.1016/j.polymer.2005.03.069.

43 Yu, J. and Qiu, Z. (2011). Isothermal and nonisothermal cold crystallization behaviors of biodegradable poly(L-lactide)/octavinyl-polyhedral oligomeric silsesquioxanes nanocomposites. *Industrial and Engineering Chemistry Research* 50: 12579–12586. https://doi.org/10.1021/ie201691y.

44 Qiao, H., Guo, J., Wang, L. et al. (2020). Effects of divinylbenzene-maleic anhydride copolymer hollow microspheres on crystallization behaviors, mechanical properties and heat resistance of poly(L-lactide acid). *Polymers for Advanced Technologies* 31: 817–826. https://doi.org/10.1002/pat.4817.

45 Zhou, W.Y., Duan, B., Wang, M., and Cheung, W.L. (2009). Crystallization kinetics of poly(L-lactide)/carbonated hydroxyapatite nanocomposite microspheres. *Journal of Applied Polymer Science* 113: 4100–4115. https://doi.org/10.1002/app.30527.

46 Li, Y., Chen, C., Li, J., and Sun, X.S. (2012). Isothermal crystallization and melting behaviors of bionanocomposites from poly(lactic acid) and TiO_2 nanowires. *Journal of Applied Polymer Science* 124: 2968–2977. https://doi.org/10.1002/app.35326.

47 Cai, Y., Yan, S., Yin, J. et al. (2011). Crystallization behavior of biodegradable poly(L-lactic acid) filled with a powerful nucleating agent: *N,N′*-bis(benzoyl) suberic acid dihydrazide. *Journal of Applied Polymer Science* 121: 1408–1416. https://doi.org/10.1002/app.33633.

48 Xu, W., Ge, M., and He, P. (2001). Nonisothermal crystallization kinetics of polyoxymethylene/montmorillonite nanocomposite. *Journal of Applied Polymer Science* 82: 2281–2289. https://doi.org/10.1002/app.2076.

49 Al-Mulla, A., Mathew, J., Yeh, S.-K., and Gupta, R. (2008). Nonisothermal crystallization kinetics of PBT nanocomposites. *Composites, Part A: Applied Science and Manufacturing* 39: 204–217. https://doi.org/10.1016/j.compositesa.2007.11.001.

50 Qu, X., Ding, H., Lu, J. et al. (2004). Isothermal and nonisothermal crystallization kinetics of MC nylon and polyazomethine/MC nylon composites. *Journal of Applied Polymer Science* 93: 2844–2855. https://doi.org/10.1002/app.20832.

51 An, Y., Li, L., Dong, L. et al. (1999). Nonisothermal crystallization and melting behavior of poly(β-hydroxybutyrate)-poly(vinyl-acetate) blends. *Journal of Polymer Science Part B: Polymer Physics* 37: 443–450. https://doi.org/10.1002/(SICI)1099-0488(19990301)37:5<443::AID-POLB4>3.0.CO;2-B.

52 Tsuji, H., Hyon, S.H., and Ikada, Y. (1991). Stereocomplex formation between enantiomeric poly(lactic acid)s. 3. Calorimetric studies on blend films cast from dilute solution. *Macromolecules* 24: 5651–5656. https://doi.org/10.1021/ma00020a026.

53 Zhao, C., Shi, W., Ma, Y. et al. (2012). Synthesis, structures and characterization of triarm PPO-PDLAPLLA block copolymers and its stereocomplex crystallization behavior. *Acta Chimica Sinica* 70: 881. https://doi.org/10.6023/A1110151.

54 He, L., Zuo, Q., Shi, Y., and Xue, W. (2014). Microstructural characteristics and crystallization behaviors of poly(L-lactide) scaffolds by thermally induced phase separation. *Journal of Applied Polymer Science* 131: n/a-n/a. https://doi.org/10.1002/app.39436.

55 Laredo, E., Newman, D., Pezzoli, R. et al. (2016). A complete TSDC description of molecular mobilities in polylactide/starch blends from local to normal modes: effect of composition, moisture, and crystallinity. *Journal of Polymer Science Part B: Polymer Physics* 54: 680–691. https://doi.org/10.1002/polb.23963.

56 do Rufino, T., C. and Felisberti, M.I. (2016). Confined PEO crystallisation in immiscible PEO/PLLA blends. *RSC Advances* 6: 30937–30950. https://doi.org/10.1039/C6RA02406H.

57 Wang, Y., Wei, X., Duan, J. et al. (2017). Greatly enhanced hydrolytic degradation ability of poly(L-lactide) achieved by adding poly(ethylene glycol). *Chinese Journal of Polymer Science* 35: 386–399. https://doi.org/10.1007/s10118-017-1904-y.

58 Shi, X., Qin, J., Wang, L. et al. (2018). Introduction of stereocomplex crystallites of PLA for the solid and microcellular poly(lactide)/poly(butylene adipate-co-terephthalate) blends. *RSC Advances* 8: 11850–11861. https://doi.org/10.1039/C8RA01570H.

59 Dell'Erba, R., Groeninckx, G., Maglio, G. et al. (2001). Immiscible polymer blends of semicrystalline biocompatible components: thermal properties and phase morphology analysis of PLLA/PCL blends. *Polymer* 42: 7831–7840. https://doi.org/10.1016/S0032-3861(01)00269-5.

60 Shibata, M., Inoue, Y., and Miyoshi, M. (2006). Mechanical properties, morphology, and crystallization behavior of blends of poly(L-lactide) with poly(butylene succinate-co-L-lactate) and poly(butylene succinate). *Polymer* 47: 3557–3564. https://doi.org/10.1016/j.polymer.2006.03.065.

61 Shi, X., Jing, Z., and Zhang, G. (2018). Influence of PLA stereocomplex crystals and thermal treatment temperature on the rheology and crystallization behavior of asymmetric poly(L-lactide)/poly(D-lactide) blends. *Journal of Polymer Research* 25: https://doi.org/10.1007/s10965-018-1467-9.

62 Tsuji, H. and Ikada, Y. (1996). Blends of isotactic and atactic poly(lactide)s: 2, Molecular-weight effects of atactic component on crystallization and morphology of equimolar blends from the melt. *Polymer* 37: 595–602. https://doi.org/10.1016/0032-3861(96)83146-6.

63 Sun, J., Yu, H., Zhuang, X. et al. (2011). Crystallization behavior of asymmetric PLLA/PDLA blends. *The Journal of Physical Chemistry B* 115: 2864–2869. https://doi.org/10.1021/jp111894m.

64 Wei, X.-F., Bao, R.-Y., Cao, Z.-Q. et al. (2014). Stereocomplex crystallite network in asymmetric PLLA/PDLA blends: formation, structure, and confining effect on the crystallization rate of homocrystallites. *Macromolecules* 47: 1439–1448. https://doi.org/10.1021/ma402653a.

65 Qiu, J., Guan, J., Wang, H. et al. (2014). Enhanced crystallization rate of poly(L-lactic acid) (PLLA) by polyoxymethylene (POM) fragment crystals in the PLLA/POM blends with a small amount of POM. *The Journal of Physical Chemistry B* 118: 7167–7176. https://doi.org/10.1021/jp412519g.

66 Ye, L., Ye, C., Xie, K. et al. (2015). Morphologies and crystallization behaviors in melt-miscible crystalline/crystalline blends with close melting temperatures but different crystallization kinetics. *Macromolecules* 48: 8515–8525. https://doi.org/10.1021/acs.macromol.5b01904.

67 Li, R., Zhao, X., Coates, P. et al. (2021). Highly reinforced poly(lactic acid) foam fabricated by formation of a heat-resistant oriented stereocomplex crystalline structure. *ACS Sustainable Chemistry & Engineering* 9: 12674–12686. https://doi.org/10.1021/acssuschemeng.1c04862.

68 Tsuji, H., Smith, R., Bonfield, W., and Ikada, Y. (2000). Porous biodegradable polyesters. I. Preparation of porous poly(L-lactide) films by extraction of poly(ethylene oxide) from their blends. *Journal of Applied Polymer Science* 75: 629–637. https://doi.org/10.1002/(SICI)1097-4628(20000131)75:5<629::AID-APP5>3.0.CO;2-A.

69 Yang, J.-M., Chen, H.-L., You, J.-W., and Hwang, J.C. (1997). Miscibility and crystallization of poly(L-lactide)/poly(ethylene glycol) and poly(L-lactide)/poly(ε-caprolactone) blends. *Polymer Journal* 29: 657–662. https://doi.org/10.1295/polymj.29.657.

70 Nakane, K., Tamaki, C., Hata, Y. et al. (2008). Blends of poly(L-lactic acid) with poly(ω-pentadecalactone) synthesized by enzyme-catalyzed polymerization. *Journal of Applied Polymer Science* 108: 2139–2143. https://doi.org/10.1002/app.27838.

71 Banpean, A. and Sakurai, S. (2021). Confined crystallization of poly(ethylene glycol) in spherulites of poly(L-lactic acid) in a PLLA/PEG blend. *Polymer* 215: 123370. https://doi.org/10.1016/j.polymer.2020.123370.

72 Xu, W., Zheng, Y., Yuan, W. et al. (2021). Polymorphic phase formation of liquid crystals distributed in semicrystalline polymers: an indicator of inter-lamellar and interspherulitic segregation. *Journal of Physical Chemistry Letters* 12: 4378–4384. https://doi.org/10.1021/acs.jpclett.1c01092.

73 Xin, S., Li, Y., Zhao, H. et al. (2015). Confinement crystallization of poly(L-lactide) induced by multiwalled carbon nanotubes and graphene nanosheets: a comparative study. *Journal of Thermal Analysis and Calorimetry* 122: 379–391. https://doi.org/10.1007/s10973-015-4695-9.

74 Tverdokhlebov, S.I., Bolbasov, E.N., Shesterikov, E.V. et al. (2015). Modification of polylactic acid surface using RF plasma discharge with sputter deposition of a hydroxyapatite target for increased biocompatibility. *Applied Surface Science* 329: 32–39. https://doi.org/10.1016/j.apsusc.2014.12.127.

75 Cai, Y.-H., Ren, L.-P., and Tang, Y. (2015). Improvement of thermal properties of poly(L-lactic acid) by blending with zinc lactate. *Journal of the National Science Foundation of Sri Lanka* 43: 247. https://doi.org/10.4038/jnsfsr.v43i3.7952.

76 Gardella, L., Furfaro, D., Galimberti, M., and Monticelli, O. (2015). On the development of a facile approach based on the use of ionic liquids: preparation of PLLA (sc-PLA)/high surface area nano-graphite systems. *Green Chemistry* 17: 4082–4088. https://doi.org/10.1039/C5GC00964B.

77 Cai, Y.-H., Zhao, L.-S., and Zhang, Y.-H. (2015). Composites based green poly(L-lactic acid) and dioctyl phthalate: preparation and performance. *Advances in Materials Science and Engineering* 2015: 1–5. https://doi.org/10.1155/2015/289725.

78 Carfì Pavia, F., Palumbo, F.S., La Carrubba, V. et al. (2016). Modulation of physical and biological properties of a composite PLLA and polyaspartamide derivative obtained via thermally induced phase separation (TIPS) technique. *Materials Science and Engineering: C* 67: 561–569. https://doi.org/10.1016/j.msec.2016.05.040.

79 Jing, Z., Shi, X., and Zhang, G. (2016). Poly(L-lactide)/four-armed star poly(L-lactide)-grafted multiwalled carbon nanotubes nanocomposites: preparation, rheology, crystallization, and mechanical properties. *Polymer Composites* 37: 2744–2755. https://doi.org/10.1002/pc.23469.

80 Zhao, L., Li, Q., Zhang, R. et al. (2016). Effects of functionalized graphenes on the isothermal crystallization of poly(L-lactide) nanocomposites. *Chinese Journal of Polymer Science* 34: 111–121. https://doi.org/10.1007/s10118-016-1732-5.

81 Girdthep, S., Sankong, W., Pongmalee, A. et al. (2017). Enhanced crystallization, thermal properties, and hydrolysis resistance of poly(L-lactic acid) and its stereocomplex by incorporation of graphene nanoplatelets. *Polymer Testing* 61: 229–239. https://doi.org/10.1016/j.polymertesting.2017.05.009.

82 Auliawan, A. and Woo, E.M. (2012). Crystallization kinetics and degradation of nanocomposites based on ternary blend of poly(L-lactic acid), poly(methyl methacrylate), and poly(ethylene oxide) with two different organoclays. *Journal of Applied Polymer Science* 125: E444–E458. https://doi.org/10.1002/app.36761.

83 Tsuji, H., Kawashima, Y., and Takikawa, H. (2007). Poly(L-lactide)/C_{60} nanocomposites: effects of C_{60} on crystallization of poly(L-lactide). *Journal of Polymer Science Part B: Polymer Physics* 45: 2167–2176. https://doi.org/10.1002/polb.21215.

84 Pan, P., Zhu, B., Dong, T., and Inoue, Y. (2008). Poly(L-lactide)/layered double hydroxides nanocomposites: preparation and crystallization behavior: PLLA/layered double hydroxides nanocomposites. *Journal of Polymer Science Part B: Polymer Physics* 46: 2222–2233. https://doi.org/10.1002/polb.21554.

85 Pan, P., Liang, Z., Cao, A., and Inoue, Y. (2009). Layered metal phosphonate reinforced poly(L-lactide) composites with a highly enhanced crystallization rate. *ACS Applied Materials & Interfaces* 1: 402–411. https://doi.org/10.1021/am800106f.

86 Li, X., Yin, J., Yu, Z. et al. (2009). Isothermal crystallization behavior of poly(L-lactic acid)/organo-montmorillonite nanocomposites. *Polymer Composites* 30: 1338–1344. https://doi.org/10.1002/pc.20721.

87 Huang, C., Bai, H., Xiu, H. et al. (2014). Matrix crystallization induced simultaneous enhancement of electrical conductivity and mechanical performance in poly(L-lactide)/multiwalled carbon nanotubes (PLLA/MWCNTs) nanocomposites. *Composites Science and Technology* 102: 20–27. https://doi.org/10.1016/j.compscitech.2014.07.016.

88 Huang, H.-D., Xu, J.-Z., Fan, Y. et al. (2013). Poly(L-lactic acid) crystallization in a confined space containing graphene oxide nanosheets. *The Journal of Physical Chemistry B* 117: 10641–10651. https://doi.org/10.1021/jp4055796.

89 Li, J., Xiao, P., Li, H. et al. (2015). Crystalline structures and crystallization behaviors of poly(L-lactide) in poly(L-lactide)/graphene nanosheet composites. *Polymer Chemistry* 6: 3988–4002. https://doi.org/10.1039/C5PY00254K.

90 Lizundia, E., Ruiz-Rubio, L., Vilas, J.L., and León, L.M. (2016). Poly(L-lactide)/ZNO nanocomposites as efficient UV-shielding coatings for packaging applications. *Journal of Applied Polymer Science* 133: n/a-n/a. https://doi.org/10.1002/app.42426.

91 Jin, X., Yu, X., Yang, C. et al. (2019). Crystallization and hydrolytic degradation behaviors of poly(L-lactide) induced by carbon nanofibers with different surface modifications. *Polymer Degradation and Stability* 170: 109014. https://doi.org/10.1016/j.polymdegradstab.2019.109014.

92 Shi, J., Wang, W., Feng, Z. et al. (2019). Multiple influences of hydrogen bonding interactions on PLLA crystallization behaviors in PLLA/TSOS hybrid blending systems. *Polymer* 175: 152–160. https://doi.org/10.1016/j.polymer.2019.05.008.

93 Tseng, W.-H., Hsieh, P.-Y., Ho, R.-M. et al. (2006). Oriented microstructures of polystyrene-b-poly(L-lactide) thin films induced by crystallizable solvents. *Macromolecules* 39: 7071–7077. https://doi.org/10.1021/ma0608929.

94 Ho, R.-M., Chen, C.-K., and Chiang, Y.-W. (2009). Novel nanostructures from self-assembly of chiral block copolymers: novel nanostructures from

self-assembly of chiral block copolymers. *Macromolecular Rapid Communications* 30: 1439–1456. https://doi.org/10.1002/marc.200900181.

95 Chen, C.-K., Lin, S.-C., Ho, R.-M. et al. (2010). Kinetically controlled self-assembled superstructures from semicrystalline chiral block copolymers. *Macromolecules* 43: 7752–7758. https://doi.org/10.1021/ma1009879.

96 Huang, C.-I., Tsai, S.-H., and Chen, C.-M. (2006). Isothermal crystallization behavior of poly(L-lactide) in poly(L-lactide)-*block*-poly(ethylene glycol) diblock copolymers. *Journal of Polymer Science Part B: Polymer Physics* 44: 2438–2448. https://doi.org/10.1002/polb.20890.

97 Cai, C., Wang, L., and Dong, C.-M. (2006). Synthesis, characterization, effect of architecture on crystallization, and spherulitic growth of poly(L-lactide)-*b*-poly(ethylene oxide) copolymers with different branch arms. *Journal of Polymer Science Part A: Polymer Chemistry* 44: 2034–2044. https://doi.org/10.1002/pola.21318.

98 Hamley, I.W., Parras, P., Castelletto, V. et al. (2006). Melt structure and its transformation by sequential crystallization of the two blocks within poly(L-lactide)-*block*-poly(ε-caprolactone) double crystalline diblock copolymers. *Macromolecular Chemistry and Physics* 207: 941–953. https://doi.org/10.1002/macp.200600085.

99 Wang, W., Wang, W., Chen, X. et al. (2009). Hydrogen bonding and crystallization in biodegradable multiblock poly(ester urethane) copolymer. *Journal of Polymer Science Part B: Polymer Physics* 47: 685–695. https://doi.org/10.1002/polb.21674.

100 Voet, V.S.D., Alberda van Ekenstein, G.O.R., Meereboer, N.L. et al. (2014). Double-crystalline PLLA-*b*-PVDF-*b*-PLLA triblock copolymers: preparation and crystallization. *Polymer Chemistry* 5: 2219. https://doi.org/10.1039/c3py01560b.

101 Yang, J., Liang, Y., and Han, C.C. (2015). Effect of crystallization temperature on the interactive crystallization behavior of poly(L-lactide)-*block*-poly(ethylene glycol) copolymer. *Polymer* 79: 56–64. https://doi.org/10.1016/j.polymer.2015.09.067.

102 Boissé, S., Kryuchkov, M.A., Tien, N.-D. et al. (2016). PLLA crystallization in linear AB and BAB copolymers of L-Lactide and 2-dimethylaminoethyl methacrylate. *Macromolecules* 49: 6973–6986. https://doi.org/10.1021/acs.macromol.6b01139.

103 Huang, S.-H., Huang, Y.-W., Chiang, Y.-W. et al. (2016). Nanoporous crystalline templates from double-crystalline block copolymers by control of interactive confinement. *Macromolecules* 49: 9048–9059. https://doi.org/10.1021/acs.macromol.6b01725.

104 Yau, M.Y.E., Gunkel, I., Hartmann-Azanza, B. et al. (2017). Semicrystalline block copolymers in rigid confining nanopores. *Macromolecules* 50: 8637–8646. https://doi.org/10.1021/acs.macromol.7b01567.

105 Yin, Y., Zhang, X., Song, Y. et al. (2015). Effect of nucleating agents on the strain-induced crystallization of poly(L-lactide). *Polymer* 65: 223–232. https://doi.org/10.1016/j.polymer.2015.03.061.

106 Cai, Y.-H., Zhang, Y.-H., and Zhao, L.-S. (2016). Evaluation of the effect of N,N'-bis(1H-benzotriazole) dodecanedioic acid acethydrazide on poly(L-lactic acid). *Polimery* 61: 773–778. https://doi.org/10.14314/polimery.2016.773.

107 Zhang, Y.-H. and Cai, Y.-H. (2016). A study on physical performance for poly(L-lactic acid) in addition to layered strontium phenylphosphonate. *International Journal of Polymeric Science* 2016: 1–6. https://doi.org/10.1155/2016/3926876.

108 Jiang, L., Shen, T., Xu, P. et al. (2016). Crystallization modification of poly(lactide) by using nucleating agents and stereocomplexation. *E-Polymers* 16: 1–13. https://doi.org/10.1515/epoly-2015-0179.

109 Wu, B., Zeng, X., Wu, L., and Li, B.-G. (2017). Nucleating agent-containing P(LLA-mb-BSA) multi-block copolymers with balanced mechanical properties. *Journal of Applied Polymer Science* 134: https://doi.org/10.1002/app.44777.

110 Cai, Y.-H., Zhao, L.-S., and Tang, Y. (2017). Thermal performance of a blend system based on poly(L-lactic acid) and an aliphatic multiamide derivative of ^{1}H-benzotriazole. *Journal of Macromolecular Science, Part B Physics* 56: 64–73. https://doi.org/10.1080/00222348.2016.1261594.

111 Song, P., Sang, L., Zheng, L. et al. (2017). Insight into the role of bound water of a nucleating agent in polymer nucleation: a comparative study of anhydrous and monohydrated orotic acid on crystallization of poly(L-lactic acid). *RSC Advances* 7: 27150–27161. https://doi.org/10.1039/C7RA02617J.

112 Naffakh, M., Fernández, M., Shuttleworth, P.S. et al. (2020). Nanocomposite materials with poly(L-lactic acid) and transition-metal dichalcogenide nanosheets 2D-TMDCs WS_2. *Polymers* 12: 2699. https://doi.org/10.3390/polym12112699.

113 Zhang, L. and Zhao, G. (2020). Poly(L-lactic acid) crystallization in pressurized CO_2: an in situ microscopic study and a new model for secondary nucleation in supercritical CO_2. *Journal of Physical Chemistry C* 124: 9021–9034. https://doi.org/10.1021/acs.jpcc.0c00049.

114 Schmidt, S.C. and Hillmyer, M.A. (2001). Polylactide stereocomplex crystallites as nucleating agents for isotactic polylactide. *Journal of Polymer Science Part B: Polymer Physics* 39: 300–313. https://doi.org/10.1002/1099-0488(20010201)39:3<300::AID-POLB1002>3.0.CO;2-M.

115 Zhao, L., Qiao, J., Shan, X. et al. (2021). Effect of 1,4-naphthalenedicarboxylic acid derivative on crystallization and performances of poly(L-lactide). *Materiale Plastice* 58: 57–68. https://doi.org/10.37358/MP.21.1.5445.

116 Zhao, L., Liu, X., Cai, Y., and Chen, W. (2021). Crystallization and properties of nucleated poly(L-lactic acid): effect of cyclopentanecarboxylic acid derivative as a new nucleating agent. *Journal of Thermoplastic Composite Materials* 89270572110019. https://doi.org/10.1177/08927057211001921.

117 Tsuji, H., Takai, H., Fukuda, N., and Takikawa, H. (2006). Non-isothermal crystallization behavior of poly(L-lactic acid) in the presence of various additives. *Macromolecular Materials and Engineering* 291: 325–335. https://doi.org/10.1002/mame.200500371.

118 Anderson, K.S. and Hillmyer, M.A. (2006). Melt preparation and nucleation efficiency of polylactide stereocomplex crystallites. *Polymer* 47: 2030–2035. https://doi.org/10.1016/j.polymer.2006.01.062.

119 Tsuji, H., Takai, H., and Saha, S.K. (2006). Isothermal and non-isothermal crystallization behavior of poly(L-lactic acid): effects of stereocomplex as nucleating agent. *Polymer* 47: 3826–3837. https://doi.org/10.1016/j.polymer.2006.03.074.

120 Xie, K., Shen, J., Ye, L. et al. (2019). Increased gt conformer contents of PLLA molecular chains induced by Li-TFSI in melt: another route to promote PLLA crystallization. *Macromolecules* 52: 7065–7072. https://doi.org/10.1021/acs.macromol.9b01188.

121 Wanamaker, C.L., Tolman, W.B., and Hillmyer, M.A. (2009). Poly(D-lactide)-poly(menthide)-poly(D-lactide) triblock copolymers as crystal nucleating agents for poly(L-lactide). *Macromolecular Symposia* 283, 284: 130–138. https://doi.org/10.1002/masy.200950917.

122 Nakajima, H., Takahashi, M., and Kimura, Y. (2010). Induced crystallization of PLLA in the presence of 1,3,5-benzenetricarboxylamide derivatives as nucleators: preparation of haze-free crystalline PLLA materials: induced crystallization of PLLA in the presence of 1,3,5-benzenetricarboxylamide. *Macromolecular Materials and Engineering* 295: 460–468. https://doi.org/10.1002/mame.200900353.

123 Ning, N., Zhang, W., Zhao, Y. et al. (2012). Nanohybrid shish kebab structure and its effect on mechanical properties in poly(L-lactide)/carbon nanotube nanocomposite fibers. *Polymer International* 61: 1634–1639. https://doi.org/10.1002/pi.4252.

124 Lai, W.-C. (2011). Thermal behavior and crystal structure of poly(L-lactic acid) with 1,3:2,4-dibenzylidene-d-sorbitol. *The Journal of Physical Chemistry B* 115: 11029–11037. https://doi.org/10.1021/jp2037312.

125 Cai, Y.-H. and Zhang, Y.-H. (2014). The crystallization, melting behavior, and thermal stability of poly(L-lactic acid) induced by N,N,N'-tris(benzoyl) trimesic acid hydrazide as an organic nucleating agent. *Advances in Materials Science and Engineering* 2014: 1–8. https://doi.org/10.1155/2014/843564.

126 Gao, T., Zhang, Z.-M., Li, L. et al. (2018). Tailoring crystalline morphology by high-efficiency nucleating fiber: toward high-performance poly(L-lactide) biocomposites. *ACS Applied Materials & Interfaces* 10: 20044–20054. https://doi.org/10.1021/acsami.8b04907.

127 Cai, Y.-H., Tang, Y., and Zhao, L.-S. (2015). Poly(L-lactic acid) with the organic nucleating agent N,N,N'-tris(1H-benzotriazole) trimesinic acid acethydrazide: crystallization and melting behavior. *Journal of Applied Polymer Science* 132: https://doi.org/10.1002/app.42402.

128 Kulinski, Z., Piorkowska, E., Gadzinowska, K., and Stasiak, M. (2006). Plasticization of poly(L-lactide) with poly(propylene glycol). *Biomacromolecules* 7: 2128–2135. https://doi.org/10.1021/bm060089m.

129 Hu, Y., Rogunova, M., Topolkaraev, V. et al. (2003). Aging of poly(lactide)/poly(ethylene glycol) blends. Part 1. Poly(lactide) with low stereoregularity. *Polymer* 44: 5701–5710. https://doi.org/10.1016/S0032-3861(03)00614-1.

130 Sungsanit, K., Kao, N., and Bhattacharya, S.N. (2012). Properties of linear poly(lactic acid)/polyethylene glycol blends. *Polymer Engineering and Science* 52: 108–116. https://doi.org/10.1002/pen.22052.

131 Guo, J., Liu, X., Liu, M. et al. (2021). Effect of molecular weight of poly(ethylene glycol) on plasticization of poly(L-lactic acid). *Polymer* 223: 123720. https://doi.org/10.1016/j.polymer.2021.123720.

132 Martino, V.P., Ruseckaite, R.A., and Jiménez, A. (2006). Thermal and mechanical characterization of plasticized poly(L-lactide-co-D,L-lactide) films for food packaging. *Journal of Thermal Analysis and Calorimetry* 86: 707–712. https://doi.org/10.1007/s10973-006-7897-3.

133 Ljungberg, N. and Wesslén, B. (2004). Thermomechanical film properties and aging of blends of poly(lactic acid) and malonate oligomers: blends of PLA and malonate oligomers. *Journal of Applied Polymer Science* 94: 2140–2149. https://doi.org/10.1002/app.21100.

134 Ljungberg, N. and Wesslén, B. (2003). Tributyl citrate oligomers as plasticizers for poly(lactic acid): thermo-mechanical film properties and aging. *Polymer* 44: 7679–7688. https://doi.org/10.1016/j.polymer.2003.09.055.

135 Thi Ngoc Diep, P., Takagi, H., Shimizu, N. et al. (2019). Effects of loading amount of plasticizers on improved crystallization of poly(L-lactic acid). *Journal of Fiber Science and Technology* 75: 99–111. https://doi.org/10.2115/fiberst.2019-0013.

7

Crystallization in PCL-Based Blends and Composites

Madhushree Hegde[1], Akshatha Chandrashekar[1], Mouna Nataraja[1], Jineesh A. Gopi[2], Niranjana Prabhu[1], and Jyotishkumar Parameswaranpillai[2]

[1]M.S. Ramaiah University of Applied Sciences, Department of Chemistry, Faculty of Mathematical and Physical Sciences, New BEL Road, MSR Nagar, Bengaluru, Karnataka, 560054, India
[2]Alliance University, Division of Chemistry, Department of Science, Anekal - Chandapura Road, Bengaluru, Karnataka, 562106, India

7.1 Introduction

Poly(ε-caprolactone) (PCL) is a saturated aliphatic polyester with hexanoate repeat units [1]. It is a semi-crystalline biodegradable polymer broadly used in various biomedical applications, including tissue engineering [2]. PCL also exhibits good mechanical properties, and it is compatible with several other polymers. The excellent processability of PCL makes it a suitable candidate for the preparation of different types of scaffolds [3]. PCL also finds applications in other fields such as drug delivery [4], microelectronics [5], adhesives [6], and packaging [7]. PCL was synthesized mainly by ring-opening polymerization of ε-caprolactone or free radical ring-opening polymerization of 2-methylene-1-3-dioxepane [8]. Different synthesis methods for the preparation of PCL are compiled by Guarino et al. [9] as shown in Figure 7.1.

The molecular weight (M_{wt}) of the PCL can be controlled with the help of the catalyst used in its preparation. The relatively lower melting temperature (59–64 °C) of PCL enables its processability at relatively lower temperatures. PCL shows a lower glass transition temperature (T_g) of −60 °C and can be used in sub-ambient temperatures [2, 10]. PCL can be synthesized with crystallinity of up to 70% [11]. The crystallinity of PCL generally decreases with an increase in the M_{wt} due to chain folding [12, 13].

7.2 Crystallinity of PCL and the Factors Affecting Crystallinity

Crystallization is a major factor controlling the material design of commodity plastics such as polyethylene (PE) and polypropylene (PP), engineering plastics

Polymer Crystallization: Methods, Characterization, and Applications, First Edition.
Edited by Jyotishkumar Parameswaranpillai, Jenny Jacob, Senthilkumar Krishnasamy, Aswathy Jayakumar, and Nishar Hameed.
© 2023 WILEY-VCH GmbH. Published 2023 by WILEY-VCH GmbH.

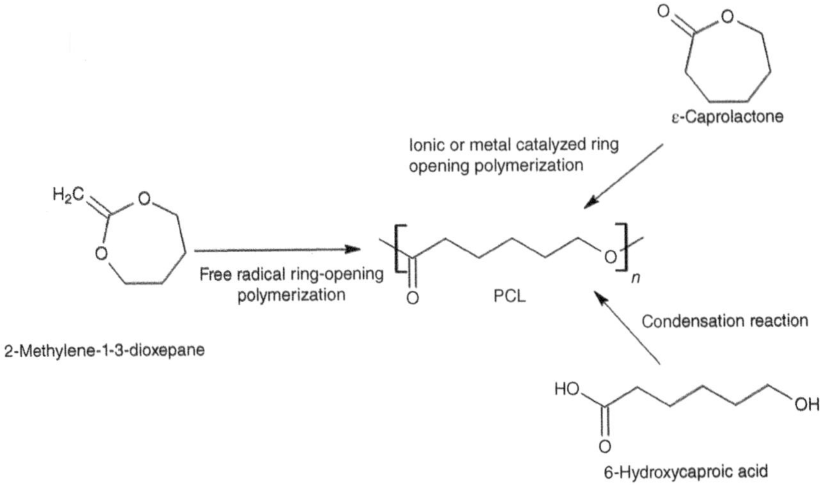

Figure 7.1 Different synthesis methods for PCL. Source: Guarino et al. [9]/John Wiley & Sons.

such as nylon 6,6, etc., and biopolymers such as polylactic acid (PLA) and PCL. The crystallinity of a polymer determines the mechanical, optical properties, and degradability. The processing conditions and the incorporation of second components into the polymer alter the crystallinity. Therefore, it is necessary to study the crystallinity of polymers and the change in crystallinity of the polymers during processing, blending with other polymers, and composite fabrication. Differential scanning calorimetry, X-ray diffraction techniques, infrared spectroscopy, and microscopic techniques such as hot stage microscopy and polarized optical microscopy are the commonly used methods to study the crystallization in polymers. The effect of crystallization characteristics on the flow properties of polymers can be studied using rheology [14]. Along with these characterization techniques, structural analysis with the help of nuclear magnetic resonance (NMR) spectroscopy offers information on the effect of changes in the structure on the crystallization behavior of polymers. Crystal structure formation is governed by the crystallization process or kinetics of crystallization, and the properties of PCL depend on the crystalline behavior and morphology. Therefore, it is essential to gather good understanding of the crystallization mechanism and morphology of PCL [15].

The access of water molecules to the bulk of PCL is an important factor in the degradation (hydrolytic cleavage) of PCL, and this access of water molecules depends on the percentage crystallinity of the polymer (spread of crystalline and amorphous domains in the PCL). The presence of amorphous regions facilitates the water molecules to enter the bulk of the polymer and increases the rate of the bulk degradation process [12]. A lower degradation rate was observed for the polymer with higher degrees of crystallinity. Moreover, PCL can be completely degraded using bacterial and fungal enzymes. Both esterase and lipase enzymes can degrade PCL very efficiently [16].

Compared to other biodegradable natural polymers and biodegradable polyesters, the degradation rate of PCL was less due to hydrophobic $-CH_2-$ repeating units in

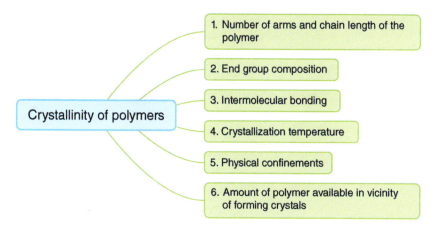

Figure 7.2 Factors affecting the polymer crystallization [9, 11, 17].

its structure [17]. The enthalpy of fusion of 100% crystalline PCL is 139.5 kJ kg^{-1}, and the crystallinity of semicrystalline PCL can be estimated from their enthalpy of fusion values [18]. The key factors that affect polymer crystallization are listed in Figure 7.2.

The crystallinity of 50%, 41%, and 33% was reported for the PCL grades with M_{wt} of 98 200, 161 000, and 224 500, respectively [13]. The reduced crystallinity with the increase in M_{wt} or increased polymer chain can be explained with the help of the unified reptation–nucleation theory. As per this theory, long chains arrange themselves into loops and cilia. These loops and cilium disturb the reptation process, which leads to hindered lamella thickening and reduced degree of crystallinity [13]. The thickness of the polymer film also has a significant effect on the spherulites morphology. If the thickness of the film is c.120 nm, then it behaves like a bulk polymer, and the spherulites were formed because of surface nucleation. Once the film thickness is less than 12 nm, the polymer crystallization is a diffusion-controlled process; on the other hand, if the film thickness is less than 4 nm, the PCL films cannot crystallize [19]. Upon reducing the film thickness, different morphological transitions are possible, such as quasi-2D spherulites, seaweed structures, and dendrites, as shown in Figure 7.3.

Zhu et al. [20] synthesized PCL with different end groups such as phenyl, naphthyl, and anthracenyl, and studied their effect on the crystallization of PCL using differential scanning calorimetry (DSC) and polarized optical microscopy (POM). An increase in the size of the end group reduces the spherulite growth rate. But the rate of crystallization of PCL with phenyl and naphthyl groups is almost same; this is due to the compatibility of naphthyl groups with PCL. The rheology studies also reported low viscosity for the naphthyl-modified PCL. The effect of block copolymer architecture on the isothermal crystallization kinetics of PCL and poly(trimethylene carbonate) di and triblock copolymers was investigated by Castillo et al. [21]. The sequence of block copolymer had a strong influence on the crystallization kinetics of the block copolymer, and the effect is dependent mainly on the nucleation step due to the topological constraints.

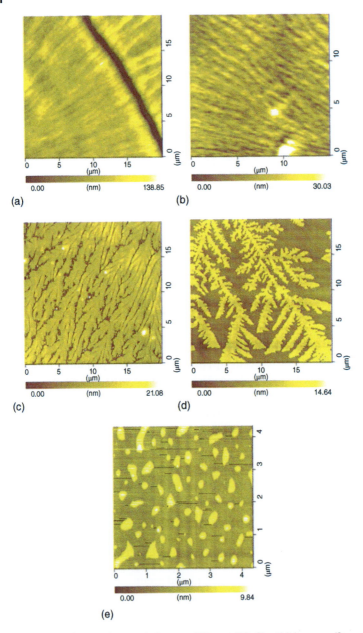

Figure 7.3 Change in morphology at different PCL film thicknesses (from (a) to (e), thickness = 120, 20, 12, 8, and 4 nm, respectively). Source: Qiao et al. [19]/Reproduced with permission from John Wiley & Sons.

7.3 Crystalline Behavior of PCL-Based Multiphase Polymer Systems

Polymer blends, copolymers, composites, interpenetrating polymer networks, etc., are examples of multiphase polymer systems, and these systems attract huge attention due to their high performance compared to neat polymer systems [22]. Various kinds of polymer crystallization in these multiphase systems and their characteristics are shown in Figure 7.4.

7.3.1 Crystallization Behavior of Blends of PCL

The polymer blending technique is an easy and effective method to enhance the various characteristics of polymers. Polymer blend systems can be categorized into miscible, partially miscible, or immiscible systems based on the miscibility between the components in the polymer blends. Based on the crystalline nature of the components, polymer blend systems can be classified into amorphous/amorphous, amorphous/crystalline, and crystalline/crystalline. The characteristics of these blends also depend on the morphology of the blends and the crystalline behavior of the component polymers. This in turn depends on the polymer blend composition, miscibility between the components in the blend, the thermal history of the polymer blend, and processing parameters [22, 23].

To improve biodegradability and reduce cost, blending PCL with starch is a good option. This blending leads to reduced stiffness and enhanced tear strength [17].

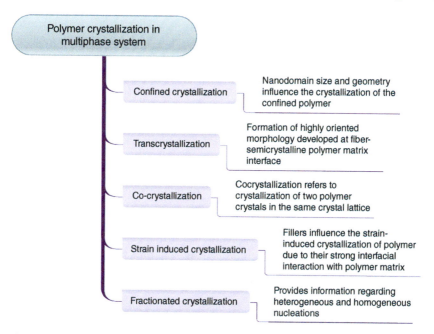

Figure 7.4 Types of polymer crystallization in multiphase systems, Source: Arif et al. [22]/with permission of Elsevier.

Ramírez-Arreola et al. [24] analyzed the effect of PCL concentration on the blends of PCL and thermoplastic starch and found that increase in the concentration of PCL increased the hydrophobicity of the blends. To improve the interaction between PCL and starch, citric acid was introduced into the blends by Ortega-Toro et al. [25] and it helped the blend to achieve better functional properties. Similarly, blending with chitosan was used to improve the antimicrobial property and tensile properties and to reduce water vapor permeability for food packaging applications, especially in the field of active food packaging [26]. Thus, different polymers can be blended into PCL depending on the required characteristics and the intended applications.

Defieuw et al. [27] prepared miscible and partially miscible blends based on PCL and amorphous chlorinated polyethylene (CPE). The crystalline behavior of these blends was analyzed using optical and electron microscopy, X-ray diffraction analysis (XRD), and DSC. The study showed that CPE with 49.1 wt% chlorine content is miscible with PCL in entire blend ratios and temperature ranges. At the same time, CPE with 42.1 wt% chlorine is showing miscibility with PCL only in certain blend ratios and temperatures. CPE with 35.6 wt% chlorine is partially miscible with PCL. In the PCL-rich blends (70–90 wt%) with CPE of 42.1% and 49.1% chlorine content, PCL crystallizes as volume-filling spherulites, and the amorphous CPE segregates inter-fibrillar. In CPE-rich blends (50 wt%), PCL crystallites as domains (not as volume-filling spherulites due to their small volume fraction), and the amorphous CPE segregates inter-spherulitically. In the partially miscible blend of PCL and CPE with 35.6 wt% chlorine, both PCL and CPE phases were crystallized.

Madbouly [28] prepared PCL/crosslinked carboxylated polyester resin (CPER) binary blends and studied non-isothermal crystallization kinetics and spherulitic growth using POM and DSC techniques. The crystallization of PCL was decreased due to the incorporation of CPER. Different theoretical models, such as modified Avrami, Ozawa, and combined Avrami–Ozawa models, were used to discuss the non-isothermal crystallization kinetics of these blends. The kinetic parameters calculated from these models established that the blending process decreased the rate of non-isothermal crystallization without a change in the mechanism of crystallization. Activation energy of the non-isothermal crystallization process of PCL and PCL/CPER blends was determined from the peak maximum of crystallization temperature (T_c) and cooling rate with the help of Kissinger equation, and this study showed that the activation energy was increased from 83 kJ mol^{-1} for neat PCL to 115 and 119 kJ mol^{-1} for the blends with 25 and 50 wt% of CPER, respectively. This increase in the activation energy is attributed to the restriction imparted by the CPER molecules on the motion of PCL chains, and this constraint made the crystallization process much harder compared to PCL.

The effect of ethylene/octene multiblock copolymer (OBC) on the crystallization behavior and melting point of PCL was studied by Lai et al. [29]. DSC cooling and heating curves of neat PCL, OBC, and their blends are shown in Figure 7.5. A minor decrease in the T_c of PCL was observed for the blends with higher OBC loadings. Similarly, the T_c of OBC was also decreased in the presence of PCL. These observations are due to the interference of one polymer chain with the crystal growth of other polymer chains. Similarly, the melting point of each component

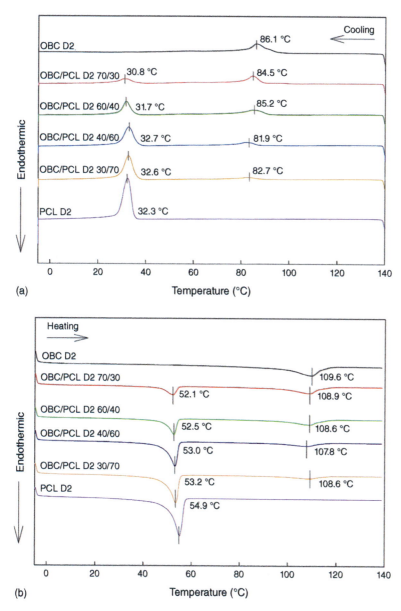

Figure 7.5 DSC (a) cooling and (b) heating profiles of OBC, PCL, and their blends. Source: Reproduced with permission from Lai et al. [29]/Elsevier /Public Domain CC BY 4.0.

in the blends also dropped in the presence of other components. Changes were observed in the crystallinity of polymers before and after blending. For example, 70% OBC content in the blend reduced the crystallinity of PCL to 30.4% from 37.4% of neat PCL, and the crystallinity of OBC changed from 9.3% to 6.7%. The decrease in the crystallinity is due to the hindrance effect of each component on the crystal growth of the other components. The blends showed excellent thermo-triggered two-way multiple-shape memory properties.

Small-angle XRD and optical microscopy techniques were used to study the effect of styrene-co-maleic anhydride (SMA) copolymers with different maleic anhydride contents on the crystallization behavior of PCL. The incorporation of SMA copolymer into PCL reduced the spherulite growth rate of PCL. This may be due to the local diffusion effects of SMA during the lamellar growth process of PCL [30]. Xiao et al. [31] studied the influence of blending poly(butylene adipate-co-terephthalate) on the crystallization characteristics of PCL and established that the maximum rate of crystallization depends on crystallization temperature as well as the nature of the polymer components. Kim et al. [32] prepared PCL/poly-L-lactic acid (PLLA) blends compatibilized with poly(L,L-lactide-co-ε-caprolactone) and observed that the incorporation of the copolymer reduced the crystallization of PLLA and PCL due to the increased compatibility of the copolymer. On contrary to this, the incorporation of block copolymer PLLA-PCL-PLLA into the PCL-PLLA blends did not change the increased crystallization rate of PLLA in presence of PCL domains [33].

7.3.2 Crystallization Behavior of Block Copolymers of PCL

The physical and chemical properties of PCL can be altered by the copolymerization technique. Copolymers of PCL have different crystallinity, solubility, and degradation properties compared to PCL homopolymers [34]. For example, the rate of hydrolysis of PCL can be modified with the help of copolymerizing caprolactone with glycolides/lactides. Copolymers of PCL can achieve desired thermo-mechanical properties and can be used in drug delivery devices, medical devices, and scaffolds. Monomers such as ethylene oxide, dilactide, valerlactone, styrene, methyl methacrylate, vinyl acetate, and diglycolides are used for the preparation of copolymers of PCL [35].

Peponi et al. [36] synthesized PCL-b-PLLA di-block copolymers and studied the effect of M_{wt} on the crystalline behavior of the block copolymer. The crystallization of both PLLA and PCL blocks is dependent on the molecular weight (both polymer blocks need minimum length to crystallize). It was found that the presence of PCL block does not influence the crystallization behavior of PLLA block. But the PLLA crystallization affects the crystallization of PCL block. When PLLA block is amorphous, PCL block crystallizes at shorter length (c. 851 g mol^{-1}), but if the PLLA block is crystalline, then the PCL block crystallizes at above c. 2000 g mol^{-1}. A series of four-armed diblock copolymers of poly(ε-caprolactone)-b-poly(D-lactic acid) (4a-PCL-b-PDLA) were synthesized by Ning et al. [37] and studied the influence of length of PCL and PDLA blocks on the crystallization behavior of the synthesized block copolymers. Studies showed that PDLA block length affects the crystallization behavior of PCL under different crystallization conditions.

Peng et al. [38] synthesized a miktoarm star copolymer (μ-PEG-PCL-PLLA) and studied the crystallization behavior of this copolymer. The crystallization of PLLA was reduced in the triblock copolymer compared to diblock copolymer polyethylene glycol (PEG)-PLLA. During crystallization process only small grains of PLLA were observed at 110 °C. On the other hand, when the temperature is reduced to 30 °C, the PCL spherulites were formed with PLLA crystalline grains as nucleus. Similarly, poly(ester-urethane)s and tri-block copolymers based on PCL and PLLA were prepared by ring-opening polymerization with varying PLLA content. The researchers observed better crystallization in triblock copolymers than in poly(ester-urethane)s [39]. Studies also showed that block copolymers have different spherulite structure compared to homopolymers. For example, a star-shaped PCL-PLLA block copolymer with an increase in the arm-length ratio of PCL-PLLA content exhibited a change in spherulite structure from ordinary spherulites to band spherulites [40].

7.3.3 Effect of Fillers on the Crystalline Behavior of PCL

In polymer composites, the crystallization of polymers is based on various factors such as type of filler and the matrix and the dispersion of the filler in the matrix. The incorporation of fillers affects several features of crystallization such as nucleation, kinetics of crystallization, spherulitic growth, crystal structure, and morphology. It is important to know the action of nanofillers as nucleating agents and their effect on the crystallization and biodegradability of PCL. The incorporation of polymer-based particles and fibers as fillers into the PCL matrix also affects the crystallization behavior of PCL. For example, well-dispersed polyamide 6 (PA6) particles in PCL promoted crystallization by acting as nucleating agents without affecting crystal structure or crystallization mechanism [41]. Similarly, the incorporation of polyvinyl chloride (PVC) fibers enhanced the non-isothermal crystallization peak temperatures and isothermal crystallization rate along with an enhancement in the strength of the composites by acting as a nucleating agent [42].

Incorporation of fillers such as nanocellulose (NC), metal nanoparticles, glass fibers, and glass microspheres into PLA matrix alters the crystallization behavior of PLA. Vignali et al. [43] prepared PCL composites using rotational molding with 3-(aminopropyl)triethoxysilane (APTES)-modified hollow glass microspheres (HGM). The surface-modified HGM acted as a nucleating agent and increased the rate of crystallization of PLA. Also, modified HGM reduced thermo-oxidative degradation of PCL. The lightweight nature of this composite with enhanced mechanical properties compared to PCL can gain attention in industries. Composites of PCL with different loadings of niobium pentoxide (Nb_2O_5) and alumina (Al_2O_3) were analyzed for their thermal properties, biodegradability, and crystallization kinetics. The incorporation of the 1% oxides increased the thermal stability of the PCL, followed by a drop in values, but better than pure PCL. On the other hand, biodegradation of PCL showed fluctuations in values with the increase in filler loading. The non-isothermal crystallization of the PCL was described well by pseudo-Avrami and Mo models [44]. Liverani et al. [45] prepared electron-spun triethoxysilane-terminated PCL (PCL-TES) – bioactive glasses (BG) fiber composites for shape memory applications. DSC studies showed that the presence of BG hindered the mobility of PCL chains, which reduced the crystallinity of PCL due to additional crosslinking caused by the hydrolysis and condensation reactions

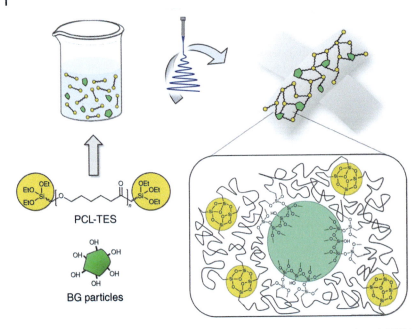

Figure 7.6 Schematic of binding in PCL-TES/BG system. Source: Liverani et al. [45] (under creative commons license).

of PCL-TES chains with BG particle surface, and this mechanism is depicted in Figure 7.6.

Recently, NC has received huge attention in the field of polymer science because of its interesting chemical and physical properties, biodegradability, and ease to functionalize [46]. Mi and team [47] investigated the effect of cellulose nanocrystal (CNC) on the crystallinity of PCL and found that CNC acts as the nucleating agent and enhanced the T_c and rate of crystallization. A similar nucleation effect could be seen on the PCL-grafted-CNC-filled PCL composites [48] and the cellulose nanoparticle-filled PCL [49].

The mechanism behind the nucleation, the interaction between the CNC and PCL, and the effect of CNC on the lamellar structure formation in PCL were studied by Lv and team [50]. They also explored the effect of acetylation on CNC and its effect on the crystallization behavior of PCL composites. The researchers found out that pristine CNC is acting as a heterogenous nucleating agent in PCL and increases the crystallization temperature of PCL. On the other hand, acetylated CNC acts as an antinucleating agent and decreases the degree of crystallinity and crystallization temperature, and this may be due to restrained nucleation kinetics and degraded lamellar structure. The effect of loading of NC on the crystallization of PCL in isothermal and non-isothermal conditions was investigated by Li et al. [51]. In both isothermal and non-isothermal conditions, the addition of NC enhanced the rate of crystallization and reduced the crystallization half-time with an increase in the NC loadings, i.e. NC acted as a nucleating agent and increased the degree of crystallinity.

The effect of both CNCs and chitin nanocrystals (ChNCs) on the crystallization behavior of PCL was studied by Li and Wu [52] who observed that CNCs acted as nucleating agent and enhanced the rate of crystallization, and on the other hand ChNCs acted as anti-nucleation agent, which decreased the rate of crystallization in PCL. This difference may be ascribed due to the difference in the interaction between the PCL and the nanocrystals. Celebi et al. [53] studied the effect of surface modification of NC by n-octadecyl isocyanate on the crystallinity of PCL. The surface modification improved the dispersion and network formation and reported enhanced dynamic mechanical properties. However, the unmodified and modified NCs reported no change in the crystallinity of the PCL. Zhang et al. [54] observed that, in the case of starch nanocrystals, acetylation as the surface modification method reduces the nucleating nature of starch nanocrystals in PCL. That means the acetylated starch nanocrystals (SNC) particles even acted as anti-nucleating agents, thus having increased spherulite size. These nucleating and antinucleating nature of nanofillers can be efficiently used to alter the crystallization behavior of PCL, which will be useful for certain applications.

In an interesting study, Eriksson et al. [55] studied the nucleating effect of PCL-grafted-CNC and non-grafted-CNC on the PCL matrix under non-shear (quiescent) conditions and shear conditions using small-angle X-ray scattering (SAXS), DSC, and rheology. The PCL-grafted-CNC reported better nucleating efficiency under quiescent conditions. While the non-grafted CNC reported better nucleation effect under the influence of shear. To enhance the interfacial adhesion between nano-lignin and PCL/PLA, Yang et al. [56] grafted nano-lignin particles with poly(lactide-ε-caprolactone) copolymer. The researchers investigated the effect of grafted lignin nanoparticles on PLA/PCL blends and observed that the crystallization ability of both PLA and PCL was increased due to the enhanced interfacial adhesion between grafted nanoparticles and PLA/PCL chains. The introduction of grafted lignin in PLA/PCL blend also resulted in enhancement in toughness.

Among the polymer nanocomposites, polymer–carbon nanocomposites gained interest in the automobile, electronic, biomedical, and space industries due to their unique characteristics [57]. Agglomeration of carbon nanofillers in the polymer matrix leads to inferior properties of the composites. To avoid the agglomeration of carbon-based nanofillers in the polymer matrix, surface functionalization or chemical modifications of carbon-based nanofillers are widely used [58]. Based on the morphology of carbon-based nanofillers, they are classified into 0D (fullerene), 1D (carbon nanotubes and carbon nanofibers), 2D (graphene), and 3D (graphite) fillers. The incorporation of these carbon-based nanomaterials affects the crystallization behavior of polymers.

The effects of graphite oxide (GO) and reduced graphene oxide (RGO) on the crystallization of PCL were analyzed by various researchers. The incorporation of GO reduced the crystallization half-time for both the isothermal and non-isothermal crystallizations of PCL. This indicates that GO is acting as nucleating agent in the crystallization of PCL. The nucleation density is also enhanced with the incorporation of GO, which is confirmed using polarized optical microscopy [59, 60].

In the study conducted by Zhang and Qiu [61], thermally reduced graphene (TRG) acted as a nucleation agent and enhanced both non-isothermal and isothermal melt crystallization of PCL. On the other hand, there was no change in crystallization mechanism and crystal structure of PCL due to the presence of TRG. RGO was found to be a nucleating agent for PCL in the injection-molded PCL-RGO composites. Also, XRD results confirmed that RGO increases the orientation degree of PCL crystals in the flow direction, which improved the mechanical properties of PCL. This is due to the hindrance to the motion of the polymer chains caused by the presence of RGO [62]. The influence of shear on epitaxial crystallization of PCL on RGO was studied, and it was found that an increase in the shear promoted the PCL chains to epitaxially crystallize on RGO surfaces and led to the formation of thicker lamellae of PCL on RGO [63]. This indicates that not only fillers, but also conditions affect crystallization of PCL in PCL-based composites.

Non-isothermal crystallization studies have been conducted by Jana and Cho [64] to investigate the effect of functionalized multiwalled carbon nanotubes (MWNTs) (f-MWNTs) on the crystallization behavior of PCL. XRD studies showed that the incorporation of f-MWNTs decreased the crystallinity of PCL. This is due to the higher number of heterogeneous nucleation sites induced by the presence of f-MWNTs, which led to more imperfections of PCL crystallites. The spherulite size was reduced by the introduction of f-MWNTs due to an increase in the nucleation density and the hindrance to the motion of the PCL chains in the composites.

Biobased reinforcing fillers, including natural fibers, can be used to develop cost-effective biocomposites. The effect of biobased fillers on the crystallization behavior was studied by different researchers, who found that the biobased fillers can alter the crystallization behavior of PCL, including the rate of crystallization, morphology, size, and amount of the crystals in the PCL. Bhagabati et al. [65] studied the effect of bamboo-root flour on the crystallization behavior of PCL matrix. The interaction between the fillers and PLA matrix led to nucleation and trans-crystallization at the surface of the filler, which is evident from the morphology studies using POM. The spherulite growth in neat PCL and PCL biocomposites at different times (POM micrographs) is given in Figure 7.7.

Surface-modified Macaiba shell [66] and fibers [67] showed nucleating effects in PCL. Also, PCL-Macaiba fiber composites showed enhanced biodegradation compared to neat PCL [68]. The effect of unmodified and surface-modified brown coir on the crystallization of PCL was investigated by Lima et al. [69] who found that the incorporation of coir enhanced the T_c to slight extent, which is 3.5 °C higher compared to neat PCL, and this change is independent of the amount of loading. There was no evidence of the change in crystallinity caused by the incorporation of coir into PCL. But the incorporation of 20% coir increased Young's modulus and tensile strength with a reduction in elongation at break. In another study, the incorporation of alfa fiber mats into PCL enhanced the crystallization capacity of the PCL matrix, and this capacity increases with higher loading of alfa fiber mats [70].

The incorporation of rice husk (RH) into PCL reported decreased crystallinity index of the PCL in the composites. The decrease in the mobility of PCL chains due to RH may lead to a reduction in packing efficiency, which leads to a less perfect

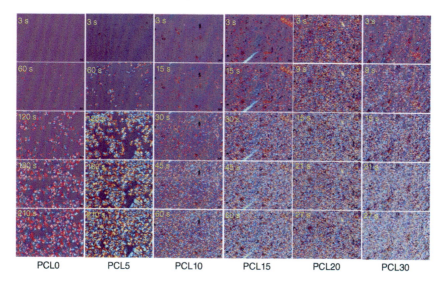

Figure 7.7 Spherulite growth in neat PCL and PCL biocomposites at different times (POM micrographs) (magnification: 50x). Source: Bhagabati et al. [65]/Royal Society of Chemistry / CC BY 3.0.

crystalline structure of PCL [71]. Madbouly [72] prepared a semi-interpenetrating network using PCL and tung oil and studied both isothermal and non-isothermal crystallization kinetics. This study showed that crystallization kinetics highly depends on the miscibility and morphology of the blends. A considerable increase in the crystallization kinetics was observed for both isothermal and non-isothermal crystallization in an immiscible blend containing 50 wt% PCL. In contrast to this, a decrease in the isothermal and non-isothermal crystallization kinetics was observed for the partially miscible blends with PCL content between 10 and 30 wt%.

7.4 Conclusion

PCL is a well-known polymer that has attracted attention from both research and industry communities due to its biocompatibility, mechanical strength, and capacity to form blends with different polymers. Being semi-crystalline in nature, the properties of PCL are extremely influenced by the degree of crystallinity, crystal structure, and size. To achieve the required properties for different applications, multiphase polymer systems of PCL are required. In this book chapter, a detailed description has been given with the help of various literature on PCL-based multicomponent systems and their difference in crystallization behavior compared to neat PCL. The difference in the crystallization behavior eventually affects the properties of the multicomponent systems. Blending PCL with other polymers is an efficient way to enhance the properties of PCL, and the properties of the blends depend on factors such as miscibility between the polymers and the crystalline nature of the individual polymers. Crystallization behavior of PCL is affected by other polymers in the blends due to the restriction imparted by the polymer chains of the second polymer on the mobility of PCL chains, changes in the spherulite growth of PCL due to local diffusion effects during the lamellar growth process of PCL, the concurrent crystallization process, etc. Copolymers of PCL have different crystallinity compared to the crystallinity of neat PCL, and copolymers show different degradation and solubility properties. A significant number of studies related to the crystallization behavior of PCL-based block copolymers are discussed in this chapter. From these studies, one can conclude that the crystalline and degradation nature of block copolymers of PCL are dependent on copolymer composition, M_{wt}, etc. In PCL composites, the fillers can induce heterogeneous nucleation in PCL. Nanoparticles with the large surface area could affect crystallization in PLA by influencing the chain mobility near the particle surface. There are several literatures available where the fillers do not have any effect or have retarding effect on the crystallization of PCL. The change in the crystallization behavior, such as crystallization kinetics and nucleation of PCL crystals, with the incorporation of biobased fillers, including natural fibers, are also discussed in the chapter.

References

1 Zhang, C. (2015). Biodegradable polyesters: synthesis, properties, applications. In: *Biodegradable Polyesters*, vol. 15 (ed. S. Fakirov), 1–14. Wiley-VCH.
2 Fernández, M.D., Guzmán, D.J., Ramos, J.R., and Fernández, M.J. (2019). Effect of alkyl chain length in POSS nanocage on non-isothermal crystallization behavior of PCL/amino-POSS nanocomposites. *Polymers* 11 (10): 1719.
3 Dwivedi, R., Kumar, S., Pandey, R. et al. (2020). Polycaprolactone as biomaterial for bone scaffolds: review of literature. *Journal of Oral Biology and Craniofacial Research* 10 (1): 381–388.
4 Chen, D.R., Bei, J.Z., and Wang, S.G. (2000). Polycaprolactone microparticles and their biodegradation. *Polymer Degradation and Stability* 67 (3): 455–459.
5 Hedrick, J.L., Magbitang, T., Connor, E.F. et al. (2002). Application of complex macromolecular architectures for advanced microelectronic materials. *Chemistry A European Journal* 8 (15): 3308.
6 Tous, L., Ruseckaite, R.A., and Ciannamea, E.M. (2019). Sustainable hot-melt adhesives based on soybean protein isolate and polycaprolactone. *Industrial Crops and Products* 135: 153–158.
7 Ikada, Y. and Tsuji, H. (2000). Biodegradable polyesters for medical and ecological applications. *Macromolecular Rapid Communications* 21 (3): 117–132.
8 Labet, M. and Thielemans, W. (2009). Synthesis of polycaprolactone: a review. *Chemical Society Reviews* 38 (12): 3484–3504.
9 Guarino, V., Gentile, G., Sorrentino, L., and Ambrosio, L. (2017). Polycaprolactone: synthesis, properties, and applications. In: *Encyclopedia of Polymer Science and Technology* (ed. H.F. Mark), 1–36. Wiley.

10 Lecomte, P. and Jérôme, C. (2011). Recent developments in ring-opening polymerization of lactones. In: *Synthetic Biodegradable Polymers* (ed. B. Rieger, A. Künkel, G.W. Coates, et al.), 173–217. Springer.
11 Woodruff, M.A. and Hutmacher, D.W. (2010). The return of a forgotten polymer – polycaprolactone in the 21st century. *Progress in Polymer Science* 35 (10): 1217–1256.
12 Sisson, A.L., Ekinci, D., and Lendlein, A. (2013). The contemporary role of ε-caprolactone chemistry to create advanced polymer architectures. *Polymer* 54 (17): 4333–4350.
13 Jenkins, M.J. and Harrison, K.L. (2006). The effect of molecular weight on the crystallization kinetics of polycaprolactone. *Polymers for Advanced Technologies* 17 (6): 474–478.
14 Speranza, V., Sorrentino, A., De Santis, F., and Pantani, R. (2014). Characterization of the polycaprolactone melt crystallization: complementary optical microscopy, DSC, and AFM studies. *Scientific World Journal* 2014: 720157.
15 Mileva, D., Tranchida, D., and Gahleitner, M. (2018). Designing polymer crystallinity: an industrial perspective. *Polymer Crystallization* 1 (2): e10009.
16 Kulkarni, A., Reiche, J., Kratz, K. et al. (2007). Enzymatic chain-scission kinetics of poly(ε-caprolactone) monolayers. *Langmuir* 23 (24): 12202–12207.
17 Thakur, M., Majid, I., Hussain, S., and Nanda, V. (2021). Poly(ε-caprolactone): a potential polymer for biodegradable food packaging applications. *Packaging Technology and Science* 34 (8): 449–461.
18 Pitt, C.G. (1990). Poly-ε-caprolactone and its copolymers. In: *Biodegradable Polymers as Drug Delivery Systems* (ed. M. Chasin and R. Langer), 71–120. Marcel Dekker Inc.
19 Qiao, C., Zhao, J., Jiang, S. et al. (2005). Crystalline morphology evolution in PCL thin films. *Journal of Polymer Science Part B: Polymer Physics* 43 (11): 1303–1309.
20 Zhu, L., Li, J., Li, H. et al. (2022). End groups affected crystallization behavior of unentangled poly(ε-caprolactone)s. *Polymer* 241: 124534.
21 Castillo, R.V., Fleury, G., Navarro, C. et al. (2017). Impact of the architecture on the crystallization kinetics of poly(ε-caprolactone)/poly(trimethylene carbonate) block copolymers. *European Polymer Journal* 95: 711–727.
22 Arif, P.M., Kalarikkal, N., and Thomas, S. (2018). Introduction on crystallization in multiphase polymer systems. In: *Crystallization in Multiphase Polymer Systems* (ed. S. Thomas, P.M. Arif, E.B. Gowd, and N. Kalarikkal), 1–16. Elsevier.
23 Kong, Y., Ma, Y., Lei, L. et al. (2017). Crystallization of poly(ε-caprolactone) in poly(vinylidene fluoride)/poly(ε-caprolactone) blend. *Polymers* 9 (12): 42.
24 Ramírez-Arreola, D.E., Robledo-Ortiz, J.R., Moscoso, F. et al. (2012). Film processability and properties of polycaprolactone/thermoplastic starch blends. *Journal of Applied Polymer Science* 123 (1): 179–190.
25 Ortega-Toro, R., Collazo-Bigliardi, S., Talens, P., and Chiralt, A. (2016). Influence of citric acid on the properties and stability of starch-polycaprolactone based films. *Journal of Applied Polymer Science* 133 (2): 42220.

26 Joseph, C.S., Prashanth, K.V.H., Rastogi, N.K. et al. (2009). Optimum blend of chitosan and poly-(ε-caprolactone) for fabrication of films for food packaging applications. *Food and Bioprocess Technology* 4 (7): 1179–1185.

27 Defieuw, G., Groeninckx, G., and Reynaers, H. (1989). Miscibility and morphology of binary polymer blends of polycaprolactone with solution-chlorinated polyethylenes. *Polymer* 30 (4): 595–603.

28 Madbouly, S.A. (2011). Nonisothermal crystallization kinetics of miscible blends of polycaprolactone and crosslinked carboxylated polyester resin. *Journal of Macromolecular Science, Physics* 50 (3): 427–443.

29 Lai, S.M., Fan Jiang, S.Y., Chou, H.C. et al. (2021). Novel two-way multiple shape memory effects of olefin block copolymer (OBC)/polycaprolactone (PCL) blends. *Polymer Testing* 102: 107333.

30 Defieuw, G., Groeninckx, G., and Reynaers, H. (1989). Miscibility, crystallization and melting behaviour, and morphology of binary blends of polycaprolactone with styrene-*co*-maleic anhydride copolymers. *Polymer* 30 (12): 2158–2163.

31 Xiao, H., Lu, W., and Yeh, J.T. (2009). Crystallization behavior of fully biodegradable poly(lactic acid)/poly(butylene adipate-*co*-terephthalate) blends. *Journal of Applied Polymer Science* 112 (6): 3754–3763.

32 Kim, C.-H., Cho, K., Choi, E.-J.C., and Park, J.-K. (2000). Effect of P(lLA-co-εCL) on the compatibility and crystallization behavior of PCL/PLLA blends. *Journal of Applied Polymer Science* 77 (1): 226–231.

33 Dell'Erba, R., Groeninckx, G., Maglio, G. et al. (2001). Immiscible polymer blends of semicrystalline biocompatible components: thermal properties and phase morphology analysis of PLLA/PCL blends. *Polymer* 42 (18): 7831–7840.

34 Espinoza, S.M., Patil, H.I., San Martin Martinez, E. et al. (2019). Poly-ε-caprolactone (PCL), a promising polymer for pharmaceutical and biomedical applications: focus on nanomedicine in cancer. *International Journal of Polymeric Materials* 69 (2): 85–126.

35 Okada, M. (2002). Chemical syntheses of biodegradable polymers. *Progress in Polymer Science* 27 (1): 87–133.

36 Peponi, L., Navarro-Baena, I., Báez, J.E. et al. (2012). Effect of the molecular weight on the crystallinity of PCL-*b*-PLLA di-block copolymers. *Polymer* 53 (21): 4561–4568.

37 Ning, Z., Jiang, N., and Gan, Z. (2014). Four-armed PCL-*b*-PDLA diblock copolymer: 1. Synthesis, crystallization and degradation. *Polymer Degradation and Stability* 107: 120–128.

38 Peng, X., Zhang, Y., Chen, Y. et al. (2016). Synthesis and crystallization of well-defined biodegradable miktoarm star PEG–PCL–PLLA copolymer. *Materials Letters* 171: 83–86.

39 Navarro-Baena, I., Kenny, J.M., and Peponi, L. (2014). Crystallization and thermal characterization of biodegradable tri-block copolymers and poly(ester-urethane)s based on PCL and PLLA. *Polymer Degradation and Stability* 108: 140–150.

40 Wang, J.L. and Dong, C.M. (2006). Synthesis, sequential crystallization and morphological evolution of well-defined star-shaped

poly(ε-caprolactone)-*b*-poly(L-lactide) block copolymer. *Macromolecular Chemistry and Physics* 207 (5): 554–562.

41 Li, M., Zhang, Y., Zhu, F. et al. (2021). Influence of PA6 particle filler on morphology, crystallization behavior and dynamic mechanical properties of poly(ε-caprolactone) as an efficient nucleating agent. *Journal of Polymer Research* 28 (12): 1–9.

42 Li, Y., Yao, S., Shi, H. et al. (2021). Enhancing the crystallization of biodegradable poly(ε-caprolactone) using a polyvinyl alcohol fiber favoring nucleation. *Thermochimica Acta* 706: 179065.

43 Vignali, A., Iannace, S., Falcone, G. et al. (2019). Lightweight poly(ε-caprolactone) composites with surface modified hollow glass microspheres for use in rotational molding: thermal, rheological and mechanical properties. *Polymers* 11 (4): 624.

44 Sousa, J.C., Costa, A.R.M., Lima, J.C. et al. (2021). Crystallization kinetics modeling, thermal properties and biodegradability of poly(ε-caprolactone)/niobium pentoxide and alumina compounds. *Polymer Bulletin* 78 (12): 7337–7353.

45 Liverani, L., Liguori, A., Zezza, P. et al. (2022). Nanocomposite electrospun fibers of poly(ε-caprolactone)/bioactive glass with shape memory properties. *Bioactive Materials* 11: 230–239.

46 Si, J., Cui, Z., Wang, Q. et al. (2016). Biomimetic composite scaffolds based on mineralization of hydroxyapatite on electrospun poly(ε-caprolactone)/nanocellulose fibers. *Carbohydrate Polymers* 143: 270–278.

47 Mi, H.Y., Jing, X., Peng, J. et al. (2014). Poly(ε-caprolactone) (PCL)/cellulose nano-crystal (CNC) nanocomposites and foams. *Cellulose* 21 (4): 2727–2741.

48 Bellani, C.F., Pollet, E., Hebraud, A. et al. (2016). Morphological, thermal, and mechanical properties of poly(ε-caprolactone)/poly(ε-caprolactone)-grafted-cellulose nanocrystals mats produced by electrospinning. *Journal of Applied Polymer Science* 133 (21): 43445.

49 Siqueira, G., Fraschini, C., Bras, J. et al. (2011). Impact of the nature and shape of cellulosic nanoparticles on the isothermal crystallization kinetics of poly(ε-caprolactone). *European Polymer Journal* 47 (12): 2216–2227.

50 Lv, Q., Xu, C., Wu, D. et al. (2017). The role of nanocrystalline cellulose during crystallization of poly(ε-caprolactone) composites: nucleation agent or not? *Composites, Part A Applied Science and Manufacturing* 92: 17–26.

51 Li, Y., Han, C., Yu, Y., and Xiao, L. (2020). Effect of loadings of nanocellulose on the significantly improved crystallization and mechanical properties of biodegradable poly(ε-caprolactone). *International Journal of Biological Macromolecules* 147: 34–45.

52 Li, J. and Wu, D. (2022). Nucleation roles of cellulose nanocrystals and chitin nanocrystals in poly(ε-caprolactone) nanocomposites. *International Journal of Biological Macromolecules* 205: 587–594.

53 Celebi, H., Ilgar, M., and Seyhan, A.T. (2021). Evaluation of the effect of isocyanate modification on the thermal and rheological properties of poly(ε-caprolactone)/cellulose composites. *Polymer Bulletin* 79: 4941–4955.

54 Zhang, G., Xu, C., Wu, D. et al. (2018). Crystallization of green poly(ε-caprolactone) nanocomposites with starch nanocrystal: the nucleation role switching of starch nanocrystal with its surface acetylation. *Industrial and Engineering Chemistry Research* 57 (18): 6257–6264.

55 Eriksson, M., Goffin, A.L., Dubois, P. et al. (2018). The influence of grafting on flow-induced crystallization and rheological properties of poly(ε-caprolactone)/cellulose nanocrystal nanocomposites. *Nanocomposites* 4 (3): 87–101.

56 Yang, W., Qi, G., Ding, H. et al. (2020). Biodegradable poly(lactic acid)–poly(ε-caprolactone)–nanolignin composite films with excellent flexibility and UV barrier performance. *Composites Communications* 22: 100497.

57 Jineesh, A.G. and Mohapatra, S. (2019). Thermal properties of polymer–carbon nanocomposites. In: *Carbon-Containing Polymer Composites* (ed. M. Rahaman, D. Khastgir, and A.K. Aldalbahi), 235–270. Singapore: Springer.

58 Roy, N., Sengupta, R., and Bhowmick, A.K. (2012). Modifications of carbon for polymer composites and nanocomposites. *Progress in Polymer Science* 37 (6): 781–819.

59 Hua, L., Kai, W.H., and Inoue, Y. (2007). Crystallization behavior of poly(ε-caprolactone)/graphite oxide composites. *Journal of Applied Polymer Science* 106 (6): 4225–4232.

60 Hua, L., Kai, W., and Inoue, Y. (2007). Synthesis and characterization of poly(ε-caprolactone)–graphite oxide composites. *Journal of Applied Polymer Science* 106 (3): 1880–1884.

61 Zhang, J. and Qiu, Z. (2011). Morphology, crystallization behavior, and dynamic mechanical properties of biodegradable poly(ε-caprolactone)/thermally reduced graphene nanocomposites. *Industrial and Engineering Chemistry Research* 50 (24): 13885–13891.

62 Wang, B., Li, Y., Weng, G. et al. (2014). Reduced graphene oxide enhances the crystallization and orientation of poly(ε-caprolactone). *Composites Science and Technology* 96: 63–70.

63 Wu, F., Jiang, L., Miao, W. et al. (2018). Effects of shear on epitaxial crystallization of poly(ε-caprolactone) on reduced graphene oxide. *RSC Advances* 8 (12): 6406–6413.

64 Jana, R.N. and Cho, J.W. (2010). Non-isothermal crystallization of poly(ε-caprolactone)-grafted multi-walled carbon nanotubes. *Composites, Part A Applied Science and Manufacturing* 41 (10): 1524–1530.

65 Bhagabati, P., Das, D., and Katiyar, V. (2021). Bamboo-flour-filled cost-effective poly(ε-caprolactone) biocomposites: a potential contender for flexible cryo-packaging applications. *Materials Advances* 2 (1): 280–291.

66 Siqueira, D.D., Luna, C.B.B., Araújo, E.M. et al. (2021). Approaches on PCL/macaíba biocomposites – mechanical, thermal, morphological properties and crystallization kinetics. *Polymers for Advanced Technologies* 32 (9): 3572–3587.

67 Siqueira, D.D., Luna, C.B.B., Araújo, E.M. et al. (2019). Biocomposites based on PCL and macaiba fiber. Detailed characterization of main properties. *Materials Research Express* 6 (9): 095335.

68 dos Santos Filho, E.A., Siqueira, D.D., Araújo, E.M. et al. (2022). The impact of the macaíba components addition on the biodegradation acceleration of poly(ε-caprolactone) (PCL). *Journal of Polymers and the Environment* 30 (2): 443–460.

69 Lima, J.C., Sousa, J.C., Arruda, S.A. et al. (2019). Polycaprolactone matrix composites reinforced with brown coir: rheological, crystallization, and mechanical behavior. *Polymer Composites* 40 (S1): E678–E686.

70 Marrakchi, Z., Oueslati, H., Belgacem, M.N. et al. (2012). Biocomposites based on polycaprolactone reinforced with alfa fibre mats. *Composites, Part A Applied Science and Manufacturing* 43 (4): 742–747.

71 Zhao, Q., Tao, J., Yam, R.C.M. et al. (2008). Biodegradation behavior of polycaprolactone/rice husk ecocomposites in simulated soil medium. *Polymer Degradation and Stability* 93 (8): 1571–1576.

72 Madbouly, S.A. (2020). Nano/micro-scale morphologies of semi-interpenetrating poly(ε-caprolactone)/tung oil polymer networks: isothermal and non-isothermal crystallization kinetics. *Polymer Testing* 89: 106586.

8

Crystallization and Shape Memory Effect

Shiji Mathew

Temple University, Department of Biology, College of Science and Technology, Philadelphia, PA 19122, USA

8.1 Introduction

Polymer crystallinity is the measure to which there are regions where polymer chains are aligned to each other. This is achieved in the presence of some level of stereoregularity. The degree of crystallinity of polymers is temperature dependent, and it governs their mechanical properties such as elastic modulus, yield stress, and impact resistance [1]. Crystalline regions in polymers are formed from the stereoregular blocks in the polymer chains. In the melted form, the polymer chains continue to remain entangled with each other in an irregular coil-like structure. In some cases, when the melted polymer is cooled down, the chains remain tangled in a disordered fashion. This is due to large degree of irregularity in their polymer chains, which is called stereorandomness, and such polymers are named amorphous polymers.

In the case of other polymers where chains have stereoregular portions, distinct ordered crystalline regions are formed where chains fold among themselves to form lamellae. Due to the presence of stereorandom blocks in the chain, complete crystallization does not occur. In addition, if, in some cases, branching has occurred, it will result in loss of stereoregularity and will end up in inhibition of crystallization. As some parts of the polymer remain uncrystallized, such polymers are known as semi-crystalline polymers [2]. The common properties of amorphous and semi-crystalline polymers are included in Table 8.1.

Shape memory effect (SME) in polymers was first discovered by Vernon and colleagues in 1941 [3]. Stimuli-responsive materials are smart materials, as they can sense their surroundings and respond accordingly [4]. Polymeric stimuli-responsive materials are otherwise called shape memory polymers (SMPs). SME has been found in many ranges of polymers such as amorphous polymers, semi-crystalline polymers, and liquid crystalline elastomers. SME can be described as the ability of a material to get deformed and fixed into a temporary shape that can be recovered only by applying an external stimulus. SMPs are highly deformable materials

---

Polymer Crystallization: Methods, Characterization, and Applications, First Edition.
Edited by Jyotishkumar Parameswaranpillai, Jenny Jacob, Senthilkumar Krishnasamy, Aswathy Jayakumar, and Nishar Hameed.
© 2023 WILEY-VCH GmbH. Published 2023 by WILEY-VCH GmbH.

Table 8.1 Differences between properties of amorphous and semi-crystalline polymers.

Amorphous polymers	Semi-crystalline polymers
High melt viscosity	Low melt viscosity
No distinct melting point	Distinct melting point
Low strength	High strength
Poor fatigue and wear resistance	High fatigue and wear resistance
Transparent	Translucent
Molecular orientation in molten phase: random	Molecular orientation in molten phase: random
Molecular orientation in solid phase: random	Molecular orientation in solid phase: presence of crystallites

Source: Reproduced with permission from Crawford and Quinn [2]/Elsevier.

that can be pre-programmed to memorize and recover from a temporary shape change and return to their original form when triggered by an external stimulus. In addition, the SME can also be generated in a reversible manner by enabling actuation behavior through macroscale deformation and processing and also by dictating the macromolecular orientation of the actuation units and the skeleton structure of geometry-determining units in the polymers. Implementation of smart movements in artificial materials can result in the development of materials that can respond to changes in environment, such as light and temperature, as a one-time event or reversibly. This invention has resulted in the development of SMPs. This chapter mainly details the definition of SMPs, the mechanism of SME, and the varied applications of SMPs in biomedical field.

8.2 Shape Memory Cycle

The whole process of representing the SME of a polymer is termed as shape memory cycle. The shape memory cycle consists of three stages, namely programming, storage, and recovery. Figure 8.1a,b shows the shape memory cycle of a thermally activated SMP in 2D and 3D, respectively. In the programming stage, the polymer is deformed by the application of thermal energy that is above its transition temperature (T_{trans}). This temperature is denoted as the deformation temperature (T_d), and at this temperature, the shape of a material can be easily manipulated to a desired shape. The storage stage is initiated with the cooling process, where the material is cooled under pre-strain constraint from T_d to a set temperature (T_s), which is below T_{trans}. During this storage stage, the deformation history is noted and fixation of a temporary shape happens [6]. In this fixation stage, the initial deformation constraint is released at T_s, and a stress-free condition is gained. The sample at this stage is referred to as a pre-deformed SMP. The last step named recovery is the unconstrained recovery where the molecular switches are opened again by heating to a temperature above T_{trans}, under stress-free condition, and the

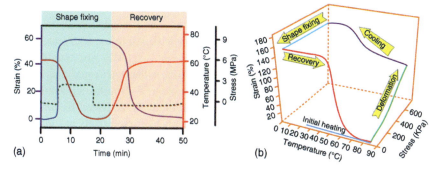

Figure 8.1 (a) 2D and (b) 3D representation of shape memory cycle of a thermally activated SMP. Source: Adapted from Thakur and Hu [5].

specimen regains its original shape. The overall time for the shape memory cycle for a material depends upon certain factors such as its properties and geometric and experimental conditions [5, 7].

8.3 Mechanism of Shape Memory Effect

The mechanism of SME is based on the fact that SMPs, when exposed to high temperatures, get soft and can be deformed into a temporary shape. Upon cooling, this temporary shape gets fixed, which can in turn regain shape when the right stimulus is applied. The polymers must possess certain essential features to show good SME. First, the polymer should have the capability to retain a permanent shape. Second, the material must be able to undergo programming, a process of deforming the material into a temporary shape at high temperatures and then cooling it before releasing the constraint. Third, upon application of the appropriate stimulus, the programmed material must regain its original shape [8] (Figure 8.2).

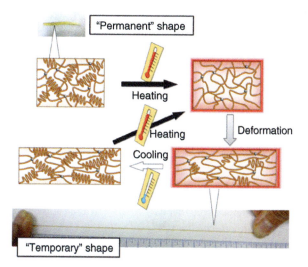

Figure 8.2 Mechanism of shape memory effect in polymers. Source: Ebara et al [9]/from Taylor & Francis Group.

8.4 Types of Shape Memory Polymers

Depending on the structure, composition, and type of stimulus, SMPs are classified into different types. Based on the type of stimulus, SMPs are classified as magnetically driven SMPs, thermally activated SMPs, solvent-activated SMPs, and light-driven SMPs. SMPs can also respond to physiological stimuli such as pH and ion concentration [10]. Table 8.2 gives examples of SMPs activated by different stimuli. Thermally activated SMPs can be triggered by heating directly or indirectly by thermal activations. While direct transfer of heat to SMPs can be done with the help of hot liquids such as water, indirect heating methods include advanced techniques such as light exposure, magnetic fields, microwaves, and electricity. The indirect heating methods are more convenient and easier to apply due to remote controlling capabilities.

Heat-responsive SMPs can be further classified into four classes as chemically cross-linked amorphous SMPs (Class I or thermosets), chemically cross-linked semi-crystalline SMPs (Class II), physically cross-linked amorphous SMPs (Class III, thermoplastics), and physically cross-linked semi-crystalline SMPs (Class IV). Examples of amorphous polymers include polystyrene and polyvinyl chloride, while examples of semi-crystalline polymers include polyether ether ketone, polyethylene terephthalate, and polytetrafluoroethylene.

Based on the shape memory functionality, SMPs are of three types: one-way, two-way, and multiple SMPs [25, 26]. Examples of one-way SMPs include heat-shrinkable tubes, toys, and labels. Two-way SMPs, also called reversible SMPs, can autonomously deform at high temperatures and revert to their original shape at lower temperatures [27]. Multiple SMPs are polymers that are capable of memorizing more than one temporary shape [28]. Figure 8.3 illustrates a schematic of one-way, two-way, and multi-SMPs.

8.5 Biomedical Applications of Shape Memory Polymers

SMPs find great application in biomedical field owing to their minimal toxicity, biocompatibility, biodegradable properties, and tunability. The commonly employed polymers include polylactide (PLA), polycaprolactone (PCL), polyglycolide (PGA), and polyurethane (PU). SMPs find promising applications in tissue engineering, bone repairs, medical stents, self-healing materials, vascular embolization, etc. The following sections explain the use of SMPs in these applications with suitable examples.

8.5.1 Tissue Engineering

Tissue engineering involves treatment of damaged tissues with biological substitutes that replace allografts and autografts. Tissue engineering applications include the use of SMPs in contraceptive implants, sutures, artificial muscles/tendons, aneurysm treatment and SMP with antibacterial functions. Only few of the SMPs

Table 8.2 Examples of thermoactivated SMPs triggered by different stimuli.

Actuation stimulus	Description	T_t	References
Magnetism driven	Nickel/zinc/ferrite particles of size 6.7–43.6 μm were incorporated into SMPs to achieve their actuation by inductive heating from magnetic field (12.2 MHz)	Temperature changed with particle content	[11]
	Nickel/manganese/gallium magnetic particles (60 μm) were added into SMPs	Above 60 °C	[12]
	Iron (II/III) oxide with various sizes were incorporated into SMPs	Temperature changed with particle content	[13–15]
Light driven	Graphene oxide incorporated into SMPs for actuation under 808 nm near infrared (NIR) radiation	Changed with irradiation time and graphene oxide (GO) content	[16]
	Diketopyrrolopyrrole-based conjugated polymer (PDPP3T) used as photothermal filler into shape memory elastomer matrix of polycaprolactone-co-polyurethane (PCL-PU)		[17]
	Polyurethane composite film containing copper sulfide nanoparticles and modified cellulose nanocrystals		[18]
	Porphyrin loaded into SMPs to achieve SME under laser irradiation	Changed with laser power and particle content	[19]
	0.08% Black phosphorous (BP) was added into the SMP to achieve shape recovery under 808 nm NIR radiation	Changed with irradiation time and BP content	
Solvent	Water-responsive poly(butanetetrol fumarate)		[20]
	Water bath was used to heat the SMP	37, 39 °C	[21, 22]
	Bound water absorbed in polymers was used to actuate shape recovery	30 °C	[23, 24]

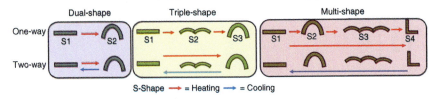

Figure 8.3 Diagram representing classification of SMPs based on shape memory functionality. Source: Adapted and modified from [7].

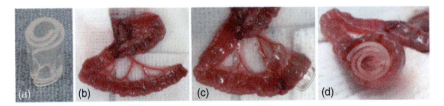

Figure 8.4 Photographs taken during operation. (a) Radially expanding SMP device. (b) SMP placed with the end of Roux limb. (c) SMP is wrapped around it (c and d). Source: Fisher et al. [33]/Reproduced with permission from Elsevier.

are employed for clinical testing and applications in tissue engineering; those include PCL glassy thermoset urethane foams, PCL crystalline copolymer networks, and PCL dimethylacrylate, which are vastly engaged in soft and hard tissue regeneration processes. Many studies have reported the applications of SMPs, which include tissue-specific musculoskeletal repair such as artificial muscles [29, 30], cardiovascular repair [31], and neural injury repairs [32].

In a study, Fisher et al. used a radially self-expanding SMP-made cylinder device that was placed extraluminally following a simple operative approach to create distraction enterogenesis in rat models. The study showed that the SMP could successfully lengthen the intestine, without causing any damage to the mucosa. Figure 8.4 depicts the photographs of self-expanding SMP and its application in enterogenesis taken during the operation.

8.5.2 Bone Engineering

In a recent study, an SMP-based foam composed of polyurethane/hydroxyapatite was developed using gas foaming method, and then its application as a scaffold substrate in bone regeneration in rabbit femoral defect model was investigated. The expanding property of SMP foam as a bone scaffold was monitored in vivo, and its self-fitting behavior was analyzed systematically. Figure 8.5 illustrates the schematic of the application of polyurethane/hydroxyapatite-based SMP foam and its application in bone repair. The results displayed that the SMP foam could be successfully implanted into defective bones with a compact shape. In addition, micro-computed tomography showed that the bone ingrowth started at the periphery of SMP foam with a constant decrease toward inside. Histological analysis revealed successful

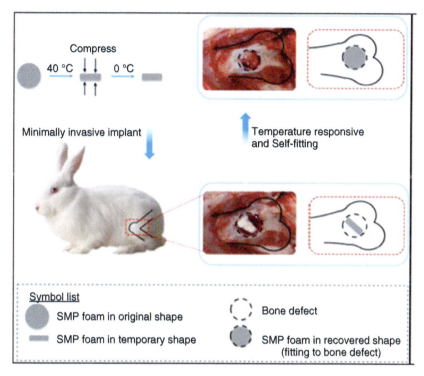

Figure 8.5 Role of SMPs in bone engineering. Application of polyurethane/hydroxyapatite-based SMP foam and its application in bone repair. Source: Xie et al. [34]/Reproduced with permission from Elsevier.

vascularization and bone remodeling, thereby proving that the developed SMP had greater potential in bone regeneration [34].

In a similar study [35], a biomimetic shape memory hyaluronic acid based cryogel scaffolds were developed. These SMP cryogel scaffolds were then loaded with primary chondrocytes, which were found to provide a more conductive microenvironment for cell adhesion, cell proliferation, and matrix biosynthesis. The study showed that these SMPs could be easily injected into the joint space using syringe for the non-surgically invasive treatment and repair of cartilage defects.

8.5.3 Medical Stents

Another important application of SMPs is in the development of stents. A stent is a tube that is inserted into narrow arteries or ducts to open them mechanically in order to maintain normal blood flow. In a recent study, 3D printed cylindrical and bifurcated stents were prepared using polyurethane-based SMPs were printed on a MarkerPi 3D printer. The shape of the stent was designed to be tubular so that the blood and bodily fluids could flow smoothly without obstruction via the inner pathway. After printing, the stent was allowed to soften by immersing it in water basin with a temperature slightly above the T_g. After softening, the stent was removed and shaped into a compact structure to form a half cylinder and immediately cooled down while applying pressure, to retain the deformed shape. So, during the cooling process, the shape was memorized so that upon heating above T_g, the original shape is regained. The printed bifurcated stents were then inserted into artificial blood vessels and expanded to get desired shapes [36]. Figure 8.6 shows the development and application of polyurethane-based stents.

Yang et al. [37] developed a thermally induced, shape-memory dual drug-eluting stent (SMDES) by cross-linking PEG-PCL copolymers (cPEG-PCL). The developed stent was able to perform a temporary shape memory phenomenon from a temporary linear form to a permanent spiral shape within a transition temperature nearer to body temperature. The stent was also designed to incorporate a controlled dual drug release system in conjugation with mitomycin C and curcumin coating to establish a shape memory dual drug eluting stent. The results showed that the stent would controllably release curcumin (the anticoagulant drug) for 14 days and mitomycin C (the antiproliferation drug) for over 70 days. This design of drug release served to reduce platelet adhesion in the early stages and also prevent in-stent restenosis of the vessel during short- and long-term therapeutic use. Moreover, the SMDES was also found to be biocompatible and biodegradable. Figure 8.7 demonstrates the SME of SMP to be applied as medical stent.

8.5.4 Drug Delivery Application

The role of SMPs in the development of drug delivery purposes has been discussed in few works [38–40]. The concept and design of drug-releasing SMPs are illustrated in Figure 8.8. According to this concept, the drug is implanted on the surface of SMPs is then programmed to attain a temporary shape. Once it is placed inside the body (physiological environment), the SMPs regain their original shape, and the drug is released in a controlled manner to targeted site.

Melocchi et al. developed an SMP to evaluate its feasibility in developing expandable gastroretentive drug delivery systems (GRDDSs), which were able to self-modify their configuration upon an external stimulus based on the water-induced SME of polyvinyl alcohol. Figure 8.9 demonstrates the graphical representation of GRDDS and its working design. The developed GRDDS was proposed to shift instinctively from temporary shape when programmed by deformation, suitable for oral administration, to an expanded original one enabling gastric retention due to proper spatial hindrance [41].

8.5.5 SMPs as Self-Healing Materials

Self-healing is a property of materials to recover from physical damage. The most common damage that can be healed in polymers is mechanical damage caused by cracking, scratching, puncture, and delamination [42]. The mechanism of self-healing in polymers can be considered to involve two interacting effects. First, a physical flow has to occur, which can close the damaged site. Next, the polymer

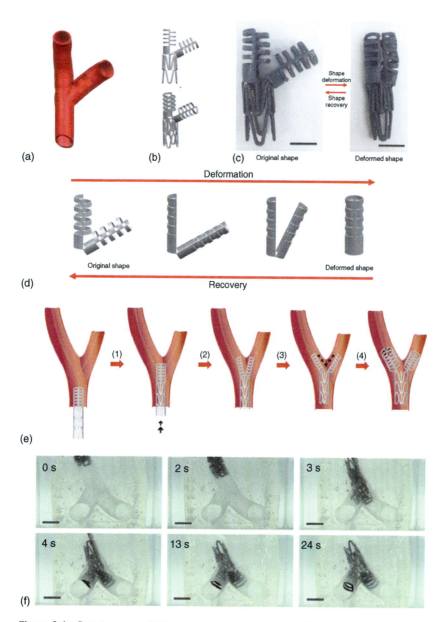

Figure 8.6 Development of bifurcated shape memory stent from shape transformable bifurcated stents. (a) 3D model of blood vessel. (b) Stent design from blood vessel model. (c) Printed bifurcated stent. (d) Deformation and recovery process of main and side branches. (e) Bifurcated stent deployment process. (f) Bifurcated stent deployment experiment. Source: Kim and Lee [36]/Springer Nature/CC BY 4.0.

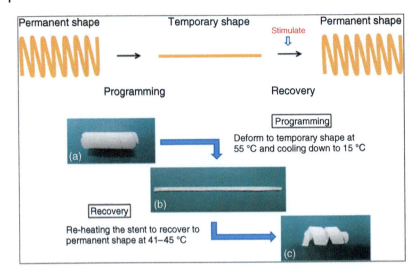

Figure 8.7 Shape memory effect of SMP for application as medical stents. Source: Yang et al. [37]/Reproduced with permission from American Chemical Society.

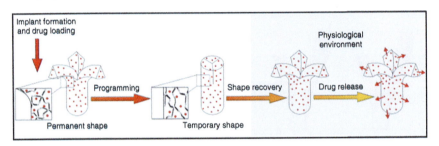

Figure 8.8 Drug delivery application concept of SMPs. Source: Reproduced with permission from Wischke et al. [40]/Elsevier.

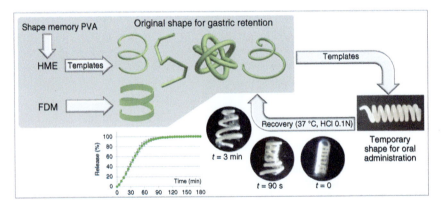

Figure 8.9 Graphical representation of shape memory polymers used as expandable drug delivery system for gastric retention developed via 3D printing and extrusion.
PVA = polyvinyl alcohol, HME = hot melt extrusion, FDM = fused deposition modelling.
Source: Melocchi et al. [41]/Reproduced with permission from Elsevier.

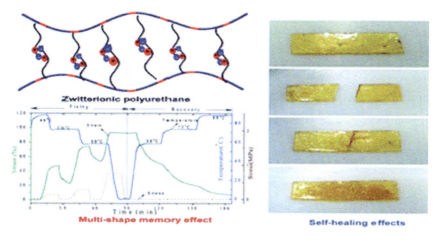

Figure 8.10 Zwitterionic multi-shape memory polyurethanes Source: Chen et al. [44]/Reproduced with permission from Royal Society of Chemistry.

network has to be rebuilt and reconnected chemically in order to safeguard the restoration of its mechanical and functional integrity [43].

Chen et al. developed zwitterion-based multi-shape memory polyurethanes (ZSMPUs) from N-methyldiethanolamine (MDEA), hexamethylene diisocyanate (HDI), and 1,3 propanesultone (PS). This novel ZSMPUs were unique such that it was able to remember four different shapes, and shape recovery decreased with increase in sulfobetaine content. Immersing the ZSMPUs in moisture-rich conditions and drying them at lower temperatures preserved the shape memory capabilities and demonstrated self-healing properties. Figure 8.10 shows the schematic of the development of ZSMPUs and their self-healing ability.

Another similar study conducted by the researchers at the University of Southern California Viterbi School of Engineering created a 3D-printed rubber shoe pads that fix themselves under photopolymerization without human intervention. The shoe pad, once cut, was found to self-heal within two hours at 60 °C. Figure 8.11 depicts

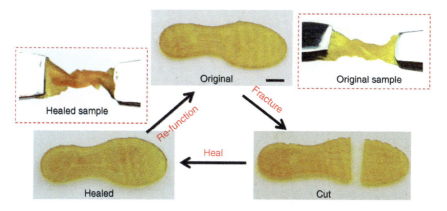

Figure 8.11 Self-healing 3D-printed shoe pad. Source: Yu et al. [45]/Springer Nature.

the schematic of the workings of 3D-printed rubber shoe pads. The sealed shoe pad was capable of sustaining 540 °C twist again [45].

8.5.6 Vascular Embolization

Endovascular aneurysms are commonly treated using transcatheter arterial embolization (TAE) [46]. Metal coils such as platinum wires are generally used to treat aneurysms in clinics, which ultimately end in accidental rupture of the aneurysm. So in order to reduce the risk, coating metal coils with SMPs has been found to be a good solution [47].

Boyle et al., designed a foam-over-wire (FOW) embolization device from SMP that was delivered in vivo and in vitro pig saccular aneurysm models to investigate the efficacy of the device, aneurysm occlusion, and acute clotting [48]. The study was designed to deliver the SMP foam to the aneurysm over a radiopaque nickel-titanium (nitinol) and platinum wire backbone, which was mechanically detached and then actuated either passively using physiological conditions or using a laser-heated injection. Figure 8.12 illustrates the schematic of the SMP FOW model design (upper panel), delivery and detachment design of SMP foam (middle panel), and laser-heated injection in the bottom panel. The results demonstrated that the device was successfully delivered and stably implanted in vitro. In addition, the FOW device was found to effectively occlude in vivo porcine aneurysms. The designed device exhibited combined filling and surface area properties of SMP

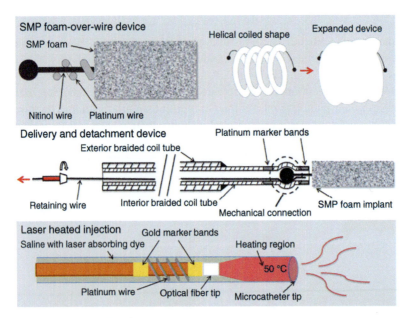

Figure 8.12 Schematic of the model of SMP foam-over-wire (upper panel), the delivery and detachment design of SMP FOW (middle panel), and the laser heating mechanism (bottom panel). Source: Boyle et al. [48]/Reproduced with permission from John Wiley & Sons.

foam along with the fluoroscopic contrast and mechanical stability of metal coils, enabling low adoption hurdle for clinicians. The FOW embolization device showed great potential for endovascular treatment of intracranial saccular aneurysms.

8.6 Conclusion

The amazing ability of SMPs to memorize and regain their shape has fetched multitude of applications in important fields of science and technology. SMPs also exhibit minimal toxicity, biocompatibility, biodegradability, and tunable properties, which find great application in biomedical field. The chapter mainly details the definition of SMPs, the mechanism of SME, and the varied applications of SMPs in biomedical field. The interventions of SMPs, such as tissue engineering scaffold, medical stent, drug delivery applications, and self-healing materials, are discussed in detail with examples.

References

1 Kong, Y. and Hay, J.N. (2002). The measurement of the crystallinity of polymers by DSC. *Polymer* 43: 3873–3878. https://doi.org/10.1016/S0032-3861(02)00235-5.

2 Crawford, C.B. and Quinn, B. (2017). Physiochemical properties and degradation. In: *Microplastic Pollutants* (ed. C.B. Crawford and B. Quinn), 57–100. Elsevier https://doi.org/10.1016/B978-0-12-809406-8.00004-9.

3 Vernon, L. B. and Vernon, H. M. (1941). Process of manufacturing articles of thermoplastic synthetic resins. US2234993A, filed 6 February 1937.

4 Zhang, X., Chen, L., Lim, K.H. et al. (2019). The pathway to intelligence: using stimuli-responsive materials as building blocks for constructing smart and functional systems. *Advanced Materials* 31: 1804540. https://doi.org/10.1002/adma.201804540.

5 Thakur, S. and Hu, J. (2017). Polyurethane: a shape memory polymer (SMP). In: *Aspects of Polyurethanes* (ed. F. Yilmaz), 53–71. INTECH https://doi.org/10.5772/intechopen.69992.

6 Jose, S., George, J.J., Siengchin, S., and Parameswaranpillai, J. (2020). Introduction to shape-memory polymers, polymer blends and composites: state of the art, opportunities, new challenges and future outlook. In: *Shape Memory Polymers, Blends and Composites, Advanced Structured Materials* (ed. J. Parameswaranpillai, S. Siengchin, J.J. George, and S. Jose), 1–19. Singapore: Springer Singapore https://doi.org/10.1007/978-981-13-8574-2_1.

7 Melly, S.K., Liu, L., Liu, Y., and Leng, J. (2020). Active composites based on shape memory polymers: overview, fabrication methods, applications, and future prospects. *Journal of Materials Science* 55: 10975–11051. https://doi.org/10.1007/s10853-020-04761-w.

8 Xiao, R. and Huang, W.M. (2020). Heating/solvent responsive shape-memory polymers for implant biomedical devices in minimally invasive surgery: current

status and challenge. *Macromolecular Bioscience* 20: 2000108. https://doi.org/10.1002/mabi.202000108.

9 Ebara, M. (2015). Shape-memory surfaces for cell mechanobiology. *Science and Technology of Advanced Materials* 16: 014804. https://doi.org/10.1088/1468-6996/16/1/014804.

10 Leng, J., Lan, X., Liu, Y., and Du, S. (2011). Shape-memory polymers and their composites: stimulus methods and applications. *Progress in Materials Science* 56: 1077–1135. https://doi.org/10.1016/j.pmatsci.2011.03.001.

11 Buckley, P.R., McKinley, G.H., Wilson, T.S. et al. (2006). Inductively heated shape memory polymer for the magnetic actuation of medical devices. *IEEE Transactions on Biomedical Engineering* 53: 2075–2083. https://doi.org/10.1109/TBME.2006.877113.

12 Scheerbaum, N., Hinz, D., Gutfleisch, O. et al. (2007). Textured polymer bonded composites with Ni–Mn–Ga magnetic shape memory particles. *Acta Materialia* 55: 2707–2713. https://doi.org/10.1016/j.actamat.2006.12.008.

13 Mohr, R., Kratz, K., Weigel, T. et al. (2006). Initiation of shape-memory effect by inductive heating of magnetic nanoparticles in thermoplastic polymers. *Proceedings of the National Academy of Sciences* 103: 3540–3545. https://doi.org/10.1073/pnas.0600079103.

14 Schmidt, A.M. (2006). Electromagnetic activation of shape memory polymer networks containing magnetic nanoparticles. *Macromolecular Rapid Communications* 27: 1168–1172. https://doi.org/10.1002/marc.200600225.

15 Shah, R.R., Davis, T.P., Glover, A.L. et al. (2015). Impact of magnetic field parameters and iron oxide nanoparticle properties on heat generation for use in magnetic hyperthermia. *Journal of Magnetism and Magnetic Materials* 387: 96–106. https://doi.org/10.1016/j.jmmm.2015.03.085.

16 Zhu, C.-H., Lu, Y., Peng, J. et al. (2012). Photothermally sensitive poly(N-isopropylacrylamide)/graphene oxide nanocomposite hydrogels as remote light-controlled liquid microvalves. *Advanced Functional Materials* 22: 4017–4022. https://doi.org/10.1002/adfm.201201020.

17 Zhang, Y., Zhou, S., Chong, K.C. et al. (2019). Near-infrared light-induced shape memory, self-healable and anti-bacterial elastomers prepared by incorporation of a diketopyrrolopyrrole-based conjugated polymer. *Materials Chemistry Frontiers* 3: 836–841. https://doi.org/10.1039/C9QM00104B.

18 Li, M., Fu, S., and Basta, A.H. (2020). Light-induced shape-memory polyurethane composite film containing copper sulfide nanoparticles and modified cellulose nanocrystals. *Carbohydrate Polymers* 230: 115676. https://doi.org/10.1016/j.carbpol.2019.115676.

19 Qian, W., Song, Y., Shi, D. et al. (2019). Photothermal-triggered shape memory polymer prepared by cross-linking porphyrin-loaded micellar particles. *Materials* 12: 496. https://doi.org/10.3390/ma12030496.

20 Guo, Y., Lv, Z., Huo, Y. et al. (2019). A biodegradable functional water-responsive shape memory polymer for biomedical applications. *Journal of Materials Chemistry B* 7: 123–132. https://doi.org/10.1039/C8TB02462F.

21 Wang, Y.-J., Jeng, U.-S., and Hsu, S. (2018). Biodegradable water-based polyurethane shape memory elastomers for bone tissue engineering. *ACS Biomaterials Science & Engineering* 4: 1397–1406. https://doi.org/10.1021/acsbiomaterials.8b00091.

22 Xie, Y., Lei, D., Wang, S. et al. (2019). A biocompatible, biodegradable, and functionalizable copolyester and its application in water-responsive shape memory scaffold. *ACS Biomaterials Science & Engineering* 5: 1668–1676. https://doi.org/10.1021/acsbiomaterials.8b01337.

23 Chae Jung, Y., Hwa So, H., and Whan Cho, J. (2006). Water-responsive shape memory polyurethane block copolymer modified with polyhedral oligomeric silsesquioxane. *Journal of Macromolecular Science, Part B* 45: 453–461. https://doi.org/10.1080/00222340600767513.

24 Huang, W.M., Yang, B., An, L. et al. (2005). Water-driven programmable polyurethane shape memory polymer: demonstration and mechanism. *Applied Physics Letters* 86: 114105. https://doi.org/10.1063/1.1880448.

25 Li, J., Duan, Q., Zhang, E., and Wang, J. (2018). Applications of shape memory polymers in kinetic buildings. *Advances in Materials Science and Engineering* 2018: 1–13. https://doi.org/10.1155/2018/7453698.

26 Zhao, Q., Qi, H.J., and Xie, T. (2015). Recent progress in shape memory polymer: new behavior, enabling materials, and mechanistic understanding. *Progress in Polymer Science* 49, 50: 79–120. https://doi.org/10.1016/j.progpolymsci.2015.04.001.

27 Li, J., Rodgers, W.R., and Xie, T. (2011). Semi-crystalline two-way shape memory elastomer. *Polymer* 52: 5320–5325. https://doi.org/10.1016/j.polymer.2011.09.030.

28 Kolesov, I.S. and Radusch, H.-J. (2008). Multiple shape-memory behavior and thermal-mechanical properties of peroxide cross-linked blends of linear and short-chain branched polyethylenes. *Express Polymer Letters* 2: 461–473. https://doi.org/10.3144/expresspolymlett.2008.56.

29 Takashima, K., Iwamoto, D., Oshiro, S. et al. (2021). Characteristics of pneumatic artificial rubber muscle using two shape-memory polymer sheets. *Journal of Robotics and Mechatronics* 33: 653–664. https://doi.org/10.20965/jrm.2021.p0653.

30 Yahara, S., Wakimoto, S., Kanda, T., and Matsushita, K. (2019). McKibben artificial muscle realizing variable contraction characteristics using helical shape-memory polymer fibers. *Sensors and Actuators A: Physics* 295: 637–642. https://doi.org/10.1016/j.sna.2019.06.012.

31 Feng, J., Shi, H., Yang, X., and Xiao, S. (2021). Self-adhesion conductive sub-micron fiber cardiac patch from shape memory polymers to promote electrical signal transduction function. *ACS Applied Materials & Interfaces* 13: 19593–19602. https://doi.org/10.1021/acsami.0c22844.

32 Wang, J., Xiong, H., Zhu, T. et al. (2020). Bioinspired multichannel nerve guidance conduit based on shape memory nanofibers for potential application in peripheral nerve repair. *ACS Nano* 14: 12579–12595. https://doi.org/10.1021/acsnano.0c03570.

33 Fisher, J.G., Sparks, E.A., Khan, F.A. et al. (2015). Extraluminal distraction enterogenesis using shape-memory polymer. *Journal of Pediatric Surgery* 50: 938–942. https://doi.org/10.1016/j.jpedsurg.2015.03.013.

34 Xie, R., Hu, J., Hoffmann, O. et al. (2018). Self-fitting shape memory polymer foam inducing bone regeneration: a rabbit femoral defect study. *Biochimica et Biophysica Acta (BBA) – General Subjects* 1862: 936–945. https://doi.org/10.1016/j.bbagen.2018.01.013.

35 He, T., Li, B., Colombani, T. et al. (2021). Hyaluronic acid-based shape-memory cryogel scaffolds for focal cartilage defect repair. *Tissue Engineering. Part A* 27: 748–760. https://doi.org/10.1089/ten.tea.2020.0264.

36 Kim, T. and Lee, Y.-G. (2018). Shape transformable bifurcated stents. *Scientific Reports* 8: 13911. https://doi.org/10.1038/s41598-018-32129-3.

37 Yang, C.-S., Wu, H.-C., Sun, J.-S. et al. (2013). Thermo-induced shape-memory PEG-PCL copolymer as a dual-drug-eluting biodegradable stent. *ACS Applied Materials & Interfaces* 5: 10985–10994. https://doi.org/10.1021/am4032295.

38 Inverardi, N., Scalet, G., Melocchi, A. et al. (2021). Experimental and computational analysis of a pharmaceutical-grade shape memory polymer applied to the development of gastroretentive drug delivery systems. *Journal of the Mechanical Behavior of Biomedical Materials* 124: 104814. https://doi.org/10.1016/j.jmbbm.2021.104814.

39 Kirillova, A. and Ionov, L. (2019). Shape-changing polymers for biomedical applications. *Journal of Materials Chemistry B* 7: 1597–1624. https://doi.org/10.1039/C8TB02579G.

40 Wischke, C., Neffe, A.T., Steuer, S., and Lendlein, A. (2009). Evaluation of a degradable shape-memory polymer network as matrix for controlled drug release. *Journal of Controlled Release* 138: 243–250. https://doi.org/10.1016/j.jconrel.2009.05.027.

41 Melocchi, A., Uboldi, M., Inverardi, N. et al. (2019). Expandable drug delivery system for gastric retention based on shape memory polymers: development via 4D printing and extrusion. *International Journal of Pharmaceutics* 571: 118700. https://doi.org/10.1016/j.ijpharm.2019.118700.

42 Blaiszik, B.J., Kramer, S.L.B., Olugebefola, S.C. et al. (2010). Self-healing polymers and composites. *Annual Review of Materials Research* 40: 179–211. https://doi.org/10.1146/annurev-matsci-070909-104532.

43 Yang, Y. and Urban, M.W. (2013). Self-healing polymeric materials. *Chemical Society Reviews* 42: 7446. https://doi.org/10.1039/c3cs60109a.

44 Chen, S., Mo, F., Yang, Y. et al. (2015). Development of zwitterionic polyurethanes with multi-shape memory effects and self-healing properties. *Journal of Materials Chemistry A* 3: 2924–2933. https://doi.org/10.1039/C4TA06304J.

45 Yu, K., Xin, A., Du, H. et al. (2019). Additive manufacturing of self-healing elastomers. *NPG Asia Materials* 11: 7. https://doi.org/10.1038/s41427-019-0109-y.

46 Lee, T.K., Kwon, J., Na, K.S. et al. (2015). Evaluation of selective arterial embolization effect by chitosan micro-hydrogels in hindlimb sarcoma rodent models using various imaging modalities. *Nuclear Medicine and Molecular Imaging* 49: 191–199. https://doi.org/10.1007/s13139-014-0316-y.

47 Sun, L., Wang, T.X., Chen, H.M. et al. (2019). A brief review of the shape memory phenomena in polymers and their typical sensor applications. *Polymers* 11: 1049. https://doi.org/10.3390/polym11061049.

48 Boyle, A.J., Landsman, T.L., Wierzbicki, M.A. et al. (2016). In vitro and in vivo evaluation of a shape memory polymer foam-over-wire embolization device delivered in saccular aneurysm models: shape memory polymer foam-over-wire embolization device. *Journal of Biomedical Materials Research. Part B, Applied Biomaterials* 104: 1407–1415. https://doi.org/10.1002/jbm.b.33489.

9

3D Printing of Crystalline Polymers

Hiriyalu S. Ashrith, Tamalapura P. Jeevan, and Hanume Gowda V. Divya

Malnad College of Engineering, Department of Mechanical Engineering, Hassan, Karnataka 573201, India

9.1 Introduction

A polymer is a substance made up of molecules that are distinguished by numerous replications of one or more atoms or clusters of atoms connected to each other to exhibit a set of properties that do not change significantly with the accumulation of one or more constitutional repeating units [1]. Crystalline polymers have a reasonably regular chain structure and a particularly favored chain configuration. Semi-crystalline polymers should be used to characterize crystalline polymers more precisely. Polymers having a smaller fraction of chain defects crystallize with a lower overall crystallinity than polymers with no chain defects. Crystalline polymers belong to an important family of engineered plastic materials used for engineering and human applications. Polyethylene (PE) and copolymers, polypropylene (PP), polyesters, polylactic acid (PLA), and nylons are some of the most widely produced polymers, which are crystalline. Examples of the most frequently used crystalline polymers in three-dimensional (3D) printing are PP, polystyrene (PS), nylon, Kevlar and Nomex, polyketones, polyamides (PA), PE, polyethylene terephthalate (PET) [2].

The 3D printing technique is one of the swiftly evolving techniques that have enormous potential for society. 3D printing is a technology in which 3D objects are rapidly manufactured directly from computer-aided design (CAD) files. 3D printing machines are versatile, as a single machine can be used to print components of various dimensions using different materials. Generally, in 3D printing techniques, the components are printed from the bottom in a layer-by-layer manner and then held together to create a 3D solid object [3]. This procedure starts with the creation of a model to be printed using computer software that interprets a standard triangle language (STL) or CAD file, geometrically slicing, followed by aligning the object for production. Using the numerical control mechanisms and the software package, the 3D printer nozzle may be controlled in three directions. The layer-by-layer manufacturing approach demonstrates the benefits of low cost, design flexibility,

Polymer Crystallization: Methods, Characterization, and Applications, First Edition.
Edited by Jyotishkumar Parameswaranpillai, Jenny Jacob, Senthilkumar Krishnasamy, Aswathy Jayakumar, and Nishar Hameed.
© 2023 WILEY-VCH GmbH. Published 2023 by WILEY-VCH GmbH.

and customizability [4]. 3D printing also minimizes the number of raw materials needed during production, employing less strain on commodity procurement, natural wealth, and the environment. Complex geometry can be easily created rapidly without using or reducing tooling and lubricants [3]. Additive manufacturing (AM) is quite popular to produce polymer components. In general, when compared with conventional plastic injection molding, 3D printing is cost-effective only for low production rates, i.e. up to 1000 parts [5]. Due to its flexibility, 3D printing applications have gradually emerged in a variety of fields such as engineering, material research, medicine, and construction [6].

3D printing technologies are broadly classified into various types according to the American Society for Testing and Materials based on layer deposition technique and type of material used. Commonly used 3D printing technologies are stereolithography (SLA), selective laser sintering (SLS), fused deposition modeling (FDM) or fused filament fabrication (FFF), laminated object manufacturing (LOM), binder jetting (BJ), and selective laser melting (SLM) [3, 7]. The scientific roadmap of AM and schematic illustration of SLA, SLS, FDM, LOM, BJ, and SLM are depicted in Figure 9.1. FDM is the most common, commercial, and inexpensive technique in which the input material in the filament form made of thermoplastic material is used to print the components [8]. FDM has also exhibited excellent ability in producing customized safety components [4]. Filaments made of plastics, metals, ceramics, and composites are successfully printed using the FDM technique. FDM can be used to print smart textiles, shape memory composites (SMCs), components for RF/microwave structures, and fiber-reinforced polymer composite structures [8–11]. However, components printed through the FDM route exhibit inferior properties than the components produced via traditional route due to the presence of porosity in the components during printing.

9.2 3D Printing Materials and Processes

9.2.1 Nylon and Polyamides

Nylon is an artificial material that has received a lot of attention due to its ease of transformation into fibers, films, and molded pieces. Nylon is biocompatible and has acceptable chemical stability and adjustable mechanical qualities. Due to this, nylon and its variants are commonly employed in sutures, catheters, dentures, and other medical devices. Nylon is a semi-crystalline material that fits into the polyamide family. Nylon is a useful thermoplastic polymer that can be formed into many forms by melting, shaping, and cooling processes. Nylon has outstanding mechanical characteristics and hence frequently utilized in the production of clothing, molded plastics, and food packaging films. It is also frequently employed in the biomedical sector for bioengineering due to its biocompatible characteristics [12].

SLS printing technology was employed to produce the solid components using polyamide nylon (PA12) in powder material. Component printed using PA12 exhibits good strength, chemical resistance, and appreciable mechanical and

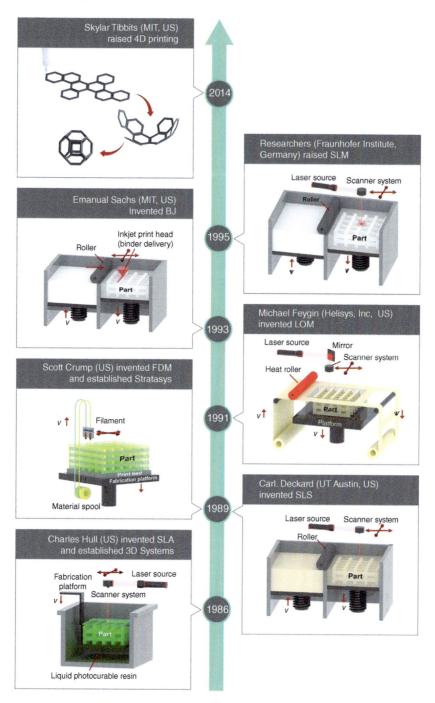

Figure 9.1 Scientific roadmap of additive manufacturing (AM) and schematic illustration of SLA, SLS, FDM, LOM, BJ, and SLM. Source: Liu et al. [7]/with permission of Elsevier/Public Domain CC BY 4.0.

thermal characteristics. Under certain conditions, PA12 is certified as food-grade plastic. The 3D printer used to print PA12 incorporates a high-accuracy galvanometer scanning and dynamic focusing system. CO_2 laser of 100 W and 0.3 mm wavelength at 12.7 ms^{-1} scanning speed was used as a source to sinter the PA12 powder at 190 °C processing temperature. A layer height of 0.1 mm was utilized during the process. The properties of the nylon powder used were as follows 0.4 g cm^{-3} bulk density; 0.95 g cm^{-3} part density; 183 °C melting point. The 3D printer used has a build volume of 300 mm × 300 mm × 300 mm [3].

Nylon 66 along with aluminum was prepared to study the crushing performance of bi-material structures. Mankati E360, an FDM-based 3D printer (Mankati Company), was used to fabricate nylon tubes. A layer of 0.2 mm thick, printing platform temperature of 45 °C, and extruder nozzle temperature of 270 °C with an infill density of 100% were used during the printing process. The remaining factors, namely withdrawal distance, nozzle moving speed, etc., were kept at default levels [4]. Ultimaker 2 and Markforged X7 systems, which work on FDM technique, were used to print spur gears using Nylon 618, Nylon 645, Onyx, and Markforged nylon proprietary materials. Default printing parameters recommended by the manufacturer with an infill percentage of 60% were used to print the components. All the materials were stored in a dry chamber during the printing process [5].

Complicated non-homogeneous dielectric material suitable for RF/Microwave applications was printed with Nylon 6 filament using FDM process. Markforged's proprietary blend Nylon 6 filament of 1.75 mm diameter was used during the printing process. A rectangular fill pattern of 0.1 mm layer thickness with 100% fill density at 275 °C was used in a 3D printer [8]. FDM technique was used to print nylon on polyamide 66 fabrics. Various printing parameters, like extruder and build plate temperature, and printing speed, significantly affected the bonding of polymer to the fabric. WANHAO Duplicator 4/4x was used to print the samples. The build volume of the printer was 22.5 cm × 14.5 cm × 15 cm with 0.4 mm diameter nozzle. Nylon on PA66 was printed using extruder temperatures of 235, 250, and 260 °C; build plate temperatures of 23 and 50 °C; and printing speeds of 18, 50, and 83 mmmin^{-1} [9].

Using the FDM process, shape memory composite (SMC) was successfully fabricated by combining a shape memory alloy (SMA) with a shape memory polymer (SMP). First, a nylon 12 filament was produced using a thermal screw extruder and PA12 powder (Arkema Corp.). Extrusion temperature of 250 °C, feeder rate of 105 rpm, and screw rate of 120 rpm were used to obtain the filament. After extrusion, the filament was placed in a water bath to maintain its initial cross section. Later Edwason Plus, Rokit Corp., made FDM 3D printer with an output speed of 200 mms^{-1}, head moving speed of 200 mms^{-1}, output temperature of 230 °C, and layer height of 0.05–0.3 mm was used to fabricate SMC [10]. Short carbon fiber (SCF)-reinforced nylon-6 thermoplastic composites (Onyx) were efficiently printed using Prusa i3 MK3s printer. A filament of 1.75 mm was printed using steel nozzles of 0.45 mm diameter. The process parameters used during the printing were as follows: extrusion temperature 275 °C; layer thickness 0.1 mm; printing speed 60 mm s^{-1}; and filament feed rate 1.6 mm s^{-1} [13]. PA6 composite filaments reinforced with carbon nanotubes, graphene nanoplatelets (GNPs), and

Kevlar were printed using FFF technique. Filabot EX2 commercial extruder was used to draw the filaments of 1.75 mm diameter using virgin PA6 at a processing temperature of 217 °C. Markforged Mark One commercial 3D printer was used to fabricate nanocomposites of 32 layers at 265 °C extrusion temperature, 0.1 mm layer thickness, and with 70% infill density [14].

Nylon 6 with various percentages of impact modifier and chain extender was printed using the FDM technique. Liestritz-Micro-27 (Germany), a twin-screw extruder was used for fabricating filaments of 2.15 mm diameter at a processing temperature of 250 °C at 100 rpm. Before processing, the Nylon 6 was desiccated for 24 hours at 80 °C to reduce the moisture content (<1%). After extrusion, the filaments were moved through a cold-water bath maintained at 7 °C. Lulzbot 3D printer (USA) with 0.50 mm nozzle diameter and borosilicate glass build plate was used to print the samples. The printing parameters, such as extrusion temperature, build plate temperature, printing speed, infill density, layer thickness, and overlap percentage, were fixed at 260 °C, 65 °C, 55 mm s^{-1}, 100%, 0.4 mm, and 5%, respectively [15].

Different grades of polyamide-12 (PA 2200 and PA12) powder were successfully printed using both SLS and multi-jet fusion (MJF) techniques. SLS printer (EOS P396) procured from EOS was used to print PA 2200, while MJF-HP 4200 procured from HP was used to print PA12. The maximum printing speed and minimum layer thickness used during printing for the SLS printer were 3.0 l h^{-1} and 0.06 mm, respectively, and 4.1 l h^{-1} and 0.08 mm, respectively, for MJF printer. Samples were printed at different orientations such as flatwise, edgewise, and upright. SLS process exhibited more error in terms of length, thickness, and width as compared to the MJF technique. SLS process resulted in lower-density components; however, the components from both the processes did not require any post-processing steps [16].

PA composites reinforced with continuous carbon and Kevlar fibers were efficaciously printed using an FDM-based printer. Filaments were procured from MarkForged®. Matrix filament was made of PA reinforced with SCFs, while continuous carbon fiber (47.98 wt%) and continuous Kevlar fiber (20.35 wt%) filaments filled with PA sizing agent were used as reinforcements. Filaments were dried before printing by a MarkForged Mark7 3D printer. Matrix filaments were drawn at a nozzle temperature of 270 °C, while a nozzle temperature of 250 °C was used to extrude continuous fiber filaments on a build plate maintained at room temperature. The specimens were made of 10 layers out of which 6 layers were made from matrix material and the remaining 4 were from fiber reinforcement [17].

Filaments of PA12 reinforced with 5 wt% hydroxyapatite and 15 wt% zirconium oxide were effectively produced by a desktop twin-screw extruder. Raw materials were mixed thoroughly, desiccated overnight in an oven at 80 °C, and pelletized before extrusion. The extrusion temperature was in the range of 180–200 °C, and extrusion speed was maintained at 150 rpm. Filaments after extrusion were made to pass through a water bath and pelletized. 3D printer filaments of 1.75 mm were extruded using an Extruder Version 1.3, Filastruder single-screw extruder. Specimens were printed using a 3D printer employing the following conditions: infill density 100%; layer height 0.2 mm; printing speed 90–150 mm s^{-1}; extrusion temperature 230 °C; and build plate temperature 110 °C [18].

FFF 3D printing process was used to print PA nanocomposites reinforced with multi-wall carbon nanotubes (MWCNTs) and carbon black (CB). PA, MWCNT, and CB were procured from Arkema (Colombes, France), Nanocyl S.A. (Sambreville, Belgium), and Nanografi Nanotechnology AS (Tallinn, Estonia), respectively. PA pellets were desiccated overnight at 80 °C before extrusion using 3D Evo Composer 450, a single-screw extruder. Extruder temperature was set in the range of 185–220 °C, and extrusion speed was set to 8.5 rpm. Craftbot Plus (Craftbot Ltd, HU) printer was used to print the samples. Extrusion temperature, nozzle diameter, layer thickness, build plate temperature, printing speed, and infill density were set to 220 °C, 0.8 mm, 0.20 mm, 60 °C, 20 mm s^{-1}, and 100%, respectively [19].

9.2.2 Polyethylene

Antifriction polymer–polymeric ultrahigh-molecular-weight polyethylene (UHMWPE) composite reinforced with grafted high-density polyethylene (HDPE-g-SMA) and PP was fabricated via the FDM process. Since UHMWPE possesses a very low melt flow rate, these were very difficult to process through traditional routes. A twin-screw extruder was employed to efficiently mix the constituents at a temperature of 210 °C. The extrudate obtained was shredded into 3–5 mm granules using a Rondol shredder. ArmPrint – 2 3D printer was employed to fabricate the composites. The materials were processed in the temperature range of 160–200 °C with 0.3 mm layer thickness and 20 mm s^{-1} deposition speed [20].

FDM technique was employed to fabricate PE composites reinforced with Martian regolith (RG) for radiation shielding purposes. Medium-density PE and commercially available basalt powder, both in powder form, were used in the as-received condition. RG in 5 and 10 wt% with PE were mixed using a mechanical stirrer (Velp Scientifica, Italy) for 15 minutes. The pallets thus formed were then used to draw the filaments of 3 mm diameter at a temperature range of 118–125 °C using PRO350EX extruder from FilaFab (UK). Filaments drawn were free from macro-defects and exhibited circular cross-sections. Ultimaker 3 printer with a 0.8 mm diameter nozzle was employed to fabricate the samples. Samples were printed using the following conditions: nozzle temperature 118 °C; build plate temperature 60–80 °C; nozzle speed 20 mm s^{-1}; infill density 100%; layer thickness 0.1 mm [21].

Biocomposites based on biopolyethylene (BioPE) and thermomechanical pulp (TMP) fibers were manufactured using the FDM 3D printing process. TMP fibers in granules of 8 mm diameter were procured from Norske Skog Saugbrugs. BioPE was procured from Braskem (Sao Paulo, Brazil), and maleic anhydride (MAPE) procured from Eastman Chemical Products, Spain, was used as a coupling agent. A Gelimat mixer was used to prepare biocomposite blends of various fiber concentrations at a processing temperature of 210 °C. Granules of approximately 10 mm diameter were prepared from blends and stored at 80 °C for 24 hours. A Noztek filament extruder was employed to extrude 3D printing filaments at 180–200 °C processing temperature. Prusa i3 3D printer was employed to print the samples [22]. Samples were printed at 210 °C nozzle temperature and 5 mm s^{-1} printing speed using a 0.4 mm diameter nozzle.

HDPE procured from LyondellBasell was used to fabricate the filament for FFF process. Twin-screw extruder was used to draw the filaments from HDPE pellets using a water cooling and winding unit. Filaments collected at 50 mm s^{-1} winding speed exhibited uniform thickness. Extruder operates at around 200 °C with a nozzle of 2.80 mm diameter and 20 rpm rotational speed. The test samples were printed with Ultimaker 2+ using 0.4 and 0.8 mm diameter circular nozzles. Extrusion temperature of around 200–260 °C, build platform temperature of 60 °C, 0.1 mm layer thickness, 100% fill density, and 25–150 ms^{-1} printing speed were adopted for printing [23]. Same printer settings were also used to print all – PE composites using PE reactor blends and UHMWPE reactor blends. A single-screw extruder of 2.30 mm nozzle diameter was used to produce PE filaments. Feed motor was directly mounted onto the print head to obtain higher extrusion pressure and defect-free filament [24].

Recycled high-density polyethylene (rHDPE) along with low-density polyethylene (LDPE) were mixed using a twin-screw micro compounder at 190 °C and 100 rpm screw speed for five minutes to produce formulated HDPE. The mixture was then extruded at a temperature of 190 °C and 0.4 mm s^{-1} using 2 mm diameter die to produce the filaments. A water bath and winding unit were utilized to obtain uniform-diameter filaments. rHDPE was collected from SWaCH (Pune, India), washed manually, dried, and shredded into flakes. The samples were printed by Julia+ Dual nozzle printer consisting of glass build plate maintained at 60 °C. Following printer settings were used during the process: 0.4–0.6 mm diameter nozzle; 230 °C nozzle extrusion temperature; 30 mm s^{-1} print head speed; 20% infill density (increased infill density increases warpage). Commonly used polyvinyl acetate-based adhesive was employed on the build plate to enhance the adhesion of HDPE [25].

GNPs in 5–30 wt% were mixed with linear low-density polyethylene (LLDPE) to manufacture the filaments suitable for 3D printing. In the first step, LLDPE/GNPs co-powders were prepared using a high-speed mixer at room temperature. A single-screw extruder was utilized to fabricate the filaments of clean LLDPE and LLDPE/GNPs nanocomposites at 30 rpm extrusion speed and 170 °C. RepRap X350pro FDM printer (Germany) was used to print the test coupons. Pro/Engineer 5.0 was used to print the model. The following printer settings were maintained during printing of samples: nozzle diameter of 0.6 mm, layer height of 350 μm, nozzle temperature of 170 °C, and infill density of 100%. The printing speed of 200–2000 mmmin^{-1} was maintained during the process. Built plate temperature was maintained at 90–100 °C to avoid or reduce problems associated with the warping of prepared filaments [26].

Core–shell filaments made of polycarbonate (PC)/ABS with HDPE/LDPE were fabricated using Rheomex 252p and Akron Extruder M-PAK150. Moisture from PC/ABS was removed by drying in an oven for 12 hours in the temperature range of 80–110 °C. The filaments after extrusion were cooled using a water bath and wound at a uniform speed to obtain constant diameter filaments. Cartesio 3D printer with 0.4 mm diameter nozzle was employed for fabricating the samples in the temperature range of 260–300 °C. Build platform temperature, layer thickness, extrusion speed, and infill density were maintained at 120 °C, 0.21 mm, 10 mm s^{-1}, and 100%, respectively. Adhesion between the build plate and the printed filament was improved using Kapton tape and a thin layer of washable adhesive [27].

9.2.3 Polyethylene Terephthalate

PET is a semi-crystalline polymer that exhibits high toughness, impact resistance, and transparency. Crystallization temperature of PET is in the range of 170–220 °C. PET is flexible and exhibits a small amount of shrinkage and wrapping during 3D printing. PET should be properly dried before using in a 3D printer since the presence of moisture may lead to the sticking of material to the nozzle [28].

Shape memory materials (SMMs) are smart materials used for manufacturing sensors and actuators used in a variety of sectors, such as medical, textile, aerospace, and automotive. Polyethylene terephthalate glycol (PETG), a type of SMP procured from YouSU 3D (China), was printed with FDM-based printer. FDM printer consists of two nozzles of 0.8 mm diameter. Build plate temperature of 50 °C was maintained during the printing of samples with 100% fill density. Printing temperature and speed were varied in the range of 200–240 °C and 10–70 mm s^{-1}, respectively [29].

SMP has drawn considerable interest in various applications such as transportation, aerospace, construction, and electric and electronic industries because of its high strength, easy processing, thermal stability, and flame retardancy. PET is a promising candidate for SMPs because it is cheaper and shows good thermal and mechanical properties. Thermoplastic SMPs filaments (of 1.75 mm diameter) based on PET copolyesters were extruded using a desktop filament extruder, Wellzoom B from China. Extruder temperature was varied in the range of 230–260 °C. Samples were printed using an FDM-based printer from HORI, China. Other details, such as printing speed, build plate temperature, and layer thickness, were not mentioned in the article [30].

PETG filament procured from Polymaker, Suzhou, China, was successfully printed without visible defects or inaccuracies using a commercially available Ultimaker 3 FDM-based printer. Samples were printed using a 0.25 and 0.4 mm diameter nozzle, and the extruder temperature used was in the range of 220–250 °C. Emissions measured during the process were in the acceptable range, especially at slow printing speeds [31].

Two types of recycled polyethylene terephthalate (rPET) materials were successfully printed using an FDM-based 3D printer. Commercially available rPET (Ultrafuse PET pellets) and shredded water bottles were used as raw materials for producing feedstock material. Water bottles were cleaned properly to remove labels, caps, and adhesives before shredding them into particles of size 5.84 mm using an open rotor scissor cut granulator. After shredding, the particles were dried for 24 hours at 38 °C using a food dehydrator to remove moisture. Samples were printed directly from rPET without converting them into filaments using a Gigabot X, a direct pellet material extrusion-based 3D printer. A constant flow of the material during printing was achieved using a compression screw and feed tube arrangement with three heat zones. The temperatures used during processing were 250, 240, and 180 °C for the bottom (near the nozzle), middle, and top zones, respectively. Samples were printed at a printing speed of 5–30 mm s^{-1} on a build plate maintained at a temperature of 100 °C [32].

PET bottles collected from the seashore were successfully turned into filaments for 3D printing process. Bottles collected were thoroughly washed before cutting and milling using a RETSCH SM2000 mill (Düsseldorf, Germany). PET flakes obtained were dried at 70 °C for 24 hours and then extruded using a single-screw extruder (Waltham, MA, USA). Filaments were extruded in the temperature range of 220–280 °C at a screw speed of 10 rpm. After extrusion, the filaments were cooled at a faster rate with the help of compressed air. FDM-based 3D printer with 0.4 mm diameter nozzle was utilized for printing samples at 50 mm s^{-1} extrusion speed and 250 °C extrusion temperature. Samples printed using rPET exhibited better properties as compared to virgin PET [33].

PETG–sepiolite (SEP) composites were successfully printed using FDM-based AEP2 printer (Rokit, Republic of Korea). SEP and PETG were procured from Sigma-Aldrich (USA) and SK Chemicals (Republic of Korea), respectively. Raw materials were dried at 65 °C for 12 hours before extrusion. A twin-screw extruder was employed to mix the constituents, followed by pelletization using a pelletizer procured from Bau-Technology, Republic of Korea. Pellets were processed at the temperature of 140–200 °C and 200 rpm screw speed. Pellets were post-processed at a temperature of 65 °C for 12 hours to remove the moisture. PETG-SEP filaments were extruded using a single-screw extruder. Filaments of 1.75 mm were extruded using a screw speed of 10 rpm at 210 °C. A layer thickness of 200 µm was maintained during sample printing with 100% fill density. Printing speed and extruder head temperature were set to 60 mm s^{-1} and 250 °C, respectively. A build plate temperature of 100 °C was used to obtain appreciable adhesion between the print platform and composite filament [34].

9.2.4 Polypropylene

FDM-based 3D printing method was adopted to print pure PP and PP embedded with 30 wt% glass fibers (GRPP). Raw materials in pellet form were used to extrude the filaments of 1.75 mm diameter at 190–220 and 200–230 °C for PP and GRPP, respectively. PP was extruded at a speed of 11 rpm, while a speed of 6.3 rpm was used to extrude GRPP. After extrusion, filaments were cooled using compressed air and an oven maintained at 30 °C to avoid the formation of voids. A commercial 3D printer, Prusa i3 with a nozzle diameter of 0.4 mm, was utilized to print the specimens. PP and GRPP specimens were printed at a nozzle temperature of 165 and 185 °C, respectively, on a glass plate/blue tape with a layer height of 0.20 and 0.35 mm and with 20%, 60%, and 100% fill density. To enhance the adhesion, the glass/blue tape build plate was replaced by a PP plate [35].

Isotactic Polypropylene (iPP) procured from Sigma-Aldrich was combined with GNPs procured from XG Sciences to prepare the composite filament for the FDM printer. Constituents were melt-mixed using a Brabender for 20 minutes at 180 °C and then shredded using a Brabender pelletizer. A Filabot extruder was used to draw the filaments of 3 mm diameter. Composite was later printed using Ultimaker

Extended 2+ and LulzBot Mini printers. Nozzle temperature, build platform temperature, and printing speed were set to 230 °C, 60 °C, and 20 mm s^{-1}, respectively [36].

PP was reinforced with varying weight percentage of CB to fabricate composite filaments for FDM 3D printing process. PP (density of 0.9 g cm^{-3}) and CB (Φ 0.1–0.2 mm) were procured from Shanghai Shien Plasticization Co., Ltd. and Xuyi Zongze carbon nano new material technology Co., Ltd., respectively. Composite filaments were processed using a single-screw extruder, which works on the hot-melt mixing principle. Base materials were dried for 24 hours at 60 °C before mixing. Filaments of 1.75 mm were drawn at 200 °C and 2 mm s^{-1} screw speed. Filaments were shredded again into pellets and fed back to extruder to achieve uniform dispersion of CB particles. Printing speed, layer thickness, nozzle temperature, nozzle diameter, and bed temperature were set to 20 mm s^{-1}, 0.1 mm, 200 °C, 0.6 mm, and 100 °C, respectively [37].

Filament for FFF-based 3D printer was successfully fabricated using rPET, PP, and PS mixture. Raw materials were cleaned with water and ethanol before shredding into small particles using a paper shredder. PET plastic salad containers (rPET) were dried for 24 hours at 120 °C under vacuum, while rPP was dried at room temperature for 24 hours. Raw materials were manually mixed in different ratios before extruding. A twin-screw extruder with eight heating zones and rotating at a constant speed of 100 rpm was used to draw the filaments. Filaments obtained were shredded using Varicut pelletizer and re-extruded at 25 rpm to achieve uniformity. Filaments were desiccated in a vacuum overnight to eliminate moisture content before printing on a FFF printer. PP/PET and PP/PS blends were printed using 260 °C nozzle temperature on PET tape surface and polyetherimide surface, respectively, maintained at 100 °C [38].

PP of different grades and PP reinforced with 40 wt% talc were successfully printed by an FDM-based 3D printer. Filaments of 1.75 mm were fabricated using a single-screw extruder having four heating zones. Filaments after extrusion were cooled using fans and wound on the spool. Samples were printed using a Roboze One 3D printer. Two types of scotch tapes were used to enhance the adhesion of PP onto the build plate. Printing speed and temperature used were in the range of 30–60 mm s^{-1} and 210–250 °C, respectively [39].

Feed spacers used for water and wastewater treatment were printed using PP powders by SLS process. Samples were prepared using an EOSINT P395 SLS machine from EOS. Powder particles of 60 µm and 0.92 g cm^{-3} density were sintered using a laser beam of 22 W at 3000 m s^{-1} scanning speed. Layer thickness and print bed temperature employed were 0.1 mm and 150–160 °C, respectively [40].

Recycled polypropylene (r-PP) was reinforced with SCF to fabricate filaments for FDM process. Filaments were extruded using a single-screw extruder of 3 mm diameter die, and extrusion temperature was in the range of 170–180 °C. Filaments were wounded on a spool at a speed of 100 m s^{-1} through a water bath. Ultimaker S3 was used to print the samples using 0.4 and 0.6 mm diameter nozzles and 100% fill density at 25 mm s^{-1} printing speed. Adhesion of the extruded filament on the build plate (maintained at 85 °C) was enhanced using Pritt glue and Kapton tape [41].

Rice husk (RH) reinforced rPP filaments were successfully extruded using a twin extruder. rPP obtained from the industrial waste was procured from Promaplast, and RH was procured from Ambala Grinder, Colombia. Particles were pulverized to a size of 250–425 μm and dried for three hours at 105 °C before extruding. After extrusion filaments were made to pass through a water bath maintained at 25 °C. Filaments extruded were pelletized and re-extruded to enhance the homogeneity. A layer height of 0.25 mm was selected during printing at a nozzle temperature of 240 °C and 100% infill density [42].

9.2.5 Polylactic Acid

Biocomposite filaments fabricated using PLA and cellulose nanofiber (CNF) were successfully printed using an FDM 3D printer. Biodegradable PLA (of $1.25\,\mathrm{g\,cm^{-3}}$ density) was obtained from Natureworks, Minnetonka, Minnesota. CNF was obtained by processing sisal raw fibers, followed by micro-grinding. PLA pellets and nano-fibers (by 1, 3, and 5 wt%) were desiccated at 80 °C in an oven before mixing. Both the constituents were dissolved in suitable chemicals, and the solutions formed were mixed using a magnetic stirrer for 12 hours. Composite slurry obtained was cast into glass molds and allowed to solidify for 48 hours at ambient temperature. Obtained composite was crushed and desiccated for 24 hours at 60 °C before extrusion at 180 °C. Samples were printed using FFF-based 3D printer utilizing 0.6 mm diameter nozzle. The remaining printing parameters namely nozzle temperature, build platform temperature, infill density, layer height, and printing speed were set to 180 °C, 60 °C, 100%, 0.2 mm, and $45\,\mathrm{mm\,s^{-1}}$, respectively [43].

PLA and PP were combined with bamboo fiber (BF) to fabricate biodegradable composite material. All the materials were used in their as-received condition without any modification. Constituents were chemically processed to prepare the composite material. Constituents were grounded and mixed using a high-speed mixer before extrusion using a twin-screw extruder with six heating zones. Extrudate obtained was pelletized and desiccated overnight at 80 °C. Filaments for 3D printing were drawn using a single-screw extruder. Samples were produced using an FDM technique at 180–200 °C nozzle temperature and 40–$60\,\mathrm{mm\,s^{-1}}$ printing speed on the build platform maintained at 40–60 °C [44].

PLA reinforced with 8 wt% of bronze was successfully printed using an FDM 3D printer. PLA/bronze filaments of 1.75 mm were directly obtained from the commercial 3D filament supplier. WANHAO Duplicator 6 was used to print the samples using 0.4 mm diameter nozzle. Nozzle and build plate temperatures were set to 190 and 60 °C, respectively [45]. PLA filaments of 1.75 mm diameter and different colors directly obtained from the commercial 3D filament supplier were printed using WANHAO Duplicator 6 desktop 3D printer. 3D printer was equipped with 0.4 mm diameter nozzle, and samples were printed at different orientations, namely horizontal, vertical, and inclined (45°) positions. Printing parameters viz printing temperature, print plate temperature, infill density, and print speed were set to 195 °C, 60 °C, 100%, and $60\,\mathrm{mm\,s^{-1}}$, respectively [46].

9.3 Characterization of 3D-Printed Crystalline Polymers

9.3.1 Mechanical Properties/Mechanical Characteristics

Mechanical characteristics of 3D-printed parts are highly influenced by printing architecture. Researchers evaluated the tensile, flexural, compressive, and impact characteristics of 3D-printed samples to make more meaningful comparisons. In all published articles, the samples were produced by bulk material standards [47].

Ambone et al. [43] carried out a systematic examination of the mechanical performance of fused filament fabricated PLA. Results illustrate that the tensile strength (T_s) and modulus (T_m) of FFF-PLA were reduced by 49% and 41%, respectively, compared to compression molded parts. Integration of a slight quantity (1 wt%) of CNFs improved the T_s and T_m of FDM-produced PLA by 84% and 63%, respectively.

Tensile testing was performed on 3D-printed PLA samples to examine mechanical behavior before and after an aging process in a salt fog environment. The outcomes demonstrated that the material crystallinity varies concerning cooling rate used during product printing. Additionally, the T_s of specimens treated to the process reduces by nearly 20% [48].

The mechanical characteristics of alkali-treated bamboo fiber (ABF) composites produced with different quantities of PLA and PP were studied. ABF of 20 wt% and maleated polypropylene (MAPP) of 5 wt% were used during composite preparation. The T_s and flexural strength rise with rise in PLA content, whereas a decrease in impact strength and a drop in elongation at break were observed. Further, 7 : 3 proportion of PP/PLA exhibited brittle characteristics [44].

Wang et al. [49] evaluated the 3D-printed polyether ether ketone (PEEK) samples for their dimensional accurateness, crystallinity, and mechanical properties. An increase in mechanical properties and crystallinity was observed at higher chamber temperatures and post-printing annealing. Various aspects like printing parameters, printer quality, material processing, and the degree of crystallinity affected the mechanical, physical, and chemical properties of FDM-printed PEEK.

Yang et al. [50] recommended a unique method for monitoring the crystallinity of 3D-printed CF/PEEK composites. Initially, samples of a CF/PEEK composite with little crystallinity and warping distortion were printed at an ambient temperature of 20 °C. Further, the recrystallization process influenced the crystallinity and mechanical properties of the composite. Similar mechanical properties were recorded for CF/PEEK composite (10 wt%) fabricated via 3D printing and injection molding processes. This study is of confident significance to encourage the claim of FDM-made PEEK composite in aerospace and orthopedic implant applications.

Yang et al. [51] analyzed the influence of 3D printing thermal processing factors on the mechanical characteristics of PEEK material. Tensile test results reveal that the mechanical properties of PEEK samples were influenced by several factors, including printing conditions, crystallinity, multiscale interfaces between printing lines, residual internal stress, and the deteriorating phenomena of polymer materials.

Chatham et al. [52] explored the blending of two immiscible, semi-crystalline polymers PET and PP in a ratio of 80 : 20 with varying concentrations of PP-*graft*-maleic anhydride (PP-g-MA) compatibilizer (up to 10 wt%). Ultimate tensile properties of the as-printed blends reveal that the uncompatibilized 80/20/0 blend had the highest stress and strain at failure.

Using rPP, a variety of composite filaments with hemp and harakeke fibers and gypsum weight percentage were generated. Best filaments concerning tensile properties contained 30% harakeke in a post-consumer PP matrix with T_s and T_m of 41 MPa and 3.8 GPa, respectively. When compared to plain PP filament, these results demonstrated 77% and 275% increases in T_s and Young's modulus, respectively [53].

Carneiro et al. [35] investigated the effect of printing circumstances, raw materials, and manufacturing procedures on the mechanical performance of a GFPP and clean PP. The tensile test results show that, compared to samples made by compression molding, the height of the layers has a slight impact on the mechanical properties of the samples, the infill degree has a dramatic and linear impact on the mechanical properties, and the loss in mechanical performance of the printed samples was approximately 20–30%, depending on the printing input variables used.

Mechanical properties of 3D printed and compression molded pure PP/glass fiber (GF) and PP/GF composites comprising maleic anhydride polyolefin (POE-g-MA) at 10, 20, and 30 wt% were investigated. Results exhibited that the addition of GF improved the modulus and strength of the composite but dropped its flexibility; conversely, the composite revealed reduced modulus and strength and improved flexibility upon adding the POE-g-MA. The samples produced via compression molding displayed better values of strength and modulus as compared to those produced via 3D printing [54].

Zuhan He et al. [55] employed a SLS process to print styrene ethylene butylene styrene (SEBS)/PP and SEBS/PP/Graphene (GE) blended tensile specimens with optimal printing parameters. SEBS/PP samples were somewhat brittle with poor tensile elongations. SEBS/PP/GE exhibits the greatest attributes in sintering samples, with a tensile strength of 2.8 MPa and elongation at break of 176%. The excellent particle fusion and reinforcement of graphene can be attributed to graphene's enhancing effect.

Morales M.A. et al. [42] manufactured composite filament from RH and rPP for 3D printing and carried out the characterization for varied RH weight percentage and raster orientation. Tensile test results exhibit increased T_s when printed at a raster orientation of 0° compared to specimens printed at 90° raster orientation, due to the poor inter-layer attachment relative to in-layer. Also, all tested materials showed similar interlayer bonding tensile strength.

Kristiawan et al. [56] investigated the consequence of a glass powder (GP) additive (2.5, 5, and 10 wt% fractions) on rPP as filaments for 3D printing applications. An increase in ultimate T_s of 38% and Young's modulus of 42% was observed with 10% GP additive in rPP-based specimens. Besides enhanced mechanical strength, the inclusion of GP decreases the bending deformation, which may be controlled by reducing the curvature, which was a concern in semi-crystalline polymer-based filaments.

The mechanical properties of 3D-printed continuous Kevlar fiber reinforced nylon composites with designed fiber layer distributions and fiber alignments were examined [6]. It has been discovered that interior-distributed fiber layers improve tensile strength by preventing the development of soft-core. The stiffness was mostly improved by the surface-distributed fiber layers. As a result, by adjusting the distribution of fiber layers, composites' mechanical characteristics may be tailored. When the strain ascribed to fiber rotation was more than 10%, the tensile stresses of composites with various fiber alignments appeared to remain constant.

9.3.2 Thermal Properties/Thermal Characteristics

The difficulty in employing semi-crystalline polymers in 3D printing is retaining the dimensional precision of the printed structure while the polymer cools, hardens, and crystallizes. As a result, the cooling and crystallization behavior of blends and homopolymers differs significantly. Various thermo-analytical techniques like thermogravimetric analysis (TGA), differential thermal analysis (DTA), and differential scanning calorimetry (DSC) are utilized to study the thermal stability and degradation temperatures of 3D-printed parts.

According to a DSC analysis, CNFs can expedite the nucleation and crystallization of 3D-printed PLA, resulting in increased crystallinity. Inclusion of CNF does not affect the thermal stability of 3D-printed PLA/CNF composites. Higher crystallinity and fewer flaws can be attributed to the improved mechanical performance of 3D-printed PLA/CNF composites [43].

DTA curves of extruded filament and 3D-printed PLA indicate no substantial deviations in melting temperature, which is about 170 °C. This aids in the setup of the fused deposition machine so that the material has adequate flowability. The TG curves of the two studied samples show that the material was thermally stable in the temperature range of 30–210 °C, without substantial changes in specimen mass [48].

TGA was used to investigate the thermal stability of the ABF/PP/PLA mix and that comprising MAPP composite [44]. Results indicate that thermal compatibility of the composites was improved by MAPP. Alkali treatment of BF improved the temperature of weight reduction, which was accredited to hemicellulose–lignin matrix partly dissolved following alkali treatment. Also, owing to the incomplete dissolution of hemicellulose–lignin in isocyanate (MDI/IPDI) treatment, thermal decomposition of isocyanate treatment of BF composite was reduced.

DSC experiments were carried out on Nylon 618 and 645, alloy 910, Onyx, and Markforged nylon at three stages: before printing, after printing, and after the nylon gear step load test. The filament crystallinity prior to printing was found to be somewhat lower than that after printing and the material from the gear tooth surface after testing. Furthermore, as compared to the other materials evaluated, Nylon 66 and Nylon 618 had considerably superior thermal performance with respect to greater glass transition temperatures, higher melting temperatures, and more crystallinity [6].

Peak melting temperatures, enthalpy of fusion, and crystallinity of PET/PP/PP-g-MA composites during the first and second heats revealed that the overall crystalline concentration in the filament was relatively like the homopolymer starting materials, with the PET phase diverting little from the as-received homopolymers. When compared to the percent crystallinity in filament form, crystallinity of the PP phase in printed structures rises. As a result, it was hypothesized that immiscible, phase-separated morphology permits printing with high crystallinity polymers [52].

Lei et al. [37] examined the thermal breakdown behavior of CB and PP composites. Thermogravimetric investigation shows that CB has high thermal conductivity and may transmit heat to the matrix. It helps/makes the composites achieve the beginning breakdown temperature. Furthermore, because CB and PP are heat resistant, a homogeneous distribution of CB in the matrix might hinder the deterioration process of composites.

Thermal stability of RH fiber, neat rPP, rPP/RH (5 wt%), and rPP/RH (10 wt%) was evaluated by TGA. The earlier degradation process for rPP/RH composites compared to rPP was disclosed by TGA due to the fiber's lignocellulosic components; however, the capability of printing was affected [42]. Based on the thermal stability analyses, adding GP improves the stability of mass changes to heat and raises the melting temperature of rPP [56].

9.3.3 Tribological Properties/Tribological Characteristics

The tribological characteristics of 3D-printed polymers are typically improved by using the optimum printing variables, including surface alterations like filament reinforcement and considering filament temperature and color [57, 58].

A step load test was conducted to determine the wear rate of 3D-printed gears. SEM was used to record distinct wear behavior and wear patterns on the gear tooth. Wear on 3D-printed gear was seen only on the pitch line, and melting of gear tooth surface sections was observed. No instances of material peeling off the tooth were noticed for the Nylon 618 printed gears, but the remaining materials showed material peeling [5].

Hanon et al. [45] examined the effect of print orientation on the tribological characteristics of a 3D-printed bronze/PLA composite. Tribological experiments were conducted in a dry environment with a reciprocating sliding motion; the introduction of bronze particles as a reinforcement for the PLA material enhanced the tribological capabilities by drastically reducing wear depth. However, because the matrix of the polymer composite was packed with hard particles, the friction remained constant.

Kichloo et al. [59] utilized a pin-on-disk tribometer to evaluate the effect of CF reinforcement on the tribological behavior of 3D-printed PETG polymer composites. Tribological research discovered that the use of CF suggestively reduced the coefficient of friction by approximately 47.3% at low speeds and 44.79% at high speeds when compared to PETG.

Influence of surface texturing on friction behavior of 3D-printed PLA was investigated by Aziz et al. [60], with three textures created by FDM technique and evaluated in dry and lubricated circumstances. Surface study in dry circumstances found that wear was mostly caused by adhesion and abrasion. The smooth surface means less adhesion damage at lower and higher speeds. Furthermore, mass plowing of material was not present. This was due to the effective layer creation between the surfaces as the temperature rose.

The impact of print direction and PLA color on tribological behavior was evaluated using a linear alternating reciprocating cylinder-on-plate friction and wear tester [46], which prints samples in horizontal, vertical, and inclined (45°) directions and with different filament colors (white, black, and gray). The findings show that tribological behavior varies owing to the variety of print orientations and filament colors. The greatest friction propensity was related to white filament, whereas maximum wear depth relates to print orientation 45 and black filament. Furthermore, sliding minimizes wear under high loads, whereas the stick-slip phenomenon was more likely to occur under low loads.

The tribological characteristics of PEEK-based composites filled with polytetrafluoroethylene (PTFE) and molybdenum disulfide (MoS_2) microparticles were investigated [61] under dry sliding friction circumstances. Loading PTFE powder into the PEEK matrix decreases the friction coefficient by up to 20 times for both the metal–polymer and ceramic–polymer tribo-pairs. Simultaneously, the resistance to wear of the composite rises by 8 times when compared to its metal counterpart, and it is 15 times greater when tested over ceramic. Table 9.1 summarizes the tests performed on 3D printed crystalline polymer and its composites.

9.4 Conclusion

For part production, all 3D printing technologies rely on a layer-wise deposition approach. Among the several 3D printing technologies, FDM is the widely used process. FDM feedstock includes both amorphous and semi-crystalline thermoplastic polymer filaments. There is an obvious prerequisite to implement more engineering and high-performance thermoplastics into the FDM material palette, in addition to the more often used synthetic plastics such as ABS or PLA, to meet the quality standards in these high-end applications. Semi-crystalline polymers have previously proven useful in a range of FDM-based applications, including medicine and electronics. It is consequently critical to properly comprehend their behavior during FDM processing.

This chapter offers a basic outline of the research on crystalline feedstock polymers for 3D printing, process parameters, and print settings utilized in 3D printing, as well as mechanical, thermal, and tribological characteristics. PLA, PP, PEEK, nylon, PET, and PA are the most studied semi-crystalline polymers in the literature, as they are common feedstock materials for FDM and SLS. However, PLA is inappropriate for high-performance applications; other polymers that can be utilized efficiently in such applications include nylon, PEEK, PP, and PA with suitable reinforcements. It is observed from the review that the crystalline filaments for 3D printing were produced using single/twin-screw extruder after being desiccated overnight at temperature of around 80 °C. The range of optimal printing process parameters adopted for printing are as follows; extruder nozzle temperature of 200–250 °C, layer thickness/height of 0.1–0.3 mm, printing speed of 25–50 mm s^{-1}, nozzle diameter of 0.4–0.8 mm, printing platform temperature of 50–70 °C, and fill density of 100%. It is also observed that various aspects like printing parameters,

Table 9.1 Tests performed on 3D-printed crystalline polymer and its composites.

References	Material	Mechanical testing	Thermal testing	Tribological testing
[43]	PLA/CNF composite	Tensile tests	TGA and DSC	None
[48]	PLA	Tensile and impact tests	TGA	None
[44]	ABF/PLA/PP	Tensile, flexural, and impact tests	TGA and DSC	None
[45]	Bronze/PLA composite	Tensile tests	None	Friction tests
[59]	PLA	Tensile and flexural tests	None	Friction and wear tests
[46]	PLA	None	None	Friction and wear tests
[49]	PEEK	Tensile, compression, and flexural tests	TGA and DSC	None
[50]	CF/PEEK composite	Tensile and three-point bending tests	None	None
[61]	PEEK/PTFE/MoS_2	None	None	Friction and wear tests
[60]	PETG	None	None	Surface texture, friction, and wear tests
[51]	PEEK	Tensile tests	DSC	None
[52]	PET/PP composite	Tensile tests	None	None
[53]	Hemp/Harakeke/PP composite	Tensile tests	None	None
[35]	GF/PP	Tensile tests	None	None
[54]	GF/PP	Tensile tests	None	None
[55]	SEBS/PP/GE	Tensile tests	None	None
[42]	RH/rPP	Tensile tests	TGA and DSC	None
[56]	GP/rPP	Tensile and impact tests	TGA and DSC	None
[37]	CB/PP	None	TGA and DSC	None
[6]	Nylon composites	Tensile tests	DSC	None
[5]	Nylon 618 and 645, alloy 910, Onyx, and Markforged nylon	None	DSC	Wear tests

quality of printer, processing of material, and degree of crystallinity influence the mechanical, thermal, and, to some extent, tribological characteristics of printed parts. Use of Scotch tapes to enhance the adhesion of polymers on the build plate is common in most of the studies.

References

1 Gedde, U. (1995). *Polymer Physics*. Springer Science & Business Media.
2 Cheng, S.Z. and Jin, S. (2002). Crystallization and melting of metastable crystalline polymers. In: *Handbook of Thermal Analysis and Calorimetry* (ed. Stephen Z.D. Cheng), vol. 3, 167–195. Elsevier Science.
3 Damanhuri, A.A.M., Fauadi, M.H.F.M., Hariri, A. et al. (2019). Emission of selected environmental exposure from selective laser sintering (SLS) polyamide nylon (PA12) 3D printing process. *Journal of Safety, Health & Ergonomics* 1: 1–6.
4 Fu, X., Zhang, X., and Huang, Z. (2021). Axial crushing of nylon and Al/nylon hybrid tubes by FDM 3D printing. *Composite Structures* 256: 113055. https://doi.org/10.1016/j.compstruct.2020.113055.
5 Zhang, Y., Purssell, C., Mao, K., and Leigh, S. (2020). A physical investigation of wear and thermal characteristics of 3D printed nylon spur gears. *Tribology International* 141: 105953. https://doi.org/10.1016/j.triboint.2019.105953.
6 Shi, K., Yan, Y., Mei, H. et al. (2021). 3D printing Kevlar fiber layer distributions and fiber orientations into nylon composites to achieve designable mechanical strength. *Additive Manufacturing* 39: 101882. https://doi.org/10.1016/j.addma.2021.101882.
7 Liu, G., Zhang, X., Chen, X. et al. (2021). Additive manufacturing of structural materials. *Materials Science and Engineering: R: Reports* 145: 100596. https://doi.org/10.1016/j.mser.2020.100596.
8 Aslanzadeh, S., Saghlatoon, H., Honari, M.M. et al. (2018). Investigation on electrical and mechanical properties of 3D printed nylon 6 for RF/microwave electronics applications. *Additive Manufacturing* 21: 69–75. https://doi.org/10.1016/j.addma.2018.02.016.
9 Sanatgar, R.H., Campagne, C., and Nierstrasz, V. (2017). Investigation of the adhesion properties of direct 3D printing of polymers and nanocomposites on textiles: effect of FDM printing process parameters. *Applied Surface Science* 403: 551–563. https://doi.org/10.1016/j.apsusc.2017.01.112.
10 Kang, M., Pyo, Y., Young Jang, J. et al. (2018). Design of a shape memory composite (SMC) using 4D printing technology. *Sensors and Actuators A: Physical* 283: 187–195. https://doi.org/10.1016/j.sna.2018.08.049.
11 Li, N., Link, G., Wang, T. et al. (2020). Path-designed 3D printing for topological optimized continuous carbon fibre reinforced composite structures. *Composites Part B: Engineering* 182: 107612. https://doi.org/10.1016/j.compositesb.2019.107612.
12 Shakiba, M., Rezvani Ghomi, E., Khosravi, F. et al. (2021). Nylon – a material introduction and overview for biomedical applications. *Polymers for Advanced Technologies* 32: 3368–3383. https://doi.org/10.1002/pat.5372.

13 Yang, D., Zhang, H., Wu, J., and McCarthy, E.D. (2021). Fibre flow and void formation in 3D printing of short-fibre reinforced thermoplastic composites: an experimental benchmark exercise. *Additive Manufacturing* 37: 101686. https://doi.org/10.1016/j.addma.2020.101686.

14 Wang, Y., Shi, J., and Liu, Z. (2021). Bending performance enhancement by nanoparticles for FFF 3D printed nylon and nylon/Kevlar composites. *Journal of Composite Materials* 55: 1017–1026. https://doi.org/10.1177%2F0021998320963524.

15 Chapman, G., Pal, A.K., Misra, M., and Mohanty, A.K. (2021). Studies on 3D printability of novel impact modified nylon 6: experimental investigations and performance evaluation. *Macromolecular Materials and Engineering* 306: 2000548. https://doi.org/10.1002/mame.202000548.

16 Mehdipour, F., Gebhardt, U., and Kästner, M. (2021). Anisotropic and rate-dependent mechanical properties of 3D printed polyamide 12-A comparison between selective laser sintering and multi jet fusion. *Results in Materials* 11: 100213. https://doi.org/10.1016/j.rinma.2021.100213.

17 Wang, K., Li, S., Wu, Y. et al. (2021). Simultaneous reinforcement of both rigidity and energy absorption of polyamide-based composites with hybrid continuous fibers by 3D printing. *Composite Structures* 267: 113854. https://doi.org/10.1016/j.compstruct.2021.113854.

18 Tuan Rahim, T.N.A., Abdullah, A.M., Md Akil, H. et al. (2015). Preparation and characterization of a newly developed polyamide composite utilising an affordable 3D printer. *Journal of Reinforced Plastics and Composites* 34: 1628–1638. https://doi.org/10.1177%2F0731684415594692.

19 Vidakis, N., Petousis, M., Tzounis, L. et al. (2021). Polyamide 12/multiwalled carbon nanotube and carbon black nanocomposites manufactured by 3D printing fused filament fabrication: a comparison of the electrical, thermoelectric, and mechanical properties. *C (Journal of Carbon Research)* 7: 38. https://doi.org/10.3390/c7020038.

20 Panin, S., Buslovich, D., Kornienko, L. et al. (2019). Structure and tribomechanical properties of extrudable ultra-high molecular weight polyethylene composites fabricated by 3D-printing. AIP Publishing LLC, p. 040011. https://doi.org/10.1063/1.5122130.

21 Zaccardi, F., Toto, E., Santonicola, M.G., and Laurenzi, S. (2022). 3D printing of radiation shielding polyethylene composites filled with Martian regolith simulant using fused filament fabrication. *Acta Astronautica* 190: 1–13. https://doi.org/10.1016/j.actaastro.2021.09.040.

22 Tarrés, Q., Melbø, J.K., Delgado-Aguilar, M. et al. (2018). Bio-polyethylene reinforced with thermomechanical pulp fibers: mechanical and micromechanical characterization and its application in 3D-printing by fused deposition modelling. *Composites Part B: Engineering* 153: 70–77. https://doi.org/10.1016/j.compositesb.2018.07.009.

23 Schirmeister, C.G., Hees, T., Licht, E.H., and Mülhaupt, R. (2019). 3D printing of high density polyethylene by fused filament fabrication. *Additive Manufacturing* 28: 152–159. https://doi.org/10.1016/j.addma.2019.05.003.

24 Schirmeister, C.G., Hees, T., Dolynchuk, O. et al. (2021). Digitally tuned multidirectional all-polyethylene composites via controlled 1D nanostructure formation during extrusion-based 3D printing. *ACS Applied Polymer Materials* 3: 1675–1686. https://doi.org/10.1021/acsapm.1c00174.

25 Gudadhe, A., Bachhar, N., Kumar, A. et al. (2020). 3D printing with waste high-density polyethylene. *Bulletin of the American Physical Society* 65: 1–10.

26 Jing, J., Chen, Y., Shi, S. et al. (2020). Facile and scalable fabrication of highly thermal conductive polyethylene/graphene nanocomposites by combining solid-state shear milling and FDM 3D-printing aligning methods. *Chemical Engineering Journal* 402: 126218. https://doi.org/10.1016/j.cej.2020.126218.

27 Peng, F., Jiang, H., Woods, A. et al. (2019). 3D printing with core–shell filaments containing high or low density polyethylene shells. *ACS Applied Polymer Materials* 1: 275–285. https://doi.org/10.1021/acsapm.8b00186.

28 Gopathi, P. and Surve, P. (2017). Possibilities and limitations of using production waste PET and PES materials in additive manufacturing (3D printing technology). Master Thesis. https://www.diva-portal.org/smash/get/diva2:1151617/FULLTEXT02.

29 Aberoumand, M., Soltanmohammadi, K., Soleyman, E. et al. (2022). A comprehensive experimental investigation on 4D printing of PET-G under bending. *Journal of Materials Research and Technology* 18: 2552–2569. https://doi.org/10.1016/j.jmrt.2022.03.121.

30 Chen, L., Zhao, H.-B., Ni, Y.-P. et al. (2019). 3D printable robust shape memory PET copolyesters with fire safety via π-stacking and synergistic crosslinking. *Journal of Materials Chemistry A* 7: 17037–17045. https://doi.org/10.1039/C9TA04187G.

31 Chýlek, R., Kudela, L., Pospíšil, J., and Šnajdárek, L. (2021). Parameters influencing the emission of ultrafine particles during 3D printing. *International Journal of Environmental Research and Public Health* 18: 11670. https://doi.org/10.3390/ijerph182111670.

32 Little, H.A., Tanikella, N.G., and J. Reich M, Fiedler MJ, Snabes SL, Pearce JM. (2020). Towards distributed recycling with additive manufacturing of PET flake feedstocks. *Materials* 13: 4273. https://doi.org/10.3390/ma13194273.

33 Ferrari, F., Esposito Corcione, C., Montagna, F., and Maffezzoli, A. (2020). 3D printing of polymer waste for improving people's awareness about marine litter. *Polymers* 12: 1738. https://doi.org/10.3390/polym12081738.

34 Kim, H., Ryu, K.-H., Baek, D. et al. (2020). 3D printing of polyethylene terephthalate glycol–sepiolite composites with nanoscale orientation. *ACS Applied Materials & Interfaces* 12: 23453–23463. https://doi.org/10.1021/acsami.0c03830.

35 Carneiro, O.S., Silva, A., and Gomes, R. (2015). Fused deposition modeling with polypropylene. *Materials & Design* 83: 768–776. https://doi.org/10.1016/j.matdes.2015.06.053.

36 Shmueli, Y., Lin, Y.-C., Zuo, X. et al. (2020). In-situ X-ray scattering study of isotactic polypropylene/graphene nanocomposites under shear during fused deposition modeling 3D printing. *Composites Science and Technology* 196: 108227. https://doi.org/10.1016/j.compscitech.2020.108227.

37 Lei, L., Yao, Z., Zhou, J. et al. (2020). 3D printing of carbon black/polypropylene composites with excellent microwave absorption performance. *Composites Science and Technology* 200: 108479. https://doi.org/10.1016/j.compscitech.2020.108479.

38 Zander, N.E., Gillan, M., Burckhard, Z., and Gardea, F. (2019). Recycled polypropylene blends as novel 3D printing materials. *Additive Manufacturing* 25: 122–130. https://doi.org/10.1016/j.addma.2018.11.009.

39 Bertolino, M., Battegazzore, D., Arrigo, R., and Frache, A. (2021). Designing 3D printable polypropylene: material and process optimisation through rheology. *Additive Manufacturing* 40: 101944. https://doi.org/10.1016/j.addma.2021.101944.

40 Tan, W.S., Chua, C.K., Chong, T.H. et al. (2016). 3D printing by selective laser sintering of polypropylene feed channel spacers for spiral wound membrane modules for the water industry. *Virtual and Physical Prototyping* 11: 151–158. https://doi.org/10.1080/17452759.2016.1211925.

41 Polline, M., Mutua, J.M., Mbuya, T.O., and Ernest, K. (2021). Recipe development and mechanical characterization of carbon fibre reinforced recycled polypropylene 3D printing filament. *Open Journal of Composite Materials* 11: 47–61. https://doi.org/10.4236/ojcm.2021.113005.

42 Morales, M.A., Atencio Martinez, C.L., Maranon, A. et al. (2021). Development and characterization of rice husk and recycled polypropylene composite filaments for 3D printing. *Polymers* 13: 1067. https://doi.org/10.3390/polym13071067.

43 Ambone, T., Torris, A., and Shanmuganathan, K. (2020). Enhancing the mechanical properties of 3D printed polylactic acid using nanocellulose. *Polymer Engineering & Science* 60: 1842–1855. https://doi.org/10.1002/pen.25421.

44 Long, H., Wu, Z., Dong, Q. et al. (2019). Mechanical and thermal properties of bamboo fiber reinforced polypropylene/polylactic acid composites for 3D printing. *Polymer Engineering & Science* 59: E247–E260. https://doi.org/10.1002/pen.25043.

45 Hanon, M.M., Alshammas, Y., and Zsidai, L. (2020). Effect of print orientation and bronze existence on tribological and mechanical properties of 3D-printed bronze/PLA composite. *The International Journal of Advanced Manufacturing Technology* 108: 553–570. https://doi.org/10.1007/s00170-020-05391-x.

46 Hanon, M.M. and Zsidai, L. (2021). Comprehending the role of process parameters and filament color on the structure and tribological performance of 3D printed PLA. *Journal of Materials Research and Technology* 15: 647–660. https://doi.org/10.1016/j.jmrt.2021.08.061.

47 Mazzanti, V., Malagutti, L., and Mollica, F. (2019). FDM 3D printing of polymers containing natural fillers: a review of their mechanical properties. *Polymers* 11: 1094. https://doi.org/10.3390/polym11071094.

48 Ambrus, S., Soporan, R., Kazamer, N. et al. (2021). Characterization and mechanical properties of fused deposited PLA material. *Materials Today: Proceedings* 45: 4356–4363. https://doi.org/10.1016/j.matpr.2021.02.760.

49 Wang, R., Cheng, K.-j., Advincula, R.C., and Chen, Q. (2019). On the thermal processing and mechanical properties of 3D-printed polyether ether ketone. *MRS Communications* 9: 1046–1052. https://doi.org/10.1557/mrc.2019.86.

50 Yang, D., Cao, Y., Zhang, Z. et al. (2021). Effects of crystallinity control on mechanical properties of 3D-printed short-carbon-fiber-reinforced polyether ether ketone composites. *Polymer Testing* 97: 107149. https://doi.org/10.1016/j.polymertesting.2021.107149.

51 Yang, C., Tian, X., Li, D. et al. (2017). Influence of thermal processing conditions in 3D printing on the crystallinity and mechanical properties of PEEK material. *Journal of Materials Processing Technology* 248: 1–7. https://doi.org/10.1016/j.jmatprotec.2017.04.027.

52 Chatham, C.A., Zawaski, C.E., Bobbitt, D.C. et al. (2019). Semi-crystalline polymer blends for material extrusion additive manufacturing printability: a case study with poly(ethylene terephthalate) and polypropylene. *Macromolecular Materials and Engineering* 304: 1800764. https://doi.org/10.1002/mame.201800764.

53 Stoof, D. and Pickering, K.L. (2017). 3D printing of natural fibre reinforced recycled polypropylene. The University of Auckland, p. 668–691. https://hdl.handle.net/10289/11095.

54 Sodeifian, G., Ghaseminejad, S., and Yousefi, A.A. (2019). Preparation of polypropylene/short glass fiber composite as fused deposition modeling (FDM) filament. *Results in Physics* 12: 205–222. https://doi.org/10.1016/j.rinp.2018.11.065.

55 He, Z., Ren, C., Zhang, A., and Bao, J. (2021). Preparation and properties of styrene ethylene butylene styrene/polypropylene thermoplastic elastomer powder for selective laser sintering 3D printing. *Journal of Applied Polymer Science* 138: 50908. https://doi.org/10.1002/app.50908.

56 Kristiawan, R.B., Rusdyanto, B., Imaduddin, F., and Ariawan, D. (2021). Glass powder additive on recycled polypropylene filaments: a sustainable material in 3D printing. *Polymers* 14: 5. https://doi.org/10.3390/polym14010005.

57 Hanon, M.M., Kovács, M., and Zsidai, L. (2019). Tribology behaviour investigation of 3D printed polymers. *International Review of Applied Sciences and Engineering* 10: 173–181. https://doi.org/10.1556/1848.2019.0021.

58 Norani, M.N.M., Abdollah, M.F.B., Abdullah, M.I.H.C. et al. (2020). Correlation of tribo-mechanical properties of internal geometry structures of fused filament fabrication 3D-printed acrylonitrile butadiene styrene. *Industrial Lubrication and Tribology* https://doi.org/10.1108/ILT-04-2020-0143.

59 Kichloo, A.F., Raina, A., Haq, M.I.U., and Wani, M.S. (2022). Impact of carbon fiber reinforcement on mechanical and tribological behavior of 3D-printed polyethylene terephthalate glycol polymer composites – an experimental investigation. *Journal of Materials Engineering and Performance* 31: 1021–1038. https://doi.org/10.1007/s11665-021-06262-6.

60 Aziz, R., Haq, M.I.U., and Raina, A. (2020). Effect of surface texturing on friction behaviour of 3D printed polylactic acid (PLA). *Polymer Testing* 85: 106434. https://doi.org/10.1016/j.polymertesting.2020.106434.

61 Panin, S., Nguyen, D.A., Kornienko, L., and Ivanova, L.. (2019). Multicomponent antifriction composites based on polyetheretherketone (PEEK) matrix. AIP Publishing LLC, p. 020267. https://doi.org/10.1063/1.5132134.

10

Crystallization from Anisotropic Polymer Melts

Daniel P. da Silva[1], James J. Holt[2], Supatra Pratumshat[3], Paula Pascoal-Faria[1], Artur Mateus[1], and Geoffrey R. Mitchell[1]

[1] Centre for Rapid and Sustainable Product Development, Polytechnic of Leiria, Rua de Portugal, Marinha Grande 2430-080, Portugal
[2] University of Reading, Department of Physics, Whiteknights, Reading RG6 6AF, UK
[3] Naresuan University, Department of Chemistry, Faculty of Science, 99 Moo 9 Phitsanulok-Nakhonsawan Road, Tambon Tapho, Muang, Phitsanulok 65000, Thailand

10.1 Introduction

The inherent chemical connectivity of polymer molecules develops a local anisotropy, and the general challenge in polymer science is to transform this local anisotropy to the global scale to influence and transform the properties of objects fabricated from that polymer. The natural configuration of polymer chains in the molten state is a random coil [1], and developing a material with extended chains over the full length of the molecule requires very specific processing. In some cases, semi-rigid molecules, which exhibit a liquid crystal phase over a limited temperature can be processed into materials with a high level of preferred orientation of the chain axes as a consequence of the long-range orientational order exhibited by such materials [2]. In order to reduce the processing temperatures of such materials, these are typically random copolymers of a variety of aromatic esters. Despite the random nature of the crystals, it has been shown that a small level of crystallinity develops through the formation of nonperiodic crystals [3]. This chapter is focused on the formation of crystals in anisotropic polymer melts, in which the crystals are generally but not exclusively chain-folded lamellar crystals. We show that the templating of crystals using anisotropy present in the polymer melt is a powerful control process for defining the morphology and properties of polymer products. We explore various material systems and polymer processing technologies to emphasize the generality of the templating process. We start with the deformation of a network to induce a critical level of anisotropy and then explore the properties of polymer melts using shear flow to impart anisotropy. We extend the concept of anisotropy to include self-assembled nanoscale networks and engineered nanoparticles such as nanotubes. We conclude by applying this approach to the emerging field of 3D

Polymer Crystallization: Methods, Characterization, and Applications, First Edition.
Edited by Jyotishkumar Parameswaranpillai, Jenny Jacob, Senthilkumar Krishnasamy, Aswathy Jayakumar, and Nishar Hameed.
© 2023 WILEY-VCH GmbH. Published 2023 by WILEY-VCH GmbH.

printing and developing the concept of morphology mapping as an extension to 3D printing.

10.2 Evaluating Anisotropy

In this work, we will evaluate the level of preferred anisotropy present in the melt phase and how this is translated to the preferred orientation present in the semi-crystalline morphology. There are many techniques, which can be employed to quantitatively evaluate the level of preferred orientation in a polymeric material. Key to a useful analysis is the identification of the structural unit on which the measurements are based, the identification of the axes, which are used to define the preferred orientation, and the scale over which the evaluation is made [4]. Polymers provide rich hierarchical structures from bonds to chains to crystals, and it is critical that we are able to define what is being measured. We have found that X-ray and neutron scattering techniques provide a powerful framework for the evaluation of preferred orientation over a range of length scales [5], and in this chapter, we will focus on their use. It is also possible to use such techniques for in-situ measurements, with the nature of the time-resolution varying between X-ray and neutron scattering. X-ray and neutron scattering are rigorous techniques [6–8], which can be performed in a quantitative manner. We will now present an overview of a framework for evaluating the level of preferred anisotropy from scattering patterns, further details are available in the literature cited. The common symmetry of deformed samples in a polymer context is a uniaxial symmetry and so we follow this approach here.

The scattering for a sample exhibiting a partial level of preferred orientation can be written as the convolution of the scattering for a perfectly aligned system $I^0(|Q|, \alpha)$ with the orientation distribution function $D(\alpha)$ [9, 10]:

$$I(|\underline{Q}|, \alpha) = I^0(|\underline{Q}|, \alpha) * D(\alpha) \tag{10.1}$$

The function $D(\alpha)$ describes the distribution of the structural units with respect to the symmetry axis of the sample.

If we write the intensity functions and the orientation distribution function in terms of a series of spherical harmonics, $I_{2n}(|Q|)$, $I^0{}_{2n}(|Q|)$, and D_{2n}, we can write this convolution as [10]:

$$I_{2n}(|\underline{Q}|) = \left\{\frac{2\pi}{(4n+1)}\right\} D_{2n} I^0_{2n}(|\underline{Q}|, \alpha) \tag{10.2}$$

where $n = 0, 1, 2, 3...\infty$. Only the even terms of each series are required due to the inversion center intrinsic to a X-ray scattering pattern for a weakly absorbing sample.

The components of each series can be obtained by:

$$I_{2n}(|\underline{Q}|) = (4n+1) \int_0^{\pi/2} I(|\underline{Q}|, \alpha) P_{2n}(\cos \alpha) \sin \alpha . d\alpha \tag{10.3}$$

and related expressions. The complete scattering pattern and equivalent functions may be recovered by:

$$I(|\underline{Q}|, \alpha) = \sum_{2n=0}^{2n=\infty} I_{2n}(|\underline{Q}|) P_{2n}(\cos \alpha) \tag{10.4}$$

10.2 Evaluating Anisotropy

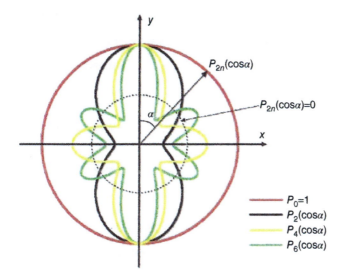

Figure 10.1 Plots in polar coordinates of the values of the first four Legendre Polynomials.

We can see that one of the advantages of using scattering techniques is that we can address all components of the orientation distribution function, whereas, the measurement of birefringence or absorption only yields values for $<P_2>$. Here we use angular brackets to indicate an average over the volume of the sample evaluated through the area of the beam and the thickness of the sample. Figure 10.1 shows a plot of the first four Legendre Polynomials in polar coordinates.

In Figure 10.2, we show the intensity recorded for a sample of isotactic polypropylene at a particular value of Q plotted in polar coordinates and we can see more clearly the process involved here. Essentially, we are identifying the fraction of the shape of the intensity in polar coordinates of each particular function shown in Figure 10.1 and then scaling that with the scattering for a perfectly aligned system. Key to the usefulness of this approach is the orthogonal nature of the Legendre Polynomials. The process is similar to the use of Fourier Components to describe peak shapes, and the convolution is the equivalent of the Stokes Theorem.

It is more convenient to work with normalized harmonics rather than the unnormalized versions inherent in Eq. (10.3). The normalized versions of the global orientation parameters $<P_{2n}>$ can be obtained using the equation [10, 11]

$$< P_{2n}^a \cos(\alpha) > = \frac{I_{2n}^a(|\underline{Q}|)}{I_0^a(|\underline{Q}|)(4n+1)P_{2n}^m(\cos \alpha)} \tag{10.5}$$

These parameters have values, which lie between −0.5 and 1, with a value of 0.0 representing an isotropic arrangement. A value of 1 corresponds to a situation where all structural elements are aligned with a common axis. A value of −0.5 corresponds to a situation where the structural units are aligned orthogonal to the common axis. In some literature $<P_2>$ is referred to as the Hermans Orientation Function, but these are essentially equivalent.

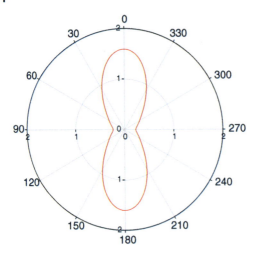

Figure 10.2 A polar plot of the intensity of the SAXS intensity for a sample of isotactic polypropylene crystallized from a sheared melt at a particular value of Q.

One final comment about this approach is that we will need to ensure that the scattering, which is used to evaluate the preferred orientation, only includes the scattering from the identified structure, such as crystals. This is usually easily achieved by examination of the Q dependence. This framework provides many advantages, which include the identification of multiple components in mixtures [10] and the deconvolution of orientation measured in poorly aligned fibers [12].

10.3 Crystallization During Deformation of Networks

Research on strain-induced crystallization in natural rubber is as old as polymer science itself [13]. The very first reference to crystallinity in strained natural rubber, involved the recently developed technique of X-ray diffraction and was due to Katz [14], and the work played a part in convincing skeptical scientists to accept Staudinger's concept of long-chain molecules. Katz showed that the crystals, which form have a high preferential alignment. Since that early start, numerous researchers have revisited the topic, bringing new methods and quantitative analysis techniques. Amongst early pioneers in this area was the work of Mitchell [15], who used quantitative static testing to establish the key features. In that work, the scattering associated with the crystals was separated from the amorphous scattering and used to estimate the level of crystallinity. The azimuthal variation of the amorphous interchain scattering was used to evaluate the level of preferred orientation present in the amorphous material. The experimental data exhibited an almost linear variation with strain until $\lambda = 4.5$, after which it exhibited a sharp upturn, eventually reaching a value for $<P_2>$ of ~0.08 at $\lambda = 7$. The lateral width of the almost perfectly aligned crystals was estimated to be 50 Å.

Figure 10.3 shows data from a recent study using the NCD-SWEET SAXS/WAXS beamline at the ALBA Synchrotron Light Source in Barcelona to record data during the uniaxial deformation of natural rubber [16]. Here we superimposed WAXS patterns taken at specific strains during the loading and unloading parts of the

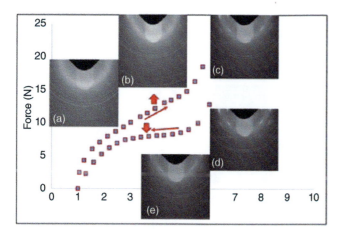

Figure 10.3 A summary of a uniaxial deformation experiment with WAXS patterns recorded at specific strain points during the loading and unloading superimposed on the plot for the force-strain curve. WAXS patterns were recorded at (a) $\lambda = 1.0$, (b) $\lambda = 4.3$, (c) $\lambda = 5.75$, (d) $\lambda = 6.0$, (e) $\lambda = 4.61$. The vertical red up arrow marks the strain at which crystalline spots were first observed and the vertical red down arrow marks the strain at which all crystalline spots were no longer observed on unloading when $\lambda < 4.0$. The WAXS patterns also show some crystalline diffraction rings, which are associated with compounds added to enable cross-linking.

cycle. The well-defined arced spots can be seen once the extension ratio λ exceeds 4.2. Upon unloading, the crystalline spots decrease in intensity and eventually disappear when $\lambda < 4.0$. In these experiments SAXS patterns were recorded and exhibit an increasing level of anisotropy and an example is shown in Figure 10.4. This scattering was located around the zero-angle point and was typical of isolated anisotropic systems. The scattering did not show any of the characteristics of stacks of chain-folded lamellar crystals. The key feature we can note from this work is the near-perfect orientation of the crystals. When the network is deformed, some shorter strands become fully aligned before the remaining chains and these chains nucleate the crystals. Small-angle X-ray scattering data were obtained and undertaken as part of the same series of experiments on natural rubber showing only evidence of isolated crystals, and there is no evidence for stacks of crystals as typically form in a strained uncross-linked melt. Figure 10.4 shows one of the SAXS patterns corresponding to Image 1c in Figure 10.3.

The study of strain-induced crystallization in natural rubber is complicated by the possibility of thermal crystallization of the rubber. Kohjiya [17] has proposed that strain-induced crystallization should be renamed as "Template Crystallization" to emphasize the unique characteristics of the process. The process occurs readily in natural rubber due to the very high stereoregularity of the polymer chains at 99.98%, a characteristic, which yet to be exhibited by synthetic cis 1,4-polyisoprene. The contribution of X-ray diffraction to the study of strain-induced crystallization of natural rubber has been comprehensively reviewed by Huneau [18] and it deals with the effects of strain rate variation and the influence of temperature on the overall process. Sotta et al. [19] have revisited Flory's theory of strain-induced crystallization

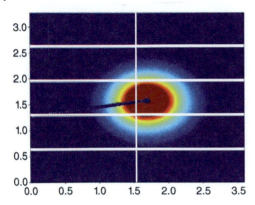

Figure 10.4 SAXS pattern for a sample deformed with 4.5.

of rubber using recently obtained data, and they conclude that this provides a good description of the process, which is driven by the relaxation of the "amorphous" part of an extended network chain. There is now a comprehensive set of new data and an excellent opportunity to move forward with this topic. Numerous authors emphasize the enhanced sensitivity of theory to measurements made under biaxial deformation and we have already made a start in this direction [20].

10.4 Sheared Polymer Melts

The rheology of the typical polydisperse polymer melt used to fabricate plastic parts remains a challenging area, but excellent progress has been made with quantitative analysis [21]. However, the basic concepts have been established for some time. In the flow conditions present during manufacturing, the longest chains become extended, allowing the majority of the chains to be in a relaxed state. As the molten material cools, if the crystallization point is reached with the long chains still extended, then these serve as row nuclei and template the crystallization so that the chain-folded lamellar crystals grow out normal to the row nuclei. The row nuclei all have a common axis of alignment and as a consequence, the initial crystals are almost perfectly aligned. Clearly, it would be most helpful to observe these very early stages of nucleation using SAXS/WAXS measurements. Pan et al. [22] have claimed to observe the SAXS of the so-called shish nuclei. These are very challenging experiments as the contrast between the nuclei and the melt is very low. We have adopted a different approach in which we make use of small-angle neutron scattering of deuterium-labeled mixtures of polyethylene, using a parallel plate shear cell to impose some anisotropy on the melt [23].

Figure 10.5 shows a schematic of the parallel plate shear cell, which was previously used for SAXS/WAXS measurements using thin mica sheets as windows [23]. One of the advantages of designing sample stages for SAXS/WAXS measurements is that the beam size is relatively small <0.5 mm, in contrast, the weaker interaction of neutrons with matter results in a larger beam to increase the signal size. For this work, we utilized the LOQ spectrometer at the UK Pulsed Neutron Facility ISIS [24] and we

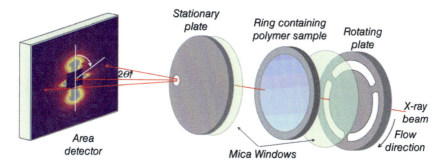

Figure 10.5 A schematic of the shear cell used for in-situ X-ray and neutron scattering studies.

Figure 10.6 The shear cell on the LOQ Instrument at ISIS in the UK. The incident neutron beam enters from the right, and the window to the detector is visible on the right.

used a beam size of 3 mm (Figure 10.6). To accommodate this beam size, we modified the original design of the shear cell and widened the windows, and replaced the lead rotating mask with one fabricated from cadmium to reduce any background scattering as the spokes passed through the incident beam. We continued to use mica as the window material. To link to existing SAXS/WAXS experiments [25] we used a commercial PE resin, which exhibited a polydisperse molecular weight with a high molecular weight fraction and formed shish-kebab or row-nucleated structures [26].

To provide contrast, we mixed this PE with a deuterated branched PE formed by deuterating a copolymer of polybutadiene. Table 10.1 shows the characteristics of the materials used in this work. We prepared two mixtures, the first for the SANS experiments was 10% of A and 90% B, and for the WAXS was 10% of A and 90% of C. In Figure 10.7, we compare the SANS patterns recorded for this mixture in the quiescent state and subject to shear. There are differences, but it is evident that the SANS pattern recorded during shear flow does not exhibit a highly anisotropic state. To analyze the complete pattern we calculate a series of spherical harmonics, which

Table 10.1 Characteristics of sample used for SANS experiments.

PE sample		F	$10^5 \, M_w$	M_w/M_n	SCB	Catalyst
A	HPE	0.10	3.45	9.6	None	Zg
B	DPE	0.90	1.80	1.06	<1.7% C_2	—
C	BPE	0.10	1.60	1.03	<1.2% C_2	—

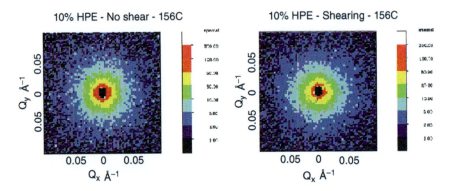

Figure 10.7 SANS patterns.

collectively represent the pattern but serve to separate the orientational behavior from the structurally dependent components (Eq. (10.3)).

Taking the approach described in Section 10.2 for the pattern shown in Figure 10.7b yields the plot shown in Figure 10.8.

And although the data is a little noisy due to the low level of anisotropy, we can observe that the harmonics $I_2(Q)$ and $I_4(Q)$ have a more or less constant value, which is independent of Q. This contrasts with the calculated harmonics of the scattering for an anisotropic Gaussian chain shown in Figure 10.9 with Rg parallel of 110 Å and Rg perpendicular of 100 Å.

The strong Q dependence of these calculated harmonics (Figure 10.9a) contrasts strongly with the observed harmonics.

A possible alternative model takes isotropic Gaussian chains with Rg of 100 Å together with a small fraction of highly anisotropic Gaussian chains, the latter with a ratio of Rg parallel/Rg perpendicular of 10, and the calculated harmonics are shown in Figure 10.9b.

This reproduces the essential features of the experimental harmonic functions by exhibiting a constant level of the harmonic as a function of Q. Of course, if we were able to record data at very low Q values then there would be a rise in the values and the development of a Q dependence, but that scattering region is not experimentally accessible on LOQ. The value of the constant value of the harmonics is directly related to the fraction of anisotropic Gaussian chains, as Figure 10.10 shows.

This suggests the molten polymer contains a small fraction of highly extended chains within a matrix of isotropic chains. This result is supported by orientation measurements made using WAXS [26].

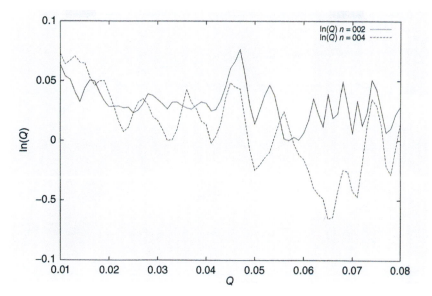

Figure 10.8 Plots of the harmonics $I_2(Q)$ and $I_4(Q)$ for the sheared sample (Figure 10.7b).

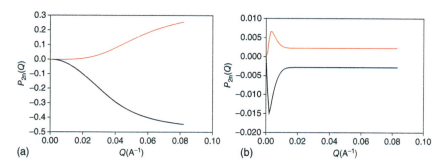

Figure 10.9 Plots of the calculated first two harmonics I_2 and I_4 for (a) a model of anisotropic Gaussian Chains with Rgparallel = 110 Å and Rgperpendicular = 100 Å and (b) a model containing 90% of isotropic Gaussian chains with Rg = 100 Å with 10% of anisotropic Gaussian chains with an anisotropy ratio of 10 in the radius of gyration.

We have explored the consequences of shear flow on the crystallization behavior of an equivalent undeuterated system. We have prepared mixture of the BPE and the high molecular weight linear HPE (A + C) as shown in Table 10.1. In order to fully appreciate the behavior of the mixture, we need to examine the properties of the components.

Figure 10.11 shows WAXS patterns for samples of BPE(a) and HPE(b) sheared at 150 °C at a shear rate of 10 s^{-1} for shear strain of 100 shear units, and then the shear was halted and the sample cooled quickly to room temperature. The right-hand curves are azimuthal sections at the peak positions of the (110) and the (200) reflections. As can be seen from the patterns, the BPE exhibits no anisotropy in the crystal peaks and similar results were obtained for shear rates up to 30 s^{-1} [25]. In contrast, the HPE showed a significant level of preferred orientation, in which the

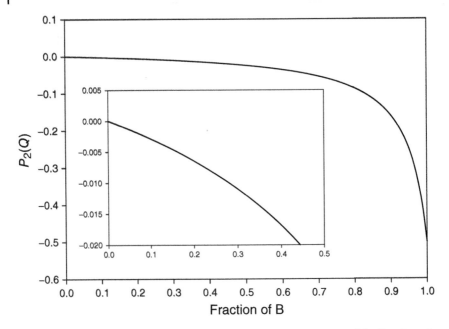

Figure 10.10 A plot of the value of the harmonic <P_2> as a function of the Fraction of highly anisotropic chains.

azimuthal position of the maximum in the crystal reflections was dependent on the shear rate [26].

Figure 10.12 is a transmission electron micrograph of a differentially etched internal surface of the sample used for Figure 10.11. It exhibits a very high level of preferred orientation. There are features running from top to bottom, which reflect the row nuclei or shish, which have directed or templated the remaining structure. We can also observe the chain-folded lamellar crystals, largely edge-on running across the micrograph. This underlines the massive amplification that takes place with respect to the level of preferred orientation upon crystallization from a deformed melt. It is noteworthy to emphasize that in this mixture there is a high molecular weight fraction, which is very responsive to the applied shear flow and the BPE whose crystallization is not influenced by the shear flow. What we find here is that all these molecular species have been templated in to a high level of preferred orientation.

10.5 Crystallization During Injection Molding

Injection molding is the most widely used technology for the fabrication of plastic parts. In this technique, molten plastic is injected at high pressure into a metal mold with a cavity, which replicates the shape of the desired product. The mold is maintained at a much lower temperature than the molten plastic so that the plastic solidifies through a glass transition or through crystallization in the case of a

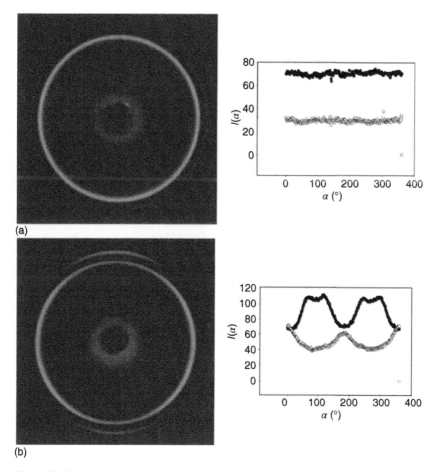

Figure 10.11 WAXS Patterns of a mixture of PE sheared as in Figure 10.7 and cooled rapidly to room temperature. The plots show the azimuthal sections for the 110 and 200 peaks.

crystallizable polymer. In fact, semi-crystalline polymers in the form of isotactic polypropylene and polyethylene are amongst the most widely used materials. Injection molding has evolved into a high-technology process, and now molds are digitally designed and the injection molding process is fully tested and explored in the virtual sense within simulations. The structure and morphology of injection molded products have been widely studied [27–29]. This shows that the morphology develops in stages, and it would be very useful to be able to follow these developments in real-time. The flow rates and the cooling/heating rates are considerable and greatly different from those available with the shear cell used in the previous section. Recently, two groups have taken on the challenge of developing experimental techniques, which allow the structure and morphology to be evaluated in real-time using X-ray scattering techniques. These two groups have followed different pathways, one involving the current authors have developed an industrially relevant injection

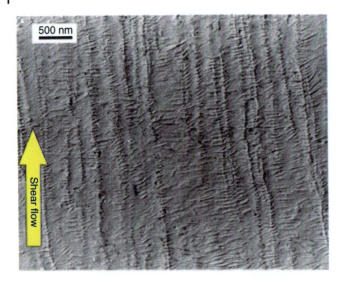

Figure 10.12 A TEM micrograph of a replica of a differentially etched internal surface of the sample (Figure 10.9).

molding system, which can be mounted on the NCD-SWEET ALBA Synchrotron Light Source [30, 31]. Figure 10.13 shows a schematic of the equipment, which is based on traditional metal mold design procedures (Figure 10.14).

Liao et al. and Zhao et al. have followed a different approach, which is focused on micro-injection molding and uses diamonds as the X-ray transparent windows in the mold cavity [32, 33]. Both of these approaches offer considerable promise for the future understanding and development of crystallization in injection molding of plastics.

10.6 Sheared Polymer Melts with Nucleating Agents

In Sections 10.4 and 10.5, we reviewed results, which underline the massive amplification available by templating crystallization using highly aligned chains as row nuclei. The success of this approach encouraged us to explore alternative approaches using nanoparticles embedded in the polymer melt. There is the possibility of adding preformed nanoparticles to the melt, and these are considered in Section 10.7. In this section, we focus on the formation of nanoparticles through self-assembly in the polymer melt, thereby sidestepping the challenges of dispersing the nanoparticles in the polymer melt. There are a number of nucleating agents, which are based on small-molecules, which self-assemble into fibrillar extended particles. One of the most widely studied of these is dibenzylidene sorbitol (DBS) [34] (Figure 10.15), which is widely used as a clarifying agent with polypropylene to stimulate a high density of nuclei to yield a large number of small crystals with a size smaller than the wavelength of visible light so as to render the final film clear and transparent and suitable for applications typically in packaging [35].

10.6 Sheared Polymer Melts with Nucleating Agents

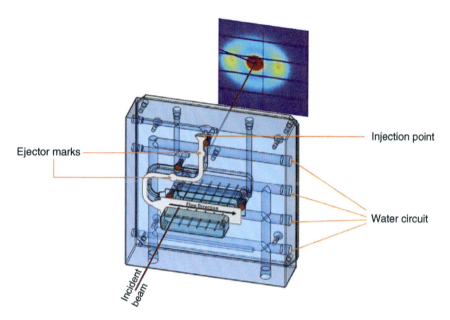

Figure 10.13 A schematic of the industrially relevant injection molding system developed by Mitchell and coworkers. Source: Andre Costa et al. [31]/MDPI.

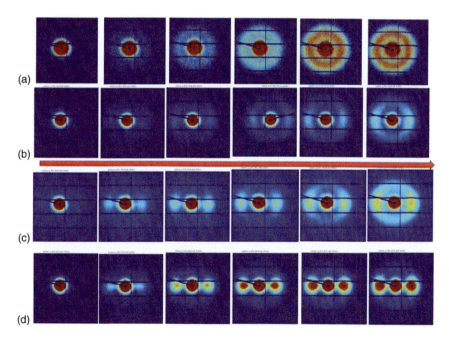

Figure 10.14 SAXS patterns for isotactic polypropylene immediately after injection in the mold. Time moves from left to right on a one second time interval. The four sequences correspond to injection temperatures of (a) 250 °C, (b) 210 °C, (c) 200 °C, and (d) 190 °C. Source: Andre Costa et al. [31]/MDPI.

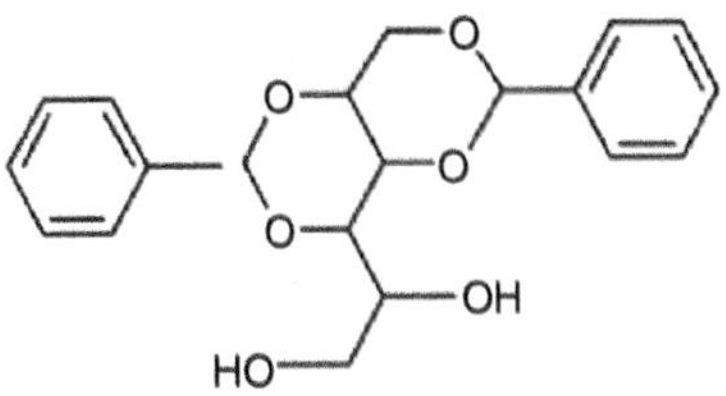

Figure 10.15 The chemical configuration of dibenzylidene sorbitol.

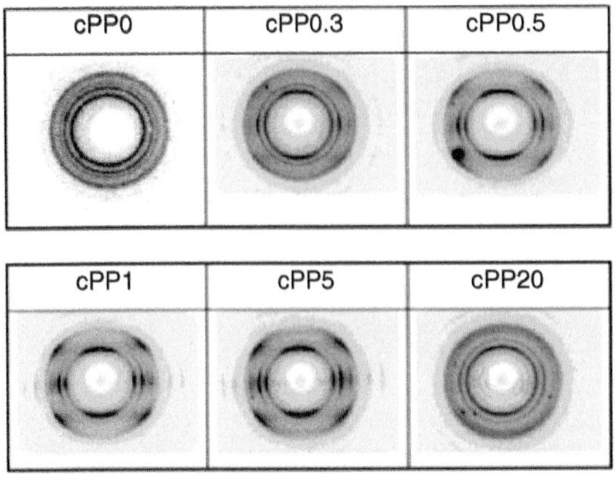

Figure 10.16 WAXS patterns of cPP with differing amounts of DBS after shearing before crystallization.

Here we review the work of Nogales et al. [36, 37], which was based upon a polypropylene copolymer (cPP) with Mw) 2.2×10^5 and Mn) 5.9×10^4 (trade name NOVOLEN 3240NC, BASF plc.), which contains 3.5% ethylene comonomer. They prepared samples of the copolymer with different fractions of DBS in the form of Millad 3905, (Milliken Chemicals) [34]. These were then sheared at 160 °C using the equipment shown in Figure 10.5 at a shear rate of $20\,s^{-1}$ for a shear strain of 120 shear units. We can see directly that the sample without DBS shows little evidence for an anisotropic structure, but increasing amounts up to 5% show increasing levels of preferred orientation, as can be evidenced by the arcing of the X-ray patterns (Figure 10.16). When 20% of DBS is added, there is strong evidence for large-scale segregation and a change in the fibrillar morphology.

Subsequently, Nogales et al. also explored the effects of DBS in polyethylene, for which it is widely reported that DBS did not function as a clarifying agent, and with poly(ε-caprolactone). Mitchell et al. [38] used SANS to clarify the phase diagram of DBS and PE. They showed that at high temperatures, the DBS was completely soluble in the PE matrix, but on cooling, the DBS phase separated through crystallization [38].

They used a linear polyethylene GX 555-2 (from Hoechst, $M_n = 24\,000$, $M_w = 171\,000$ as determined by Rapra Technology, Shawbury, Shrewsbury, Shropshire, United Kingdom, SY4 4NR). The polymer is understood to be produced by a metallocene catalyst. The additive, DBS (trade name Millad 3905, Milliken Chemicals) [39], was mixed with the polyethylene, 1% by weight, by a solution/precipitation method, involving the initial dissolution of both components in hot xylene, followed by precipitation in an excess of isopentane chilled to near its freezing point with liquid nitrogen. Samples were subjected to shearing in the melt

10.6 Sheared Polymer Melts with Nucleating Agents

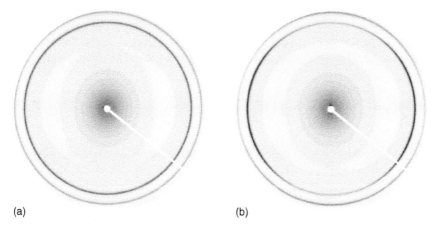

(a) (b)

Figure 10.17 Wide angle X-ray patterns of polyethylene specimens sheared in the melt and subsequently crystallized; (a) without and (b) with added sorbitol-based nucleating agent. The flow direction during the shear flow stage is vertical. These distinct crystalline features correspond to the 110 and 200 Bragg reflections. *Note*: the outer 200 reflection has been slightly cropped at upper left by the detector, so comparisons should be made of the right and bottom parts of the picture.

and then cooling to room temperature. These samples with and without DBS were studied on the A2 beamline at the HASYLAB in Hamburg. Figure 10.17 shows the WAXS patterns.

Figure 10.18 shows a TEM micrograph obtained from a replica of a differentially etched internal surface [5] for a sample of polyethylene with 1% of DBS. The micrographs show several troughs, which represent where the DBS was before etching. In Figure 10.18, we can see the actual ends of DBS fibrils, the flat end may be the result of breakage during shear flow. In the area alongside the DBS trough, the polyethylene lamellae run out normal to the DBS fibril surface. In contrast, the polyethylene lamellae in the region to the left of the ends of the DBS fibrils have

Figure 10.18 A TEM micrograph of a differentially etched internal surface of a sheared sample of polyethylene with 1% DBS.

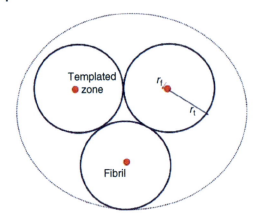

Figure 10.19 The basic elements of the templating model proposed by Wangsoub et al. Source: Adapted from Wangsoub and Mitchell [41].

a different orientation. This clearly demonstrates the templating effect of the DBS fibrils on the lamellae growth.

The influence of the addition of DBS to polypropylene and polyethylene has been extended to polycaprolactone and to the use of some halogenated DBS derivatives. The great contrast between the PCL and the DBS enabled Wangsoub et al. [40] to determine that the DBS fibrils were crystalline and that they exhibited a lower solubility in the PCL, increasing the number density of fibrils and yield with shear, resulting in the highly aligned structure shown.

Wangsoub et al. [40] developed a quantitative model of templating with DBS fibrils (Figure 10.19). In the case of highly aligned fibrils, consideration of the templating process reduces to a 2-d matter. The number density of fibrils is given by

$$N = (f-f_c)/\pi r_f^2 \tag{10.6}$$

where f is the fraction of the DBS present in the polymer and f_c is the upper limit of solubility of the DBS derivative in the polymer matrix. We consider all fibrils to have the same radius r_f. Wangsoub and Mitchell [41] used SAXS to measure the fibril radius and found that under some conditions, the distribution of fibril radii approaches the monodisperse state. If we associate a templated zone of radius r_t with each fibril the fraction of templated material, f_t, is given by:

$$f_t = (f - f_c)\frac{\pi r_t^2}{\pi r_f^2} \tag{10.7}$$

This is a relatively simple model without any distribution of fibril dimensions, and the templated zone size does not account for a random spatial distribution of fibrils. However, we do have an estimate of the fibril radius of ~7.5 nm from the SAXS measurements, and we have estimated for polyethylene that the templated zone has a radius ~100 nm. This suggests that the value of $(f - f_c)$ needs to be greater than 0.014 to achieve full templating. Clearly, the solubility of the sorbitol derivative in the polymer matrix is critical in determining the actual amount of additive required to achieve this. Wangsoub et al. [40] previously estimated that the solubility limit of DBS in PCL was ~0.01. They reported that a chlorine derivative of DBS exhibited a reduced solubility limit in PCL and led to enhanced templating of Cl-DBS.

Figure 10.20 The SAXS pattern recorded for the 3% Cl-DBS/PCL system after shearing at 80 °C at a shear rate of 10^{-1} s This pattern was obtained at the Diamond Light Source beamline I22. Source: Redrawn from Mitchell 2013.

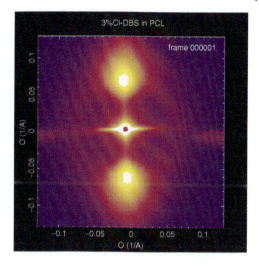

In essence, the level of direction was more or less perfect, as can be observed in Figure 10.20.

The orientation parameter $<P_2>_L$ describing the preferred orientation of the lamellae with respect to the flow axis is given by the fraction of material templated:

$$< P_2>_L = -0.5(f - f_c)\frac{\pi r_t^2}{\pi r_f^2} \qquad (10.8)$$

The factor −0.5 takes account of the fact that the lamellae are arranged normal to the flow axis. Equation (10.8) shows that efficacy of templating or directing process is directly related to the number density of the DBS fibrils. For a given fraction of DBS, the number density can be optimized by making the nanofibrils as thin as is possible. The processes described in this section are effective as a consequence of the lateral scale of the fibrils. Some have proposed that the interaction between the DBS and the polymer matrix can be described as atomic-level epitaxy. From the universality of the process with differing polymers and differing derivatives of sorbitol, we suggest that this may be more related to topology. We note that Mitchell and Olley [42] have reported the possibility of achieving orthogonal templating by using an additional agent to severely limit the length/breadth of the DBS fibril and this opens up new possibilities, which may prove invaluable for some applications.

10.7 Sheared Polymer Melts with Nanoparticles

The development of nanocomposites is a major area of activity in polymers. The goal is to fully disperse small fractions of engineered nanoparticles within a polymeric matrix so as to lead to enhanced properties. For a semi-crystalline polymer, the nanoparticles can only be dispersed within the amorphous phase. Carbon nanotubes, Halloysites, and other mineral particles as well as graphene nanoflakes have been employed. Avalos-Belmontes and coworkers [43] have reviewed the nucleating

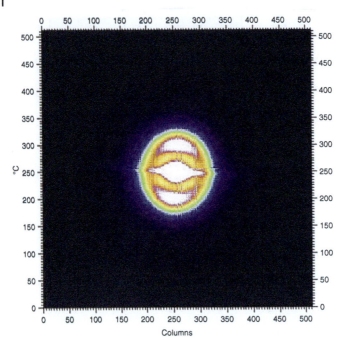

Figure 10.21 SAXS Pattern recorded for a sample of PCL/graphene sheared for 1000 shear units at 1/s at 100 °C and cooled to room temperature. Source: Ana Tojeira and Mitchell [44]/Springer Nature.

behavior of CNTs in isotactic polypropylene. Ana Tojeira and Mitchell [44] have shown that a modest shear rate of $1\,s^{-1}$ is sufficient to induce a preferred orientation of 14 nm graphene nanoplatelets, which then template the crystallization of the PCL matrix. Figure 10.21 shows the SAXS taken during shear flow.

10.8 3D Printing Using Extrusion

3D printing is one of the emerging technologies under the additive manufacturing umbrella, which forms part of the fourth Industrial revolution I 4.0 [45]. In this concept, products are manufactured from a digital definition such as a computer aided design file without the use of specialist tooling such as molds or dies. For that reason, it is sometimes referred to as direct digital manufacturing (DDM), which is a more precise description of the process. It involves a variety of technologies such as stereolithography, fused deposition modeling, and selective laser melting, and addresses both plastics, ceramics, and metals. The basic concept of fused deposition modeling in which a preformed thermoplastic filament is fed through rollers into hot extruder, and the plastic appears as a molten strand as it emerges from the extruder and is deposited where required on a build platform and the process is repeated layer by layer. Figure 10.22 (left) shows a schematic of the Fused deposition modeling.

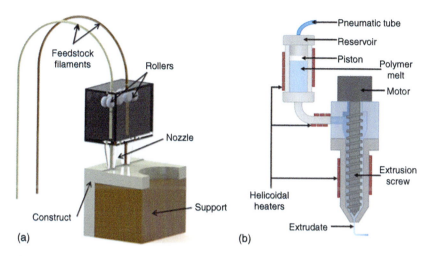

Figure 10.22 (a) A schematic of the fused deposition modeling 3D printing process (b) a schematic of the fused granular deposition 3D printing. Source: da Silva et al. [46]/MDPI/Public Domain CC BY 4.0.

We will now describe some work, which is focused on 3D printing but uses a pellet-fed extruder system, as shown schematically in Figure 10.22 (right). This offers several distinct advantages over the filament-based methodology. First and foremost, it is not necessary to prepare filament of each material, a process, which may involve handling a kilogram or more of material. The fused granular printing involves working with small pellets or fragments of polymers and is hence much more suited to materials development, where the process of producing filaments may be quite severe. Secondly, it allows great control and definition of the extrusion process, including both the extrusion die and the extrusion pressure. In some cases, we have employed an industrial scale extruder, which facilitates the addition of a sensor to provide feedback on the extrusion pressure and enables greater pressures and temperatures to be used.

Of course in the end it is a polymer processing system and all of the lessons learned in the earlier sections will and must apply. Tojeira and Mitchell [44] were the first to realize that the extrusion process could under certain conditions lead to the deposition of material on the build platform, which exhibited a high level of preferred orientation of the chain-folded lamellar crystals, which form on cooling following the printing process [47]. This led to the development of a 3D printer, which could be mounted on the NCD-SWEET ALBA Synchrotron Light Source SAXS/WAXS beamline to evaluate the development of the morphology of the polymer during the 3D printing process; this is described in Section 10.8.1.

10.8.1 In-Situ Studies of Polymer Crystallization During 3D Printing

Figure 10.23 shows the equipment specifically designed for the in-situ 3D printing of thermoplastics on the NCD-SWEET beamline at the ALBA Synchrotron Light Source in Barcelona, Spain. This equipment is described in detail elsewhere [46],

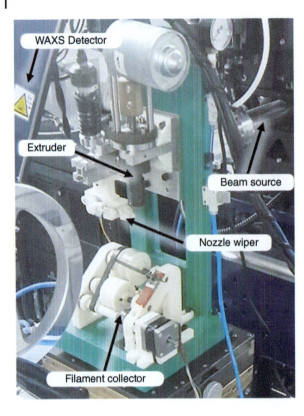

Figure 10.23 The 3D Fused Granular Deposition Printer mounted on the NCD-SWEET Beam-line at ALBA Synchrotron Light. Source: da Silva et al. [46]/MDPI.

but in essence, it provides the opportunity to follow the progression of development of the semi-crystalline morphology in the extruded filament as it emerges from the extrusion die and cools in a quasi-static manner.

Now, although the extruded filament is moving, the environment is more or less fixed. As a consequence, we can move the beam down the extruded filament and this in effect represents the time evolution. In Figure 10.24, we show a set of SAXS patterns recorded at ALBA using the equipment shown for a sample of PCL extruded at 170 °C and with a fast write speed. The write speed is the speed of relative displacement of the print head with respect to the build platform. The pattern on the extreme left corresponds to the point closest to the extruder die, and the pattern is

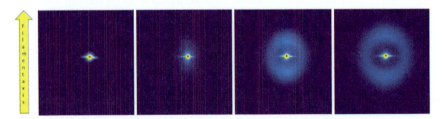

Figure 10.24 SAXS patterns recorded in-situ from an extruded filament of PCL. The position of the beam is closest to the extruder on the left moving in 0.1 mm increments to the left. Source: da Silva et al. [46]/MDPI.

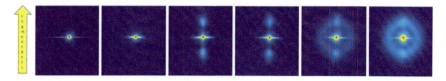

Figure 10.25 SAXS patterns recorded in-situ from an extruded filament of PCL. The position of the beam is closest to the extruder on the left moving in 0.1 mm increments to the left. The print parameters are the same as used in Figure 10.19, but the write speed is fast. Source: da Silva et al. [46]/MDPI.

featureless as the polymer density has no structure or variations in electron density on the scale probed. The next pattern to the right shows weak scattering corresponding to the early stages of crystallization and it appears to be highly anisotropic, but as we move on to successive patterns to the right, a more isotropic structure emerges and becomes the dominant feature. In the final pattern on the right-hand side, the scattering is more or less completely isotropic, and the initial highly anisotropic scattering is submerged in the mass of isotropic scattering.

Figure 10.25 shows a second set of patterns obtained using the same procedures as above and with PCL but with a faster write speed, all other conditions being the same. The sequence of patterns is very different from those shown in Figure 10.24.

The sequence starts the same, although the horizontal streak is more extended and probably relates to nanovoids in the extrudate. The next three patterns to the right show a development of a highly anisotropic structure, which is typical of a highly aligned stack of chain-folded lamellar arranged so that the growth of the lamellar is normal to the extrusion axis. As we have observed in Sections 10.4 and 10.5, this is very typical of row-nucleated structures, where nucleation provides a common global axis. The last two patterns to the right show the development of an isotropic crystalline morphology, which overlays the initial anisotropic structure. These two different morphologies exhibit different mechanical properties by at least a factor of 2–3× [47]. We can quantify the development of anisotropy in the crystalline morphology using the SAXS patterns as described in Section 10.2, and the results are shown in Figure 10.26. The advantage of performing time-resolved in-situ measurements can be clearly seen in these plots.

Figure 10.27 (left) shows three SAXS patterns from the 3D printing of LDPE with increasing write speed from left to right. The fastest write speed provides the highest level of preferred orientation as observed with PCL. Figure 10.27 (right) shows a 3D plot of the global orientational parameter $<P_2>$ as a function of the extrusion speed and write speed for printing LDPE.

10.9 Morphology Mapping

Silva et al. [46] have introduced the term "Morphology Mapping" to describe the possibility of depositing using fused granular deposition 3D printing, polymeric materials in different parts of the product with differing properties. The different properties

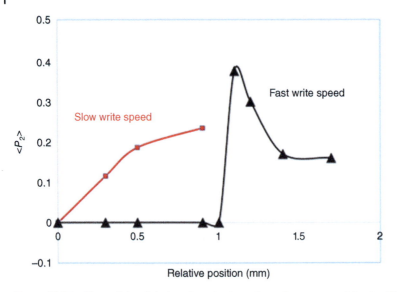

Figure 10.26 Plots of the global preferred orientation <P_2> measured for the SAXS patterns shown in Figures 10.25 and 10.26. Source: da Silva et al. [46]/MDPI/Public Domain CC BY 4.0.

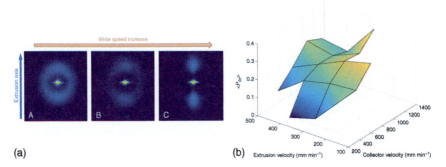

Figure 10.27 (a) SAXS Patterns measured during 3D printing with an increasing write speed moving left to right. (b) A plot of the global orientation parameter <P_2> for the normals to the chain-folded lamellar crystals evaluated using the patterns shown in Figures 10.24 and 10.25 and other data (left) and related patterns as a function of the extrusion and write speed. Source: da Silva et al. [46]/MDPI.

are defined by the print parameters selected at the point of deposition, which can be varied more or less instantaneously during printing. This arises directly from the transfer of the directing anisotropy imposed in the melt phase to the semi-crystalline morphology.

10.10 Discussion

In this chapter, we have demonstrated the powerful link between a level of anisotropy in the melt phase and the subsequent morphology of the semi-crystalline state. As observed, crystallization is directed by this anisotropy in the melt to yield a high level of common crystal orientation. Without such direction or templating, the crystals adopt an isotropic morphology. There is a clear commonality evident in these different processes. The process of strain-induced crystallization in natural rubber appears to generate the highest level of common crystal orientation. We attribute this to bother the highly extended state of the nucleating chains and that crystallization occurs through the templated crystallization of long crystalline chains as chain folding does not take place in the conditions studied. The systems, which contain DBS and derivatives dispersed in the polymer matrix also show very high common alignment for both the chain-folded lamellar crystals and the highly acicular nanoparticles that serve as row nuclei. The high aspect ratio (length to breadth ratio) of the nanoparticles will lead to high common alignment with minimum shear rate and applied shear strain. The high common alignment arises through the directing surface of the DBS nanoparticles, and we deduce that through symmetry of that surface, the lamellar crystals grow normal to the nanoparticle. Similar considerations apply to the system with graphene nanoplatelets. The generation of row nuclei in a polymer melt is a greater challenge, not least as the random coil configuration has to be unraveled in contrast to the rigid DBS nanoparticles, and there are greater opportunities for relaxation again in contrast to the DBS nanoparticles. As a consequence, the resultant high level of anisotropy present in the semi-crystalline polymer morphology is lower and similar arguments apply to the work on 3D printing. The development of morphology mapping as part of 3D printing of semi-crystalline polymers is an exciting and rapidly growing area.

Acknowledgments

This work was supported by the Fundação para a Ciência e Tecnologia (FCT) through the project references: MIT-EXPL/TDI/0044/2021, UID/Multi/04044/2013; PAMI-ROTEIRO/0328/2013 (N° 022158), Add.Additive – add additive manufacturing to Portuguese industry POCI-01-0247-FEDER-024533 and UC4EP PTDC/CTM-POL/7133/2014).

We acknowledge the critical role of large-scale facilities in this work, including the ALBA Synchrotron Light Source in Barcelona, the Daresbury Synchrotron Radiation Source in the UK, the ESRF in France, DESY in Germany, Elettra in Italy, and the STFC ISIS Pulsed Neutron Source in the UK, and we thank the scientists involved with each beamline for their involvement in this work. The experiments on 3D printing were performed at the NCD-SWEET beamline at ALBA Synchrotron with the collaboration of ALBA staff. We thank EPSRC and BP Chemicals for CASE Award for JJH (University of Reading), Naresuan University for a PhD scholarship for SP, and the Faculty of Science at Naresuan for funding a scientific visit.

References

1 de Gennes, P.G. (1979). *Scaling Concepts in Polymer Physics*. Ithaca: Cornell University Press.

2 Donald, A.M., Windle, A.H., and Hanna, S. (2006). *Liquid Crystalline Polymers*. Cambridge, UK: Cambridge University Press. ISBN: ISBN 9780521580014.

3 Spontak, R.J. and Windle, A.H. (1992). Crystallite morphology in thermotropic random copolymers: application of transmission electron microscopy. *Journal of Polymer Science Part B: Polymer Physics* 30: 61–69. https://doi.org/10.1002/polb.1992.090300106.

4 Mitchell, G.R. and Windle, A.H. (1988). Orientation in liquid crystal polymers. In: *Developments in Crystalline Polymers* (ed. D.C. Bassett). Dordrecht: Springer https://doi.org/10.1007/978-94-009-1341-7_3.

5 Mohan, S.D., Olley, R.H., Vaughan, A.S., and Mitchell, G.R. (2016). Evaluating scales of structure in polymers. In: *Controlling the Morphology of Polymers: Multiple Scales of Structure and Processing* (ed. A. Tojeiria and G.R. Mitchell), 29–68. Springer. ISBN: ISBN 978-3-319-39320-9.

6 Roe, R.-J. (2000). *Methods of X-ray and Neutron Scattering in Polymer Science*. Oxford: Oxford University Press. ISBN: ISBN 0-19-511321-7.

7 Chu, B. and Hsiao, B.S. (2001). Small-angle X-ray scattering of polymers. *Chemical Reviews* 101: 1727–1762.

8 Wignall, G.D. and Melnichenko, Y.B. (2005). Recent applications of small-angle neutron scattering in strongly interacting soft condensed matter. *Reports on Progress in Physics* 68: 1761–1810.

9 Deas, H.D. (1952). The diffraction of X-rays by a random assemblage of molecules with a partial alignment. *Acta Crystallographica* 5: 542.

10 Mitchell, G.R., Saengsuwan, S., and Bualek-Limcharoen, S. (2005, 2005). Evaluation of preferred orientation in multi-component polymer systems using X-ray scattering procedures. *Progress in Colloid and Polymer Science* 130: 149–159.

11 Lovell, R., Mitchell, G.R., and Cryst, A. (1981). Molecular orientation distribution derived from an arbitrary reflection. *Acta Crystallographica Section A* 37: 135–137.

12 Edwards, M.D., Mitchell, G.R., Mohan, S.D., and Olley, R.H. (2010). Development of orientation during electrospinning of fibres of poly(ε-caprolactone). *European Polymer Journal* 46: 1175–1183.

13 Staudinger, H. (1920). Über polymerisation. *Berichte der Deutschen Chemischen Gesellschaft* 53: 1073–1085. https://doi.org/10.1002/cber.19200530627.

14 Katz, J.R. (1925). Röntgenspektrographische Untersuchungen am gedehnten Kautschuk und ihre mögliche Bedeutung für das Problem der Dehnungseigenschaften dieser Substanz. *Naturwissenshaften* 19: 410.

15 Mitchell, G.R. (1984). A wide angle X ray scattering study of the development of molecular orientation in cross linked natural rubber. *Polymer* 25: 1562–1572.

16 Silva, D., Pinheiro, J., Abdulghani, S. et al. (2022a). Controlling morphological development during additive manufacturing: a route to the mapping of properties. *Materials Proceedings* 8: 116. https://doi.org/10.3390/materproc2022008116.

17 Kohjiya, S. (2017). Crystallisation of natural rubber its unique feature. *Elastomere und Kunststoffe Elastomers and Plastics* 10: 38.

18 Huneau, B. (2011). Strain-induced crystallisation of natural rubber – a review of X-ray diffraction investigation. *Rubber Chemistry and Technology, American Chemical Society* 84 (3): 425–452. https://doi.org/10.5254/1.3601131. hal-01007326.

19 Sotta, P. and Albouy, P.-A. (2020). Strain-induced crystallization in natural rubber: Flory's theory revisited. *Macromolecules* 53 (8): 3097–3109. https://doi.org/10.1021/acs.macromol.0c00515.

20 Lamolinara, B., da Silva, D.P., Gameiro, F. et al. (2022). Self-contained equipment for the quantitative biaxial deformation of soft elastic materials (Eng, submitted).

21 Keller, A. and Kolnaar, H.W.H. (2006). Flow-induced orientation and structure formation. In: *Materials Science and Technology* (ed. R.W. Cahn, P. Haasen, and E.J. Kramer). https://doi.org/10.1002/9783527603978.mst0210.

22 Pan, Y., Li, N., Wang, J. et al. (2021). The role of expansion angle and speed on shish-kebab formation in expansion pipe extruder head explored by high temperature rapid stretch. *Polymer Testing* 104: 107387. https://doi.org/10.1016/j.polymertesting.2021.107387.

23 Nogales, A., Thornley, S.A., and Mitchell, G.R. (2004). Shear cell for in situ WAXS SAXS and SANS experiments on polymer melts under flow fields. *Journal of Macromolecular Science, Part B Physic* 43: 1161–1170.

24 Heenan, R.K., Penfold, J., and King, S.M. (1997). SANS at pulsed neutron sources: present and future prospects. *Journal of Applied Crystallography* 30: 1140.

25 An, Y., Holt, J.J., Mitchell, G.R., and Vaughan, A.S. (2006). Influence of molecular composition on the development of microstructure from sheared polyethylene melts: molecular and lamellar templating. *Polymer* 47: 5643–5656.

26 Pople, J.A., Mitchell, G.R., and Chai, C.K. (1996). In-situ time-resolving wide-angle X-ray scattering study of crystallization from sheared polyethylene melts. *Polymer* 37: 4187–4191. https://doi.org/10.1016/0032-3861(96)00264-9.

27 Spoerer, Y., Androsch, R., Jehnichen, D., and Kuehnert, I. (2020). Process induced skin-core morphology in injection molded polyamide. *Polymers* 12: 894. https://doi.org/10.3390/polym12040894.

28 Pantani, R., Coccorullo, I., Speranza, V., and Titomanlio, G. (2005). Modeling of morphology evolution in the injection molding process of thermoplastic polymers. *Progress in Polymer Science* 30 (12): 1185–1222. https://doi.org/10.1016/j.progpolymsci.2005.09.001.

29 Speranza, V., Liparoti, S., Pantani, R., and Titomanlio, G. (2019). Hierarchical structure of iPP during injection molding process with fast mold temperature evolution. *Materials* 12: 424. https://doi.org/10.3390/ma120304243.

30 Costa, A., Gameiro, F., Potencio, A. et al. (2022a). In situ time-resolving small-angle X-ray scattering study of the injection moulding of isotactic polypropylene parts. *Materials Proceedings* 8: 30. https://doi.org/10.3390/materproc2022008030.

31 Costa, A., Gameiro, F., Potencio, A. et al. (2022b). Evaluating the injection moulding of plastic parts using in situ time-resolved small-angle X-ray scattering techniques. *Polymers* submitted.

32 Liao, T., Zhao, X., Yang, X. et al. (2021, 2021). In situ synchrotron small angle X-ray scattering investigation of structural formation of polyethylene upon micro-injection molding. *Polymer* 215: 123390. doi.org/10.1016/j.polymer.2021.123390.

33 Zhao, Z., Liao, T., Yang, X. et al. (2022). Mold temperature- and molar mass-dependent structural formation in micro-injection molding of isotactic polypropylene. *Polymer* 248: 124797. https://doi.org/10.1016/j.polymer.2022.124797.

34 Okesola, B.O., Vieira, V.M., Cornwell, D.J. et al. (2015, 2015). 1,3:2,4-dibenzylidene-D-sorbitol (DBS) and its derivatives--efficient, versatile and industrially-relevant low-molecular-weight gelators with over 100 years of history and a bright future. *Soft Matter* 11 (24): 4768–4787. https://doi.org/10.1039/c5sm00845j.

35 Balkaev, D., Neklyudov, V., Starshinova, V. et al. (2021). Novel nucleating agents for polypropylene and modifier of its physical-mechanical properties. *Materials Today Communications* 2021: 101783. https://doi.org/10.1016/j.mtcomm.2020.101783.

36 Nogales, A., Olley, R.H., and Mitchell, G.R. (2003a). Directed crystallisation of synthetic polymers by low-molar-mass self-assembled templates. *Macromolecular Rapid Communications* 24: 496–502.

37 Nogales, A., Mitchell, G.R., and Vaughan, A.S. (2003b, 2003). Anisotropic crystallization in polypropylene induced by deformation of a nucleating agent network. *Macromolecules* 36: 4898–4906.

38 Mitchell, G.R., Pratumshat, S., and Olley, R. (2019). Polyethylene and the nucleating agent: dibenzylidene sorbitol, a neutron scattering study. *AMM* 890: 199–204. https://doi.org/10.4028/www.scientific.net/amm.890.199.

39 Wypych, G. and Wypych, A. (2016). *Databook of Nucleating Agents*. Elsevier eBook. ISBN: 9781927885130.

40 Wangsoub, S., Davis, F.J., Mitchell, G.R. et al. (2008). Enhanced templating in the crystallization of poly(ε-caprolactone) using 1,3:2,4-di(4-chlorobenzylidene) sorbitol. *Macromolecular Rapid Communications* 29: 1861–1865.

41 Wangsoub, S. and Mitchell, G.R. (2009). Shear controlled crystal size definition in a low molar mass compound using a polymeric solvent. *Soft Matter* 5: 525.

42 Mitchell, G.R. and Olley, R.H. (2018). Orthogonal templating control of the crystallisation of poly(ε-caprolactone). *Polymers (Basel)* 10 (3): 300. https://doi.org/10.3390/polym10030300.

43 Fernández-García, M., Avalos-Belmontes, F., Ramos-deValle, L.F. et al. (2016). Effect of different nucleating agents on the crystallization of Ziegler-Natta isotactic polypropylene. *International Journal of Polymer Science* 687–9422. https://doi.org/10.1155/2016/9839201.

44 Tojeira, A. and Mitchell, G.R. (2016). Controlling morphology in 3-D printing. In: *Controlling Controlling the Morphology of Polymers: Multiple Scales of Structure and Processing* (ed. G. Mitchell and A. Tojeira). Springer. ISBN: 978-3-319-39320-9.

45 Gibson, I., Rosen, D., and Stucker, B. (2015). *Additive Manufacturing Technologies 3D Printing, Rapid Prototyping, and Direct Digital Manufacturing*. Springer. ISBN: 978-1-4939-2113-3.

46 da Silva, D.P., Pinheiro, J., Abdulghani, S. et al. (2022b). Property mapping of LDPE during 3D printing: A study on morphological development. *Polymers* in press.

47 da Silva, D.P., Pinheiro, J., Abdulghani, S. et al. (2022). Changing the paradigm-controlling polymer morphology during 3D printing defines properties. *Polymers* 14 (9): 1638. https://doi.org/10.3390/polym14091638.

11

Molecular Simulations of Polymer Crystallization

Yijing Nie and Jianlong Wen

Jiangsu University, Research School of Polymeric Materials, School of Materials Science and Engineering, 301 Xuefu Road, Zhenjiang 212013, China

11.1 Introduction

Up to now, though some theories have been proposed to explain the formation process of polymer crystals [1–9], microscopic mechanisms of polymer crystallization are still not fully understood. The main problem is that it is hard to directly detect the details of structural changes during polymer crystallization, especially those in the early nucleation process, owing to the limitation of observation scale in experiments. However, molecular simulations provide an effective way to solve the above problem [10, 11]. Using molecular simulations, the conformational and structural changes during polymer crystallization can be detected at molecular scale, which is beneficial for the complete understanding of the mechanisms of polymer crystallization.

Till now, many researchers have carried out outstanding research work on polymer crystallization using molecular simulations, and some exciting simulation results have been found [10–16]. In this chapter, we mainly focused on the establishment of polymer simulation systems, molecular simulations of polymer crystallization at quiescent state, and molecular simulations of flow-induced polymer crystallization.

11.2 Establishment of Polymer Simulation Systems

Nowadays, there are two kinds of molecular simulations that can be used to investigate polymer crystallization: Monte Carlo (MC) simulations and molecular dynamics (MD) simulations. Hu and coworkers used dynamic MC simulations to systematically explore the microscopic mechanisms of crystallization of polymers [3, 4, 6, 10, 17–26]. Nie's group also applied MC simulations to detect nanofiller-induced crystallization of polymers [16, 27–39] and stereocomplex crystallization of polymer blends [40–45]. Graham and Olmsted used kinetic MC

Polymer Crystallization: Methods, Characterization, and Applications, First Edition.
Edited by Jyotishkumar Parameswaranpillai, Jenny Jacob, Senthilkumar Krishnasamy, Aswathy Jayakumar, and Nishar Hameed.
© 2023 WILEY-VCH GmbH. Published 2023 by WILEY-VCH GmbH.

simulations to study flow-induced nucleation in semicrystalline polymers [46, 47]. Rutledge's group [12, 48–55], Li's group [56–59], Yamamoto's group [60–62], Sommer's group [63–65], Luo's group [66–68], and Jabbari-Farouji's group [69, 70] used MD simulations to detect microstructural changes during polymer crystallization.

It should be noted that both of the two kinds of simulations are effective ways to reveal the mechanisms of crystallization of polymers, but some differences exist. In MC simulations, coarse-grained polymer chains are usually used [6, 10, 15, 46, 47, 71, 72], and in this condition, the simulation results of the corresponding MC simulations can be used to uncover universal microscopic mechanisms of polymer crystallization (namely, this method does not care about specific polymers). However, in MD simulations, by setting the correct force field and establishing the relevant molecular chain model (coarse-grained chain model or all-atom chain model), the crystallization of a specific polymer can be simulated [12, 48, 60, 61].

11.2.1 MC Simulations

In general, in order to speed up simulation efficiency, lattice polymer chain model is always used in MC simulations [3, 4, 6, 10]. In this condition, lattice polymer chains only move on lattice sites of lattice space based on the micro-relaxation mode (one bead can jump from one site into another neighboring empty site, and partial sliding along local chain sections is also allowed [3, 4, 6, 10]. In order to realize the polymer chain movements during simulations, each micro-relaxation step is determined based on the conventional Metropolis sampling algorithm. Hu et al. introduced two different energies in their simulation work: E_p (the change of potential energy corresponding to one pair of bonds that are nonparallel-packed, which can drive polymer crystallization) and E_c (the change of potential energy corresponding to the connection of two consecutive bonds that are noncollinear, which reflects the flexibility of chains [3, 4, 6, 10]). Sometimes, some other parameters, such as B (the change of potential energy corresponding to one pair of different components, which reflects mixing interaction) and E_f (the barrier of kinetic energy corresponding to one pair of bonds that are parallel-packed, which reflects the frictional resistance to sliding diffusion of chains in crystals) were also introduced [27, 36]. Then, the change in potential energy corresponding to the micro-relaxation step can be written as follows:

$$\Delta E = \Delta c \cdot E_c + \Delta p \cdot E_p + \Delta b \cdot B + \sum_i f(i) \cdot E_f \quad (11.1)$$

where Δc corresponds to the net number of noncollinear bond pairs on chains, Δp corresponds to the net number of nonparallel bond pairs, Δb corresponds to the net number of pairs of contacts between different components, and $\sum_i f(i)$ corresponds to the sum of bonds that are parallel-packed in the local sliding diffusion path.

11.2.2 MD Simulations

In general, three different kinds of polymer chain models are usually used in MD simulations: united atom chain model, coarse-grained polymer model, and all-atom chain model [73, 74]. United atom chain model can be used for the simulations

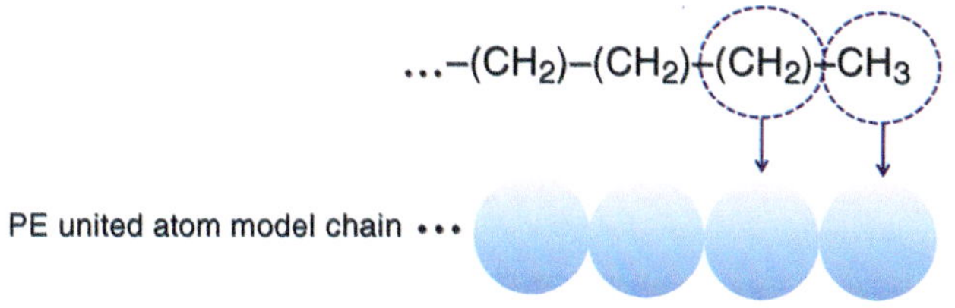

...—(CH$_2$)—(CH$_2$)—(CH$_2$)—CH$_3$

PE united atom model chain ...

Figure 11.1 Schematic diagram of a PE chain based on the UA model.

of polyethylene (PE) crystallization [12, 14, 38]. There are two different kinds of coarse-grained polymer models. One kind of coarse-grained polymer model can be used for simulations of coarse-grained polymer crystallization, and then universal microscopic mechanisms of polymer crystallization can be revealed (similar to the MC simulation). For instance, Lame's group proposed a finite-extensible nonlinear elastic (FENE) Lennard-Jones coarse-grained polymer model to realize simulation of polymer crystallization process [75, 76]. The other kind of coarse-grained polymer model can be used to simulate crystallization of one specific polymer. Meyer and Müller-Plathe proposed a coarse-grained chain model to simulate poly(vinyl alcohol) (PVA) crystallization, in which one bead corresponds to one PVA monomeric unit [77–79]. Then, Luo and Sommer further presented a patch code to perform the coarse-grained PVA chain model based on the LAMMPS package [80].

11.2.2.1 United Atom Chain Model

In the united atom chain model, one methylene was considered as one united atom unit, as shown in Figure 11.1 [56, 61]. Although the united atom model cannot fully show the details of all PE chemical structures, it can effectively reflect the crystallization behaviors of PE molecules [61]. Compared with the simulations based on the all-atom model, those based on the united atom model exhibit much higher simulation efficiency. The simulation of PE crystallization based on the united atom chain model is usually realized by using the force field of Paul et al., the detailed descriptions of which can be seen in Table 11.1 [81].

11.2.2.2 Coarse-Grained Polymer Model

For the coarse-grained polymer model used by Lame's group [76], chains are formed by beads reflecting some structural units. The FENE potential is used to reflect the interactions of bonded beads (namely, covalent bonds), and all other interactions between two beads of different chains or between two nonbonded beads of a same chain are described by a Lennard-Jones potential, as shown in Table 11.2 [76].

In the PVA coarse-grained polymer model, one bead denotes one monomeric unit [77–80]. The corresponding force field can be seen in Table 11.3 [80].

11.3 Polymer Crystallization at Quiescent State

11.3.1 Crystal Nucleation

The crystallization process of polymers can be divided into two stages: crystal nucleation and growth [1–4, 6, 10]. The classical thermodynamic nucleation

Table 11.1 Force field and potential parameters for simulation of PE based on the united atom model.

Interaction	Equation	Parameters
Bond	$E = K_l(l - l_0)^2$	$K_l = 350 \text{ kcal mol}^{-1}$, $l_0 = 1.53 \text{ Å}$
Angle	$E = K_\theta(\theta - \theta_0)^2$	$K_\theta = 60 \text{ kcal/(mol deg}^2)$, $\theta_0 = 109°$
Torsional	$E = \sum_{n=1}^{3} K_n[1 - \cos(n\emptyset)]$	$K_1 = 0.81 \text{ kcal mol}^{-1}$, $K_2 = -0.43 \text{ kcal mol}^{-1}$, $K_3 = 1.62 \text{ kcal mol}^{-1}$
Non-bonded	$E(r) = 4\varepsilon[((\sigma/r)^{12} - (\sigma/r)^6]$, $r \leq r_c$	$\varepsilon = 0.112 \text{ kcal mol}^{-1}$, $\sigma = 4.01 \text{ Å}$

Source: Adapted from Paul et al. [81].

Table 11.2 Force field and potential parameters for simulation based on the coarse-grained polymer model used by Lame's group.

Interaction	Equation	Parameters
Bond	$V_{FENE}(r) = -0.5kR_0^2 \ln\left[1 - \left(\frac{r}{R_0}\right)^2\right] + 4\varepsilon\left[\left(\frac{\sigma_F}{r}\right)^{12} - \left(\frac{\sigma_F}{r}\right)^6\right]$	$k = 30\varepsilon_u/\sigma_u^2$, $R_0 = 1.5\sigma_u$, $\varepsilon = 1\varepsilon_u$, $\sigma_F = 1.05\sigma_u$
Non-bonded	$E(r) = 4\varepsilon\left[\left(\frac{\sigma}{r}\right)^{12} - \left(\frac{\sigma}{r}\right)^6\right] - 4\varepsilon\left[\left(\frac{\sigma}{r_c}\right)^{12} - \left(\frac{\sigma}{r_c}\right)^6\right], r \leq r_c$	$\varepsilon = 1\varepsilon_u$, $r_c = 2.5\sigma$

Source: Adapted from Morthomas et al. [76].

theory [4, 11] states that there is a barrier of free energy for nucleation caused by the competition between the body free energy reduction and the surface free energy increase. This nucleation barrier must be overcome in order for a nucleus to grow into a stable crystal. When crystals grow in a mode of parallel stacking of polymer stems, two typical conditions occur on the stem-end surfaces: that is, intramolecular chain-folding and intermolecular fringed-micelle (Figure 11.3). For the chain-folding process, chains can fold back into crystals, and then a low local conformational enthalpy penalty appears; for the fringed-micelle mode, the cilia are confined to the ends of stems, resulting in higher conformational entropy penalty. The surface free energy of the PE folded-chain nucleus is estimated to be one third of that of the PE fringed-micelle nucleus [6]. In this condition, the nucleation barrier and the critical size of the folded-chain nucleus are much smaller than those of the fringed-micelle nucleus. That is, polymer crystallization at quiescent state prefers to choose the mode of intramolecular chain-folding nucleation [6, 10].

Table 11.3 Force field and potential parameters for simulation based on the coarse-grained PVA model.

Interaction	Equation	Parameters
Bond	$E = \frac{1}{2}K_l(l-l_0)^2$	$K_l = 2704\ kT/\sigma^2,\ l_0 = 0.5\sigma$
Angle[a]		
Non-bonded	$E(r) = \varepsilon[((\sigma_0/r)^9 - (\sigma_0/r)^6],\ r \le r_c$	$\varepsilon = 112\ kT,\ \sigma_0 = 0.89\sigma,\ r_c = 1.02\sigma$

a) The angular potential belongs to a tabulated potential, as illustrated in Figure 11.2 Three minima correspond to *trans–trans* state at 180°, *trans–gauche* state at near 120°, and *gauche–gauche* state at near 100°, respectively.

Source: Adapted from Luo and Sommer [80].

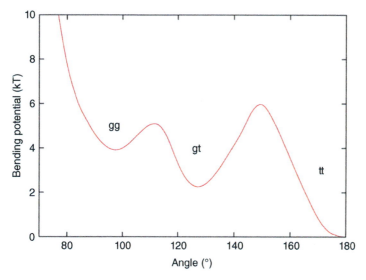

Figure 11.2 Angular potential of coarse-grained PVA model. The minimum "tt" at 180° corresponds to two successive *trans* torsion on the atomistic backbone. The minima "gg" and "gt" correspond to *gauche–gauche* and *gauche–trans* sequences, respectively [80]. Source: Reproduced with permission from Luo et al. [80]/Elsevier.

However, the intermolecular fringed-micelle nucleation mode can also occur in some cases, such as crystallization of polymer systems with large strains.

11.3.2 Intramolecular Nucleation Model

Hu proposed an intramolecular nucleation model to describe the mechanism of polymer crystal nucleation [6, 10, 11]. For the nucleation of one chain in dilute solution, it can be imagined that chain-folding can reduce the contact with solvents, thus significantly reducing the surface free energy barrier for nucleation. Besides, chain-folding effectively leads to an increase in the contact of parallel packing, thus effectively enhancing the gain of body free energy. Thus, single polymer chains prefer to choose the mode of the adjacent chain-folding to form crystal nucleus. Using

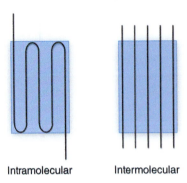

Figure 11.3 Illustration of chain-folding principle of polymer crystallization. The fast path of crystal growth will select the intramolecular mode of secondary crystal nucleation, which leads to lamellar crystals. The latter then thickens slowly toward the stable extended-chain crystals [10]. Source: Reproduced with permission from Hu [10]/Elsevier.

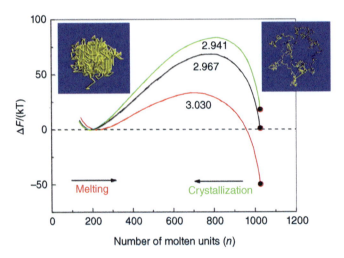

Figure 11.4 Free energy curves of single 1024-mer upon melting and crystallization at three reduced temperatures around the equilibrium melting point, obtained from dynamic Monte Carlo simulations [82]. Source: Hu [82]. Copyright (2003) AIP Publishing. Two snapshots indicate the crystalline and amorphous states of a single 1024-mer, respectively [10]. Source: Reprinted with permission from Hu [10]/Reproduced with permission from AIP Publishing LLC.

dynamic MC simulations of lattice polymer model, even the free energy of crystallization and melting of single chains can be calculated, as shown in Figure 11.4 [82].

11.4 Nanofiller-Induced Polymer Crystallization

11.4.1 Nanofiller-Induced Homopolymer Crystallization

Nanofillers can be used to help polymers improve their physical properties [83–85]. For semi-crystalline polymers, some nanofillers can play as agents to induce the heterogeneous nucleation of polymers and cause the crystalline structure changes of polymers [16, 27, 28, 83]. Li's group detected that the presence of carbon nanotubes (CNTs) in polymer solutions can cause the formation of a new nanohybrid

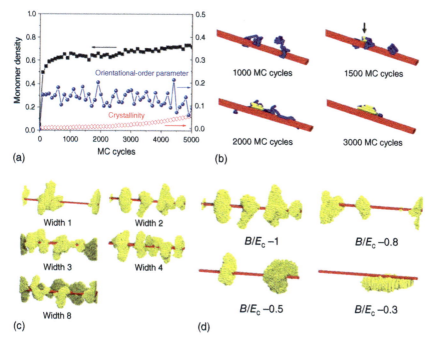

Figure 11.5 (a) Changes of monomer density, orientational-order parameters of amorphous segments in the interface, and crystallinity of the polymer filled with the filler with the width of 2 during crystallization. (b) Snapshots of the 10th and 32nd chains in the filled polymer (red strip represents the filler with the width of 2) during nucleation at different MC cycles. Blue cylinders represent amorphous bonds, and yellow cylinders represent crystalline bonds. (c) Snapshots of crystalline morphologies of polymer nanocomposites containing fillers with different widths formed at 10 000 MC cycles. (d) Snapshots of crystalline morphologies of polymer nanocomposites with different polymer-filler interactions formed at 10 000 MC cycles [27]. Source: Reproduced with permission from Nie et al. [27]/Elsevier.

shish-kebab (NHSK) structure [86, 87]. They further proposed a new theory (geometric confinement) to explain the mechanisms of the appearance of the NHSK structure [86, 87]. The CNTs of small radii can impose confinement effect on conformations of interfacial segments, making them align along the CNT's long axis. MD simulations were used to reveal the molecular details of this geometric confinement effect [88–90]. Dynamic MC simulations were also performed to reveal the mechanism of the formation of the NHSK structure [27, 28, 33]. As seen in Figure 11.5a, monomer density and orientational-order parameters of amorphous segments along the long axis of nanofiller in interfacial regions increase faster than polymer crystallinity at the early stage of crystallization, indicating the presence of the adsorption and orientation processes before the NHSK structure formation. As illustrated in Figure 11.5b, the oriented segments in interfacial regions are involved in heterogeneous nucleation at the early stage of crystallization [27, 33].

Both experimental and simulation results showed that there are some factors influencing the NHSK formation, such as the diameters of one-dimensional

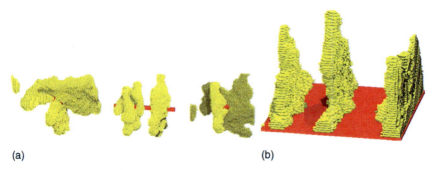

Figure 11.6 (a) Snapshots of crystalline morphologies of polymers filled with nanofillers of different dimensions (from left to right: nanofiller of zero-dimension, nanofiller of one-dimension, and nanofiller of two-dimension). The nanofillers are red, and the crystalline bonds are yellow [33]. Source: Reproduced with permission from Nie et al. [33]/Elsevier. (b) Snapshot of crystalline morphology of the grafted system. The yellow cylinders represent the crystalline bonds of free chains, and the green cylinders represent the crystalline bonds of the cyclic grafted chains. Only the crystalline bonds are shown [31]. Source: Reproduced with permission from Nie et al. [31]/American Chemical Society.

nanofillers and polymer-filler interfacial interactions [27, 87, 89, 91]. As illustrated in Figure 11.5c, the perfect NHSK structures cannot form in polymer solutions containing one-dimensional nanofillers with relatively large lateral sizes [27]. In addition, it can also be seen in Figure 11.5d that the perfect NHSK structures exist in polymer solutions with high interfacial interactions [27].

Dynamic MC simulations were further used to study crystallization of polymers containing nanofillers of different dimensions [33]. As shown in Figure 11.6a, the NHSK structures form in the system filled with the one-dimensional nanofiller but cannot form in the systems filled with the zero-dimensional nanofiller and the two-dimensional nanofiller [33]. The nanofillers with different dimensions have the different capability to cause the interfacial segment orientation [33, 92]. The simulation results also showed that the one-dimensional nanofiller exhibits the strongest ability to induce nucleation, while the zero-dimensional nanofiller has the weakest ability [33].

It is important to note that the crystal orientation of polymer chains in the system filled with two-dimensional nanofillers can also be precisely controlled through proper design. In experiments, shear flow is an effective way to improve polymer segment orientation [83, 91, 93–95]. The oriented segments can then be further evolved into oriented crystals or shish-kebabs. Nie's group reported that the two-dimensional nano-filler modified by oriented loops, as shown in Figure 11.6b, can cause the appearance of crystals with uniform orientation [31]. This simulation work provides a new idea for the crystal orientation control of polymers.

Chain topology also affects polymer crystallization. Some research groups focused on the crystallization of cyclic polymers [96, 97]. Sommer's group found that, given the same thermal history condition, compared with the corresponding linear counterparts, crystallization temperature, crystallinity, stem length, melting point, and latent heat of melting/crystallization of cyclic polymers were significantly higher

[64]. However, Muthukumar's group applied Langevin dynamic simulations to explore crystallization behavior of cyclic polymers. They observed that the melting point of the cyclic polymers was lower [98]. In general, the reason for the above different results can be attributed to a combination of two factors: kinetic factor correlated with diffusion and thermodynamic factor affected by different subcooling [99, 100]. Nie's group investigated the crystallization of cyclic polymer and linear polymer nanocomposites by using dynamic MC simulation [37]. They found that the melting and crystallization temperatures are higher and crystallization rates are faster in the filled cyclic polymers compared with the filled linear polymers [37].

On the basis of the Thomson–Gibbs equation [101], the relationship between the melting points of cyclic polymer systems with different chain lengths and their linear analogs was established [37]:

$$\frac{T_{mL}}{T_{mC}} = \left[1 + \frac{T^0_{mL} Qk}{\Delta h^0_f N}\left(C - \frac{3\ln N}{2} - \frac{3}{2N}\right)\right] \quad (11.2)$$

where T_{mL} and T_{mC} are the melting temperatures of linear and cyclic polymers, respectively, T^0_{mL} corresponds to the equilibrium melting point for linear polymers, Δh^0_f corresponds to the difference in equilibrium melting enthalpy, Q/N corresponds to the chain number, and k is the Boltzmann's constant [37].

11.4.2 Nanofiller-Induced Copolymer Crystallization

Block and random copolymers filled with nanofillers display different crystallization behaviors compared with filled homopolymers, due to the distribution of comonomer units on chains.

11.4.2.1 Nanofiller-Induced Block Copolymer Crystallization

Microphase separation occurs in block copolymers due to the competition between the repulsive interactions of different blocks and the connectivity of chemical bonds [102–104]. Crystalline morphology of crystalline-amorphous diblock copolymers is dominated by the competition between microphase separation and crystallization, leading to the appearance of more complex structures, compared with homopolymers [104–106].

How nanofillers affect block copolymer crystallization is an interesting topic. Experimental work showed that the NHSK structures appear in block copolymers reinforced by CNT bundles [107].

In addition, a new nanohybrid epitaxial brush (NHEB) structure has been found in PVCH-PE-PVCH triblock copolymer, which consists of CNTs in the center and disk-like crystal lamellae formed by PE block segments existing on the CNT surfaces surrounded by random coils of polyvinyl cyclohexane (PVCH) block segments [108]. Using dynamic MC simulations, the formation mechanisms of the NHEB structure were further detected [34, 35]. The nanofiller can adsorb amorphous segments in crystallizable blocks into the interface regions and make them orient along the filler's long axis. Then, the adsorbed block segments grow into small, oriented crystallites based on heterogeneous nucleation, leading to the appearance of the NHSK

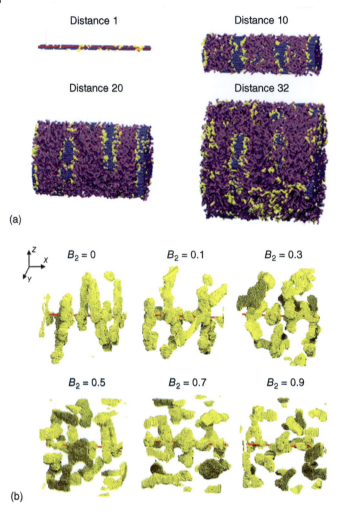

Figure 11.7 (a) Snapshots of morphologies of block copolymers within different distances from the filler surface. Red strip denotes the nanofiller, blue cylinders denote the crystalline stems, yellow cylinders denote the amorphous segments of the crystallizable blocks, and purple cylinders denote the non-crystallizable segments of the non-crystallizable blocks. Source: Reproduced with permission from Nie et al. [34]/Elsevier. (b) Snapshots of crystalline morphologies of block copolymers with different repulsive interactions. Here, the nanofiller and the crystals were drawn. Source: Reproduced with permission from Nie et al. [35]/John Wiley and Sons.

structure. It can be seen in Figure 11.7a that three different structural features exist: the center nanofiller, the lamellar crystals, and the non-crystallizable blocks existing on the outer layer [34, 35]. It was also detected that the increase in repulsive interaction can promote the microphase separation, leading to an increase in the degree of extension of local crystallizable segments near block junctions and thus the appearance of more crystals based on homogeneous nucleation, as shown in Figure 11.7b [34, 35].

11.4.2.2 Random Copolymer Nanocomposite Crystallization

Compared with homopolymers, the melting point of random copolymers will decrease due to the breaking of the chain sequence regularity. In addition, the presence of comonomers can hinder crystallization process, reduce crystallization rate, and degrade crystalline morphology [109–111].

The crystallization of random copolymers was also studied based on dynamic MC simulations [36]. Similar to homopolymer and block copolymer nanocomposites, the fillers could promote random copolymer crystallization. Crystallization behavior and crystalline morphology in random copolymer nanocomposites are also correlated with the non-crystallizable comonomer content. Increasing comonomer content slows down nucleation rate and reduces final crystallinity. The NHSK structure could appear in the random copolymer composites containing low comonomer contents [36]. However, it is hard for the NHSK structure to form in the systems with high comonomer contents [36].

11.4.3 Crystallization of Polymers Grafted on Nanofillers

In experiments, polymer chains are usually grafted on filler surfaces to enhance the polymer-filler interactions [112, 113]. These grafted polymers on nanofillers exhibit complex crystallization behaviors [114, 115]. The crystallization behaviors of grafted polymers will be influenced by grafting density [116, 117], chain length of grafted polymers [118, 119], interfacial interaction [32, 117], and filler amount [120]. Dynamic MC simulations were used to probe the crystallization of grafted polymers [29, 32, 116–119, 121].

11.5 Effect of Grafting Density

Their simulation work showed that the nucleation induction periods are reduced due to the increase in grafting densities [29, 117]. The crowding effect of chains is relatively weak in grafted polymers of low grafting densities, and grafted chains can be nucleated according to intramolecular chain-folding mode [29, 117]. However, grafted chains in the systems of high grafting densities can be nucleated based on intermolecular fringed-micelle mode, because they have a high degree of deformation owing to the strong crowding effect. As seen in Figure 11.8, the stems are parallel to the filler surface in grafted polymers of a low grafting density, but they are perpendicular to the filler surface in grafted polymers of a high grafting density [29, 117].

11.6 Effect of Chain Length

Chain length of grafted polymers can also affect chain conformations, thus influencing the corresponding crystallization behaviors [32, 119, 122]. The increase in molecular weight of polymers grafted onto nanofillers can cause an improvement in crystallization temperature [122] and also changes in nucleation mode and stem orientation [118], as shown in Figure 11.9

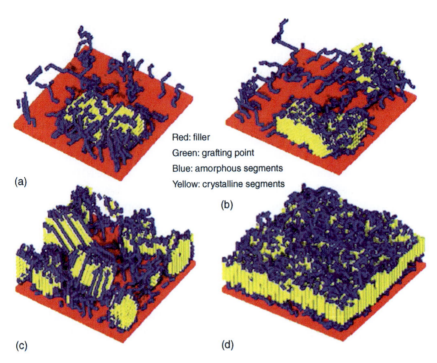

Figure 11.8 Snapshots of morphologies of grafted systems containing (a) 49, (b) 64, (c) 128, and (d) 256 chains [29]. Source: Reproduced with permission from Nie et al. [29]/Elsevier.

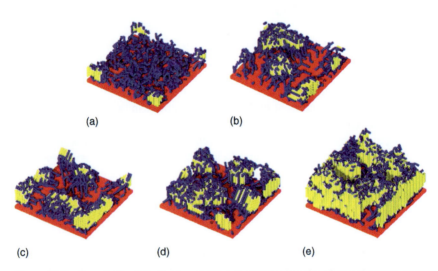

Figure 11.9 Snapshots of the final crystalline morphologies of grafted polymer systems with respective chain lengths of (a) 12, (b) 16, (c) 20, (d) 30, and (e) 62. Source: Reproduced with permission from Nie et al. [118]/Elsevier.

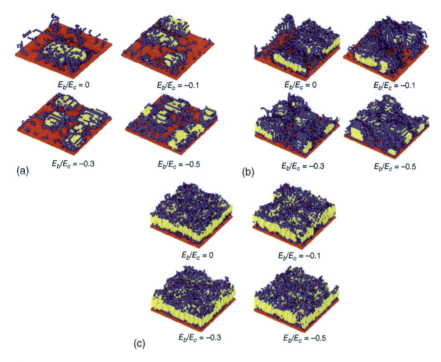

Figure 11.10 Snapshots of morphologies of grafted systems of (a) low, (b) medium, and (c) high grafting densities, respectively, with different interactions [32]. Source: Reproduced with permission from Nie et al. [32]/Elsevier.

11.7 Effect of Interfacial Interactions

The crystallization of grafted polymers with different interfacial interactions was investigated based on dynamic MC simulations, as seen in Figure 11.10 [32]. In the systems of low grafting density, heterogeneous nucleation of grafted chains can be promoted by the increase in attractive interfacial interaction, but the stem orientation does not change. In the systems of medium grafting density, the increased grafting density will result in the enhancement of the crowding effect; however, the filler can adsorb chains into the interface regions due to the enhanced interfacial interaction. In this way, the adsorption effect can make up for the loss of conformational entropy, and then the nucleation ability can be improved. In the systems of high grafting density, since the crowding effect is very strong, the crystallization is only determined by the crowding effect.

11.8 Stereocomplex Crystallization of Polymer Blends

Due to resource crises and environmental pollution, renewable polymer materials such as polylactic acid (PLA) attract much attention nowadays [123, 124]. But so far, PLA has not been able to replace traditional plastics on a large scale due to the presence of some shortcomings, such as relatively poor thermal stability and mechanical properties [125, 126]. Therefore, the enhancement of thermal stability of PLA is an interesting research topic. Ikada et al. reported that a new crystallite (stereocomplex crystallite, SC) appears in the blend of poly (L-lactic acid) (PLLA) and poly (D-lactic acid) (PDLA), which has a higher melting point (T_m) than that of homocrystallites (HC) formed from PLLA or PDLA chains alone [127, 128]. In other words, the appearance of SC in PLA can effectively improve the thermal stability of PLA. Subsequently, some experimental work has shown that the formation of SC by intermolecular packing of adjacent PLLA and PDLA segments is closely related to the hydrogen bonds between the segments of PLLA and PDLA [129, 130].

Many different methods have been proposed to facilitate SC formation, such as shearing or stretching [131–133], synthesizing polymers with star topology [134], adding another component that can interact with PLLA and PDLA to form hydrogen bonds [135], and adding nucleating agents or nanofillers [136, 137]. Although the above methods can effectively enhance the formation of SC, the microscopic mechanism of SC formation promoted by the above methods has not been fully revealed. Fortunately, molecular simulation is a powerful way to detect microscopic changes at the molecular scale.

Hu's group proved that the formation of SC could be facilitated by stretching through dynamic MC simulation [19]. Subsequently, Nie's group systematically studied the mechanisms of SC formation and found some interesting results [40–45].

11.8.1 Simulation Details

Experimental work has demonstrated the existence of intermolecular hydrogen bonds between the helical chains of enantiomers in the crystal structure [129, 130]. The energy reduction caused by the packing of a PLLA bond and a PDLA bond will be higher than that caused by the packing of two bonds of the same type. To reflect the difference, two different interaction parameters were introduced: E_{p1} is used to describe changes in potential energy corresponding to a pair of nonparallel-packed bonds of the same polymer, which can drive the HC formation, and E_{p2} is used to describe changes in potential energy corresponding to a pair of nonparallel-packed bonds formed by different polymer segments, which can drive the SC formation [40–45, 138]. Then, the value of E_{p2}/E_c should be set to a value bigger than that of E_{p1}/E_c, as shown in Figure 11.11 [42]. In the simulation process, there are three kinds of parallel packings of polymer bonds: packings of A bonds, packings of B bonds, and packings of bonds of different types. Then, three kinds of crystals can appear in the simulation process: crystals mainly composed of packing of A bonds, crystals mainly composed of packing of B bonds, and crystals mainly composed of packing of A and B bonds. The appearance of the first and second crystals is driven by $E_{p1}/E_c = 1$, while the appearance of the third crystal is driven by $E_{p2}/E_c = 1.2$. Previous simulation work shows that the third crystal shows a higher melting

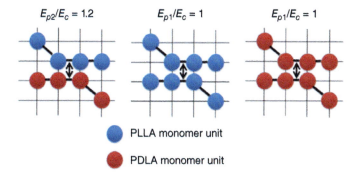

Figure 11.11 Sketch map of E_{p2}/E_c for packing of one A bond and one B bond, and E_{p1}/E_c for the packing of A bonds or B bonds. Blue and red spheres are monomer units in A and B chains, respectively [42]. Source: Reproduced with permission from Nie et al. [42]/John Wiley and Sons.

temperature than the first and second crystals [42]. Then, the first and second crystals can be treated as HCs, and the third crystal can be treated as SCs.

11.8.2 Effects of Different Methods

11.8.2.1 Effect of Chain Length

Experimental work has shown that the SC content was reduced due to the increase in PLA molecular weights [139, 140]. Dynamic MC simulations were performed to detect the SC formation mechanisms under the effect of molecular weights [43]. The formation of SC is influenced by chain length and crystallization temperature. As shown in Figure 11.12a [43], the SC content increases gradually with the increase in crystallization temperature in the blends with relatively long chain lengths. The SC content first increases with the improvement in temperature and then becomes saturated at high temperatures in the blends with short chain lengths. This difference could be owing to the different degrees of supercooling in the blends with different chain lengths. Because of the low melting points of the systems composed of short chains, the degree of supercooling for the SC formation should be lower than that of the systems composed of long chains.

At high temperatures, the degree of supercooling in the systems composed of short chains would not be enough to further induce the formation of SC, and then the SC content would not depend on temperature. For the systems composed of long chains, the degree of supercooling is still enough to improve the formation of SC at high temperatures, and then the SC content can continue to increase at high temperatures.

The formation of SC in blends is determined by three factors, namely, segment miscibility, segment mobility, and nucleation mode. The miscibility and segment mobility in the systems of short chain lengths are higher (as seen in Figure 11.12b,c) [43]. Due to the high miscibility of the short-chain system, segments have more chances to encounter other different types of chains in adjacent areas. In addition, chain segments in the systems composed of short chains tend to be nucleated according to intermolecular packing mode (as shown in Figure 11.12d), and then in the

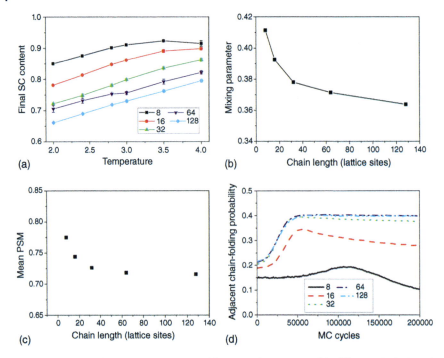

Figure 11.12 (a) Relationship between SC content in systems with different chain lengths and crystallization temperature. (b) Relationship between mixing parameter in initial relaxed systems and chain length. (c) Relationship between mean PSM of amorphous beads at the early stage of crystallization and chain length at $T = 4.0$. (d) Variations of chain-folding probability in crystals during crystallization at $T = 4.0$ [43]. Source: Reproduced with permission from Nie et al. [43]/Elsevier.

crystallization process, more chains choose alternative packing [43]. In short, those above factors contribute to the formation of more SCs in the short-chain blends.

11.8.2.2 Effect of Stretching

Through dynamic MC simulation, Hu's group demonstrated that stereocomplex formation in blends can be effectively promoted by stretching [19]. They explained that the stretching of chains can lead to the occurrence of crystal nucleation based on intermolecular fringed-micelle nucleation mode, and then more SCs can be formed [19]. Those simulation findings can be used to explain the mechanisms of the flow (stretching or shearing) and improved stereocomplex formation in PLLA/PDLA blends.

11.8.2.3 Effect of Nanofillers

Nie's group also investigated the stereocomplex formation in blends containing nanosheets [40]. The nanosheets could promote the SC formation, as seen in Figure 11.13a [40]. The increases in filler content and interface interaction both promote the formation of SC. Two factors determine the enhancement of the SC formation. Firstly, the miscibility is higher in the blends with higher filler content

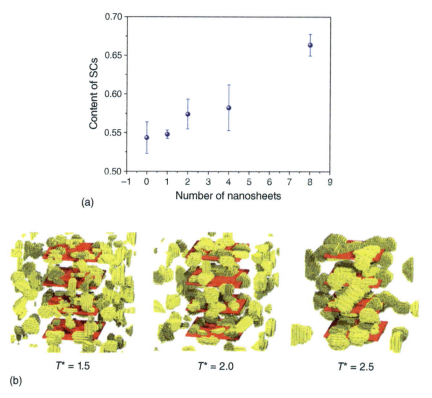

Figure 11.13 (a) Final SC content in different systems. (b) Snapshots of crystalline morphologies of system containing four nanosheets at different temperatures. Red planes are the nanosheets, and yellow cylinders are the stems. Source: Reprinted with permission from Nie et al. [40]. Copyright (2019) Royal Society of Chemistry.

or higher interfacial interactions. Secondly, more segments crystallize by intermolecular packing in blends with higher filler content or interfacial interaction. Obviously, the increase in the segment miscibility and the degree of intermolecular packing are conducive to the formation of SCs. The increase in temperature can promote stereocomplex crystallization. At higher temperatures, the degree of supercooling of the stereocomplex crystallization is higher than that of the HC crystallization. Besides, the homogeneously nucleated crystallites can be formed at low temperatures (Figure 11.13b) [40]. Since the corresponding segments in those homogeneously nucleated crystallites are almost unaffected by the nanosheet, the SC formation in those crystallites could not be affected by the nanosheet.

11.8.2.4 Effect of Chain Topology

The formation of SC in cyclic polymer blends and cyclic block copolymers was also studied [45]. The melting temperature and chain-folding ability of the cyclic polymer blends are higher, and the segment mobility and miscibility are lower, compared with the linear polymer blends. Thus, the stereocomplexation ability of the cyclic polymer blends is weaker.

But cyclic diblock copolymers have stronger stereocomplexation ability than linear diblock copolymers. This particular phenomenon may be due to the synergy between the chain topology of cyclic polymers and the structure of block copolymers. The chain topology of cyclic polymers limits segment movements into a smaller region, and then the segments of different blocks can have more chance to meet each other. In this case, the cyclic block copolymers have better segment miscibility than the linear ones. The enhancement of stereocomplexation ability of cyclic diblock copolymers is just mainly dominated by the improvement of miscibility.

11.8.2.5 Effect of Chain Structure

Dynamic MC simulation was used to explore the mechanisms of stereocomplex crystallization in multi-block copolymers [41]. The simulation results displayed that the increase in the number of blocks leads to an increase in the SC content at low block numbers. But the growth process of the SC content becomes saturated at high block numbers, as shown in Figure 11.14a [41]. The SC formation of the multi-block copolymers is controlled by two key factors: the segment miscibility of different blocks and the relative size of crystal thickness to block length. When block length is greater than crystal thickness, the segment miscibility of different blocks can be improved by the increase in block number, leading to the improvement in the SC contents. In the multi-block copolymers with the same block length and crystal thickness, one block forms one stem (as shown in Figure 11.14b), and then different blocks tend to alternate in parallel packing [41]. Then, the block copolymers reach the upper limit of the SC formation capacity, and the further increase of the block number could not lead to an increase in the SC fraction.

The formation of SCs in the asymmetric diblock copolymers was also investigated [44]. The difference in volume fraction of beads in different blocks is larger in the asymmetric diblock copolymers that have higher degree of asymmetry, resulting in the lower segment miscibility and SC content. In short, for the asymmetric diblock copolymers, miscibility is the most important factor affecting the SC formation [44].

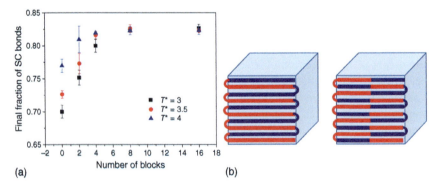

Figure 11.14 (a) SC fraction at different crystallization temperatures in blends and copolymers of different block numbers. (b) Sketch map of SC in A-B octablock copolymer (left) and A-B sixteen-block copolymer (right), respectively. The red and blue lines denote the stems of different blocks [41]. Source: Reproduced with permission from Nie et al. [41]/Royal Society of Chemistry.

11.9 Flow-Induced Polymer Crystallization

Flow-induced polymer crystallization is a typical nonequilibrium phase transition, which is different from polymer crystallization in quiescent state [94, 95]. Understanding the mechanisms of flow-induced polymer crystallization can guide the design of polymer crystalline morphology of semi-crystalline polymers to achieve the regulation of physical properties. Although different research groups, such as Alfonso's group [141, 142], Fu's group [83, 91], Kanaya's group [143, 144], Kornfield's group [145, 146], Han's group [147, 148], Hsiao's group [149, 150], Hashimoto's group [151, 152], Li's group [94, 153], Li's group [154, 155], Matsuba's group [156, 157], Rastogi's group [93, 158], and Winter's group [159, 160] have proposed different models or explanations based on the corresponding experimental results, the molecular details of flow-induced crystallization of polymers, especially flow-induced nucleation are still not fully understood.

In addition, how stretching affects the crystalline structures of semi-crystalline polymers is also an interesting topic, and some controversies remain. Peterlin stated that stretching breaks crystals, creating holes or cracks, and then polymer chains slip and become straight-chain conformations [161]. However, Flory suggested that because lamellar crystals are subject to too many topological constraints, it is difficult for them to move relative to each other, and chains cannot slip directly. Thus, Flory proposed that the folded-chain crystals melt under local stress and then recrystallize to produce extended-chain microfibers [162]. Many groups, such as Strobl's group [163, 164] and Men's group [165, 166], systematically studied the crystalline structure changes of polyolefins during stretching, and some interesting results were observed.

11.9.1 Flow-Induced Polymer Nucleation

Li's group has published several reviews of flow-induced crystallization of polymers, which gave a complete description of the corresponding multiscale and multistep process [94, 167, 168]. Based on the corresponding experimental results, they proposed that the processes of flow-induced crystallization of polymers contain the changes of the multiscale structures (from segment conformation to the whole-chain deformation) and the multistep processes (from conformational transition to the formation of shish-kebab crystals), as shown in Figure 11.15 [94].

Rutledge's group [12, 48–55] and Yamamoto's group [60–62] were among the first to study the crystallization of oriented or stretched polymers using MD simulations. The formation of the oriented crystals and the acceleration of the crystallization process were detected, as shown in Figure 11.16 [48].

Using all-atom MD simulations, Li's group demonstrated that density fluctuation appears due to the coupling effect of conformational and orientational orderings, which can further promote the flow-induced nucleation [59]. A *CO* parameter was used to reflect the synergy of the conformational and orientational orders. They found that the locations that display high density (Figure 11.17b) and high *CO* value

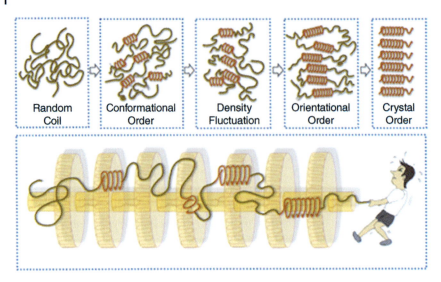

Figure 11.15 Flow-induced polymer nucleation containing multiscale and multistep ordering [94]. Source: Reproduced with permission from Li's group [94]/American Chemical Society.

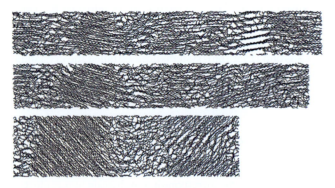

Figure 11.16 Snapshots of the system of 20 C400 chains that were deformed at 425 K and then quenched to 250, 300, and 350 K (from top to bottom). The snapshots were obtained after the systems showed an initial plateau in the alignment order parameter [48]. Source: Reproduced with permission from Rutledge et al. [48]/Elsevier.

(Figure 11.17c) can evolve into crystal nuclei (Figure 11.17a), demonstrating that density fluctuation is beneficial for flow-induced nucleation [59].

Using MD simulations, Nie's group observed that conformational transition from *gauche*-conformations to *trans*-conformations and segment orientation occur during deformation [38]. Those highly oriented segments of *trans*-conformations can be aggregated in local areas, which could be treated as the nucleation precursors [38]. This finding of the role of the aggregation of highly oriented segments coincides with that of Hashimoto and coworkers [151].

Hu's group also reported some exciting simulation results about the shish-kebab formation [26]. As shown in Figure 11.18a, the shear-induced oriented crystals (the

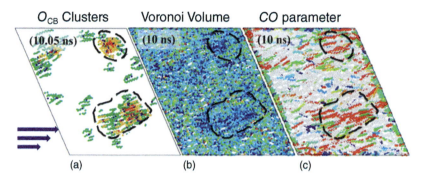

Figure 11.17 A slice of the simulation box with thickness equal to 30 nm at $T_s = 500$ K. The O_{CB} clusters (a), Voronoi volume (b), and CO parameter (c) were displayed, respectively [59]. Source: Reproduced with permission from Li's group [59]/AIP Publishing.

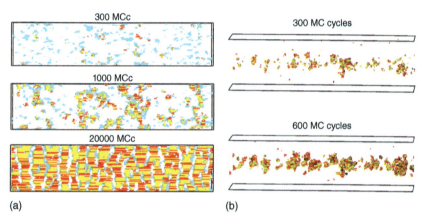

Figure 11.18 (a) Snapshots of precursors and crystals in the middle layer at different MC cycles. Blue regions are deformed segments, yellow cylinders are crystalline stems of long chains, and red cylinders are crystalline stems of short chains. (b) Snapshots of nuclei at 300 and 600 MC cycles, respectively [26]. Source: Reproduced with permission from Nie et al. [26]/American Chemical Society.

yellow and red ones) first form in the aggregation domains of deformed subchains (the blue ones), demonstrating that the precursor formation of shish can be regarded as a synergy of chain stretching and chain segregation of long chain segments. Figure 11.18a displays the shish crystal structures (the row structures of lamellar crystals) at 20 000 MC cycles [26], which are similar to the micro-shish-kebab structures found by Barham and Keller [169]. Another controversial argument is whether short chains can join the shish formation during bulk polymer crystallization induced by shearing. Figure 11.18b shows the crystal structures of the system at 300 and 600 MC cycles [26]. Some small crystal clusters appear at first, which orient along the shearing direction. They then grow to form a string of crystals arranged along the shearing direction, which are connected by tie molecules. At the same time, it can be seen that the oriented crystals also contain crystalline stems formed by short chains, indicating that the short chains with small deformation are

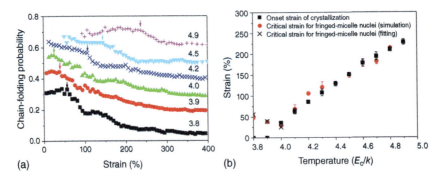

Figure 11.19 (a) Chain-folding probability with the increase of strain in small crystallites during stretching at different temperatures. The arrows denote the strains at which nucleation mode changes. (b) Comparison between the onset crystallization strains and the critical strains for the change of nucleation mode at different temperatures [25]. Source: Reproduced with permission from Nie et al. [25]/Elsevier.

also involved in the shish formation, which is in line with the findings of Kimata et al. [146]. Since the separation in the bulk polymers is incomplete, the short chains should take part in the shish formation so that the local segregated domains composed of the oriented long chain segments can be connected. However, the segregation in polymer solutions with low viscosity can be more complete, and then shish is mainly formed by the long chains.

The presence of flow can also affect the nucleation mode. Hu and coworkers applied MC simulations to study the stretch-induced crystal nucleation [25]. It can be seen in Figure 11.19a that the chain-folding probabilities are nearly constant at small strains, and then decrease above critical strain values during stretching [25]. This means that folded-chain crystal nuclei are more likely to form during nucleation under low-strain conditions; under high-strain conditions, the fringed-micelle nucleus formation is dominant, as shown in Figure 11.19b [25]. In short, stretching can cause a transition from the folded-chain nucleus formation to the fringed-micelle nucleus formation.

In polymer nanocomposites, the nanofillers influence the flow-induced polymer nucleation [91, 170]. The orientational orders of the interfacial bonds along the filler axes and the stretching direction and the orientational orders of the CNT axes and all bonds along the stretching direction in the early stage of crystallization are shown in Figure 11.20a, respectively [38]. The CNT axes display the highest orientational orders. The interfacial segments exhibit slightly higher orientational orders along the CNT axes compared with those along the stretching direction. Besides, both of them are higher than those of all bonds along the stretching direction. These findings show that deformation induces the orientation of the CNTs, thereby causing the local orientation of segments in the interface domains along the CNT axes or stretching direction.

Figure 11.20b depicts the precursor bead fractions in the interface and the entire system under different strains [38]. Under the same strain, the interfacial precursor bead fraction is higher than the entire precursor bead fraction, suggesting that

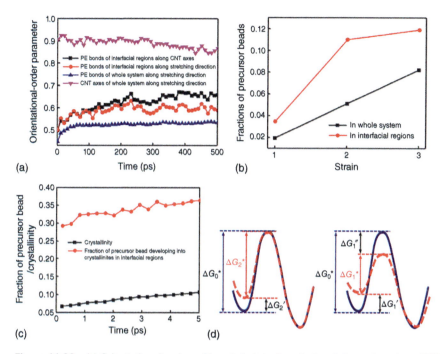

Figure 11.20 (a) Orientational orders of bonds in interface during deformation along CNT axes and stretching direction, and those of all bonds and CNT axes along stretching direction at the early stage of crystallization. (b) Fractions of precursor beads in interface and in whole system at different strains. (c) Fraction of precursor beads forming crystals in interface of stretched system at the strain of 3 during crystallization. (d) Free energy curves for heterogeneous nucleation of segments in interface (the right panel) and homogeneous nucleation of segments in non-interfacial regions (the left panel), respectively [38]. Source: Reproduced with permission from Nie et al. [38]/Elsevier.

there is a higher possibility of precursor formation for the interfacial segments. Two orientation effects exist that can affect the interfacial segment orientation. First, CNTs can cause the segments to orient along the CNT axes [16, 27, 28, 171–173]; second, stretching can also cause segments to orient along the stretching direction [174–178]. Therefore, the interfacial segment orientation should be larger than the non-interfacial segment orientation, resulting in a greater possibility of precursor formation for the interface segments. Figure 11.20c shows the proportional evolution of the precursor beads in the interface (the strain of 3) at the crystallization stage [38]. Under the same strain conditions, the fraction of precursor beads forming crystals was higher than the crystallinity, indicating that the nanofillers could induce the crystallization of the precursors composed of the highly oriented segments at the interface regions on the basis of heterogeneous nucleation.

Li and coworkers stated that both the reduction of conformational entropy (ΔS_c) and the reduction of orientation entropy (ΔS_o) influence nucleation process [179, 180]. The formation of the precursors formed by highly oriented segments can lead to the ΔS_o. Then, the ΔS_c and the ΔS_o cause the free energy change of the state prior to crystallization ($\Delta G' = -T(\Delta S_c + \Delta S_o)$), as shown in Figure 11.20d

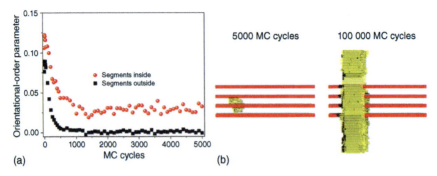

Figure 11.21 (a) Orientational orders of amorphous bonds inside and outside the filler network along the filler axis, respectively. (b) Snapshots of crystals in the corresponding system; red strips are fillers, and yellow cylinders are crystalline stems [30]. Source: Reproduced with permission from Nie's group [30]/American Chemical Society.

[38]. Since the interfacial segments exhibit higher orientation compared with the non-interfacial segments, leading to the higher absolute value of ΔS_0 of the oriented interfacial segments. Namely, $\Delta G'$ for the heterogeneous nucleation of segments in interface ($\Delta G_1'$) should be higher compared with $\Delta G'$ for the homogeneous nucleation ($\Delta G_2'$). As shown in Figure 11.20d, the peak value of the free energy curve during nucleation ($\Delta G_1''$) would decrease due to the heterogeneous nucleation [38]. Then, the barrier of free energy for the heterogeneous nucleation of segments in interface (ΔG_1^*) should be smaller compared with that for the homogeneous nucleation of segments in non-interfacial regions (ΔG_2^*).

In experiments, a number of nanoparticles, rather than a single nanoparticle, are introduced into polymers to form a filler network. Nie's group found that the relaxation and conformations of oriented chains would be restricted by the filler networks, and then the inside segments could exhibit higher degree of segment orientation and lower conformational entropy, as shown in Figure 11.21a [30]. Furthermore, the inside chains prefer to crystallize on the filler surfaces on the basis of heterogeneous nucleation, and the NHSK structure could finally form, as shown in Figure 11.21b [30].

11.9.2 Stretch-Induced Crystalline Structure Changes

Simulation work displayed that the stretch-induced structural changes in crystals depend on some factors, such as stretching direction, strain rate, and temperature [52, 181]. Yamamoto performed simulations to detect the micro-structural changes of the crystallized systems during the stretching along the transverse direction and along the fiber axis [181]. It was found that the stretching along the fiber axis can cause large reorientation of the crystalline tilted chain segments and lead to a specific yielding correlated with chain slips in crystals. In this condition, the crystallinity is almost unchanged at high strains, as shown in Figure 11.22a [181]. Differently, the transverse stretching can cause large-scale reconstruction of structures, including melting and recrystallization processes, as shown in Figure 11.22b.

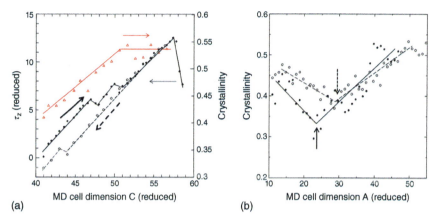

Figure 11.22 (a) Stress–strain curves during stretch (•) and contraction (○) from C~56 along the Z-axis at $T = 6.0$. Crystallinity (△) is also shown. (b) Crystallinity during transverse stretching at $T = 6.0$ (•) and $T = 5.0$ (○). Crystallinity minima are marked using solid and dotted arrows [181]. Source: Reproduced with permission from Yamamoto et al. [181]/Elsevier.

Rutledge's group also focused on the structural changes of semicrystalline polymers under tensile deformation [49, 51, 52]. In their MD simulations, the effects of strain rate and temperature were investigated in detail [52]. The cavitation in non-crystalline regions dominates the structural changes accompanied by little change in crystals, when the crystallized system is stretched using a fast strain rate at a high temperature, as shown in Figure 11.23a [52]. However, when the system was stretched using a slow extension rate at a high temperature, repeated melting/recrystallization events were observed, as shown in Figure 11.23b [52]. At the low temperature, only cavity formation mechanism was detected for both fast and slow strain rates.

Hu's group also detected the stretch-induced crystal structure changes [23]. They probed the chain-folding probability changes and crystallinity during stretching. It can be seen in Figure 11.24a that at high strains (the post-growth stage), the crystallinity levels off but the chain-folding probability decreases, suggesting that the stretching at high strains can extend the folded-chain segments with no crystallinity reduction [23]. To confirm whether the melting-recrystallization process occurs in the simulations, the evolutions of the crystallinity of two adjacent chains during stretching were also provided, as shown in Figure 11.24b [23]. The crystallinity of one chain (the 501st chain) decreases at high strains due to extending the folded-chain segments, while the crystallinity of its neighboring chain (the 207th chain) increases gradually, demonstrating the occurrence of the exchange of crystallinity based on a melting-recrystallization process. As further seen in Figure 11.24c, at the strain of 275%, there are many folded parts in the 501*st* chain, but at the strain of 400%, the folded chain parts in the crystals are extended to form highly oriented amorphous segments, and the crystallinity decreases, that is, strain-induced crystal melting occurs [23]. Those newly generated highly

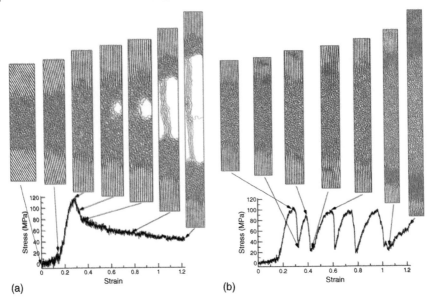

Figure 11.23 (a) Snapshots of PE and stress–strain curve with a high strain rate from a representative configuration (denoted case 2). (b) Snapshots of PE and stress–strain curve with a slow strain rate starting with the same representative configuration as in Figure 11.23a [52]. Source: Reproduced with permission from Rutledge's group [52]/American Chemical Society.

oriented amorphous segments will recrystallize together with other chains in neighboring locations to form a bundle of oriented crystallites based on the mode of intermolecular nucleation, thus increasing the crystallinity of other chains in the neighboring locations (such as the 207th chain). This "exchange" of crystallinity between chains at high strains is also the reason why the crystallinity of the polymer system remains constant at high strains.

11.10 Summary

Both MD and MC simulations are very useful for the investigations on polymer crystallization. During simulations, the structural changes of polymer systems can be probed at the molecular scale, and then the corresponding microscopic mechanisms can be thus revealed. However, the corresponding MD and MC simulations are mainly carried out based on the united atom chain model and the coarse-grained polymer model. If multi-scale simulations can be introduced to combine the chemical structure characteristics of polymers with high computational efficiency, the results of the simulations may gain more recognition in the industry.

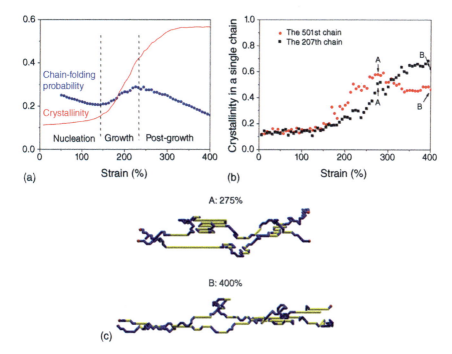

Figure 11.24 (a) Chain-folding probability and crystallinity during deformation at the temperature of 4.5. (b) crystallinity in two typical chains (the 501st and 207th chains) during deformation. The arrows show the instants at which the snapshots are shown in Figure 11.24c. (c) Snapshots of the 501st (the upper one in the middle region) and 207th (the lower one in the middle region) chains changing from folded-chain to extended-chain conformations at the strains of 275% and 400%, respectively, as marked by the arrows in Figure 11.24b. Amorphous segments were shown as blue cylinders, and crystalline ones were shown as yellow ones. Red spheres represent the chain ends restricted in the YZ boundary planes [23]. Source: Reproduced with permission from Nie et al. [23]/Elsevier.

References

1 Flory, P.J. (1953). *Principles of Polymer Chemistry*. Ithaca, New York: Cornell University Press.
2 Wunderlich, B. (1976). Macromolecular physics. In: *Crystal Nucleation, Growth, Annealing*, vol. 2, 1–34. New York: Academic Press.
3 Hu, W.B. (2013). *Polymer Physics: A Molecular Approach*. Vienna: Springer-Verlag.
4 Hu, W.B. (2013). *Principles of Polymer Crystallization*. Beijing: Chemical Industry Publisher.
5 Hoffman, J.D. and Lauritzen, J.I. (1961). Crystallization of bulk polymers with chain folding-theory of growth of lamellar spherulites. *Journal of Research of The National Bureau of Standards A Physical Chemistry* 65A: 297–336.

6 Hu, W.B. (2007). Intramolecular crystal nucleation. In: *Lecture Notes in Physics: Progress in Understanding of Polymer Crystallization* (ed. G. Strobl and G. Reiter), 47–63. Springer-Verlag: Berlin.

7 Strobl, G. (2000). From the melt via mesomorphic and granular crystalline layers to lamellar crystallites: a major route followed in polymer crystallization? *European Physical Journal E* 3: 165–183.

8 Allegra, G. (1977). Chain folding and polymer crystallization-statistical-mechanical approach. *Journal of Chemical Physics* 66: 5453–5463.

9 Sadler, D.M. (1987). New explanation for chain folding in polymers. *Nature* 326: 174–177.

10 Hu, W.B. (2018). The physics of polymer chain-folding. *Physics Reports* 747: 1–50.

11 Hu, W.B. (2022). Polymer features in crystallization. *Chinese Journal of Polymer Science* 40: 545–555.

12 Ko, M.J., Waheed, N., Lavine, M.S., and Rutledge, G.C. (2004). Characterization of polyethylene crystallization from an oriented melt by molecular dynamics simulation. *Journal of Chemical Physics* 121: 2823.

13 Jabbarzadeh, A. and Tanner, R.I. (2010). Flow-induced crystallization: unravelling the effects of shear rate and strain. *Macromolecules* 43: 8136–8142.

14 Sliozberg, Y.R., Yeh, I.C., Kröger, M. et al. (2018). Ordering and crystallization of entangled polyethylene melts under uniaxial tension: a molecular dynamics study. *Macromolecules* 51: 9635–9648.

15 Hu, W.B. and Frenkel, D. (2005). Polymer crystallization driven by anisotropic interactions. *Advances in Polymer Science* 191: 1–35.

16 Liu, R., Nie, Y., Ming, Y. et al. (2021). Simulations on polymer nanocomposite crystallization. *Polymer Crystallization* 4: e10214.

17 Guo, Y.Q., Luo, W., Wang, J.P., and Hu, W.B. (2022). Dynamic Monte Carlo simulations of strain-induced crystallization in multiblock copolymers: effects of dilution. *Soft Matter* 18: 3376–3383.

18 Guan, X.C., Wang, Y.H., Wang, J.P. et al. (2019). Effects of short-chain branches on strain-induced polymer crystallization. *Polymer International* 68: 225–230.

19 Guan, X.C., Wang, J.P., and Hu, W.B. (2018). Monte Carlo simulation of strain-enhanced stereo-complex polymer crystallization. *Journal of Physical Chemistry B* 122: 10928–10933.

20 Zha, L.Y., Wu, Y.X., and Hu, W.B. (2016). Multi-component thermodynamics of strain-induced polymer crystallization. *Journal of Physical Chemistry B* 120: 6890–6896.

21 Nie, Y.J., Gao, H.H., Wu, Y.X., and Hu, W.B. (2016). Effect of comonomer sizes on the strain-induced crystal nucleation of random copolymers. *European Polymer Journal* 81: 34–42.

22 Zhang, M.M., Zha, L.Y., Gao, H.H. et al. (2014). How polydispersity of network polymers influences strain-induced crystal nucleation in a rubber. *Chinese Journal of Polymer Science* 32: 1218–1223.

23 Nie, Y.J., Gao, H.H., and Hu, W.B. (2014). Variable trends of chain-folding in separate stages of strain-induced crystallization of bulk polymers. *Polymer* 55: 1267–1272.

24 Nie, Y.J., Gao, H.H., Wu, Y.X., and Hu, W.B. (2014). Thermodynamics of strain-induced crystallization of random copolymers. *Soft Matter* 10: 343–347.

25 Nie, Y.J., Gao, H.H., Yu, M.H. et al. (2013). Competition of crystal nucleation to fabricate the oriented semi-crystalline polymers. *Polymer* 54: 3402–3407.

26 Nie, Y.J., Zhao, Y.F., Matsuba, G., and Hu, W.B. (2018). Shish-Kebab crystallites initiated by shear fracture in bulk polymers. *Macromolecules* 51: 480–487.

27 Nie, Y., Zhang, R., Zheng, K., and Zhou, Z. (2015). Nucleation details of nanohybrid shish-kebabs in polymer solutions studied by molecular simulations. *Polymer* 76: 1–7.

28 Nie, Y., Hao, T., Wei, Y., and Zhou, Z. (2016). Polymer crystal nucleation with confinement-enhanced orientation dominating the formation of nanohybrid shish-kebabs with multiple shish. *RSC Advances* 6: 50451–50459.

29 Hao, T., Zhou, Z., Nie, Y. et al. (2016). Molecular simulations of crystallization behaviors of polymers grafted on two-dimensional filler. *Polymer* 100: 10–18.

30 Nie, Y., Hao, T., Gu, Z. et al. (2017). Relaxation and crystallization of oriented polymer melts with anisotropic filler networks. *Journal of Physical Chemistry B* 121: 1426–1437.

31 Nie, Y., Gu, Z., Zhou, Q. et al. (2017). Controllability of polymer crystal orientation using heterogeneous nucleation of deformed polymer loops grafted on two-dimensional nanofiller. *Journal of Physical Chemistry B* 121: 6685–6690.

32 Hao, T., Zhou, Z., Nie, Y. et al. (2017). Effect of the polymer-substrate interactions on crystal nucleation of polymers grafted on a flat solid substrate as studied by molecular simulations. *Polymer* 123: 169–178.

33 Gu, Z., Yang, R., Yang, J. et al. (2018). Dynamic Monte Carlo simulations of effects of nanoparticle on polymer crystallization in polymer solutions. *Computational Materials Science* 147: 217–226.

34 Liu, R., Zhou, Z., Liu, Y. et al. (2019). Epitaxial orientation and localized microphase separation prior to formation of nanohybrid shish-kebabs induced by one-dimensional nanofiller in miscible diblock copolymers with selective interaction. *Polymer* 166: 72–80.

35 Liu, R., Zhou, Z., Liu, Y. et al. (2019). Competition between interfacial interaction and microphase separation in crystallization of filled block copolymers. *Journal of Polymer Science Part B: Polymer Physics* 57: 1516–1526.

36 Liu, R., Yang, L., Qiu, X. et al. (2020). One-dimensional nanofiller induced crystallization in random copolymers studied by dynamic Monte Carlo simulations. *Molecular Simulation* 46: 669–677.

37 Liu, R.L., Zhou, Z.P., Liu, Y. et al. (2020). Differences in crystallization behaviors between cyclic and linear polymer nanocomposites. *Chinese Journal of Polymer Science* 38: 1034–1044.

38 Nie, Y., Yang, J., Liu, Z. et al. (2022). Precursor formation and crystal nucleation in stretched polyethylene/carbon nanotube nanocomposites. *Polymer* 239: 124438.

39 Yang, J., Liu, Z., Zhou, Z. et al. (2022). Studying the effects of carbon nanotube contents on stretch-induced crystallization behavior of polyethylene/carbon nanotube nanocomposites using molecular dynamics simulations. *Physical Chemistry Chemical Physics* 24: 16021–16030.

40 Gu, Z., Xu, Y., Lu, Q. et al. (2019). Stereocomplex formation in mixed polymers filled with two-dimensional nanofillers. *Physical Chemistry Chemical Physics* 21: 6443–6452.

41 Qiu, X., Liu, R., Nie, Y. et al. (2019). Monte Carlo simulations of stereocomplex formation in multiblock copolymers. *Physical Chemistry Chemical Physics* 21: 13296–13303.

42 Nie, Y., Liu, Y., Liu, R. et al. (2019). Dynamic Monte Carlo simulations of competition in crystallization of mixed polymers grafted on a substrate. *Journal of Polymer Science Part B: Polymer Physics* 57: 89–97.

43 Xu, Y., Wu, H., Yang, J. et al. (2020). Molecular simulations of microscopic mechanism of the effects of chain length on stereocomplex formation in polymer blends. *Computational Materials Science* 172: 109297.

44 Xu, Y., Yang, J., Liu, Z.F. et al. (2021). Stereocomplex crystallization in asymmetric diblock copolymers studied by dynamic monte carlo simulations. *Chinese Journal of Polymer Science* 39: 632–639.

45 Zhu, Q., Zhou, Z.P., Hao, T.F., and Nie, Y.J. (2022). Significantly improved stereocomplexation ability in cyclic block copolymers. *Chinese Journal of Polymer Science* 41: 432–441.

46 Graham, R.S. and Olmsted, P.D. (2009). Coarse-grained simulations of flow-induced nucleation in semicrystalline polymers. *Physical Review Letters* 103: 115702.

47 Graham, R.S. (2019). Understanding flow-induced crystallization in polymers: a perspective on the role of molecular simulations. *Journal of Rheology* 63: 203–214.

48 Lavine, M.S., Waheed, N., and Rutledge, G.C. (2003). Molecular dynamics simulation of orientation and crystallization of polyethylene during uniaxial extension. *Polymer* 44: 1771–1779.

49 Lee, S., Rutledge, G.C., Lee, S., and Rutledge, G.C. (2011). Plastic deformation of semicrystalline polyethylene by molecular simulation. *Macromolecules* 44: 3096–3108.

50 Yi, P., Locker, C.R., and Rutledge, G.C. (2013). Molecular dynamics simulation of homogeneous crystal nucleation in polyethylene. *Macromolecules* 46: 4723–4733.

51 Kim, J.M., Locker, R., and Rutledge, G.C. (2014). Plastic deformation of semicrystalline polyethylene under extension, compression, and shear using molecular dynamics simulation. *Macromolecules* 47: 2515–2528.

52 Yeh, I.C., Andzelm, J.M., and Rutledge, G.C. Mechanical and structural characterization of semicrystalline polyethylene under tensile deformation by molecular dynamics simulations. *Macromolecules* 48: 4228–4239.

53 Nicholson, D.A. and Rutledge, G.C. (2016). Molecular simulation of flow-enhanced nucleation in n-eicosane melts under steady shear and uniaxial extension. *Journal of Chemical Physics* 145: 244903.

54 Nicholson, D.A. and Rutledge, G.C. (2020). Flow-induced inhomogeneity and enhanced nucleation in a long alkane melt. *Polymer* 200: 122605.

55 Nicholson, D.A. and Rutledge, G.C. (2019). An assessment of models for flow-enhanced nucleation in an n-alkane melt by molecular simulation. *Journal of Rheology* 63: 465–475.

56 Yang, J., Tang, X., Wang, Z. et al. (2017). Coupling between intra-and inter-chain orderings in flow-induced crystallization of polyethylene: a non-equilibrium molecular dynamics simulation study. *Journal of Chemical Physics* 146: 014901.

57 Xie, C., Tang, X., Yang, J. et al. (2018). Stretch-Induced coil-helix transition in isotactic polypropylene: a molecular dynamics simulation. *Macromolecules* 51: 3994–4002.

58 Nie, C., Peng, F., Xu, T. et al. (2021). Biaxial stretch-induced crystallization of polymers: a molecular dynamics simulation study. *Macromolecules* 54: 9794–9803.

59 Tang, X., Yang, J., Tian, F. et al. (2018). Flow-induced density fluctuation assisted nucleation in polyethylene. *Journal of Chemical Physics* 149: 224901.

60 Yamamoto, T. (2009). Computer modeling of polymer crystallization–Toward computer-assisted materials' design. *Polymer* 50: 1975–1985.

61 Yamamoto, T. (2019). Molecular dynamics simulation of stretch-induced crystallization in polyethylene: emergence of fiber structure and molecular network. *Macromolecules* 52: 1695–1706.

62 Koyama, A., Yamamoto, T., Fukao, K., and Miyamoto, Y. (2002). Molecular dynamics simulation of polymer crystallization from an oriented amorphous state. *Physical Review E* 65: 050801.

63 Luo, C. and Sommer, J.U. (2011). Growth pathway and precursor states in single lamellar crystallization: MD simulations. *Macromolecules* 44: 1523–1529.

64 Xiao, H., Luo, C., Yan, D., and Sommer, J.U. (2017). Molecular dynamics simulation of crystallization cyclic polymer melts as compared to their linear counterparts. *Macromolecules* 50: 9796–9806.

65 Luo, C., Kröger, M., and Sommer, J.U. (2016). Entanglements and crystallization of concentrated polymer solutions: molecular dynamics simulations. *Macromolecules* 49: 9017–9025.

66 Jiang, S., Lu, Y., and Luo, C. (2022). State transitions and crystalline structures of a single polyethylene chain: MD simulations. *Journal of Physical Chemistry B* 126: 964–975.

67 Ma, R., Xu, D., and Luo, C.F. (2022). Effect of crystallization and entropy contribution upon the mechanical response of polymer nano-fibers: a steered molecular dynamics study. *Chinese Journal of Polymer Science* https://doi.org/10.1007/s10118-022-2817-y.

68 Chen, R. and Luo, C. (2022). Stretching effect on intrachain conformational ordering of polymers: a steered molecular dynamics simulation. *Polymer* 254: 125106.

69 Jabbari-Farouji, S., Rottler, J., Lame, O. et al. (2015). Plastic deformation mechanisms of semicrystalline and amorphous polymers. *ACS Macro Letters* 4: 147–150.

70 Jabbari-Farouji, S., Lame, O., Perez, M. et al. (2017). Role of the intercrystalline tie chains network in the mechanical response of semicrystalline polymers. *Physical Review Letters* 118: 217802.

71 Nie, Y., Gu, Z., Wei, Y. et al. (2017). Features of strain-induced crystallization of natural rubber revealed by experiments and simulations. *Polymer Journal* 49: 309–317.

72 Nie, Y. (2015). Strain-induced crystallization of natural rubber/zinc dimethacrylate composites studied using synchrotron X-ray diffraction and molecular simulation. *Journal of Polymer Research* 22: 1–10.

73 Zhang, W. and Larson, R.G. (2018). Direct all-atom molecular dynamics simulations of the effects of short chain branching on polyethylene oligomer crystal nucleation. *Macromolecules* 51: 4762–4769.

74 Luo, C. (2020). A challenging topic of computer simulations: Polymorphism in polymers. *Polymer Crystallization* 3: e10109.

75 Zhai, Z., Morthomas, J., Fusco, C. et al. (2019). Crystallization and molecular topology of linear semicrystalline polymers: simulation of uni- and bimodal molecular weight distribution systems. *Macromolecules* 52: 4196–4208.

76 Morthomas, J., Fusco, C., Zhai, Z. et al. (2017). Crystallization of finite-extensible nonlinear elastic Lennard-Jones coarse-grained polymers. *Physical Review E* 96: 052502.

77 Meyer, H. and Müller-Plathe, F. (2001). Formation of chain-folded structures in supercooled polymer melts. *Journal of Chemical Physics* 115: 7807–7810.

78 Meyer, H. and Müller-Plathe, F. (2002). Formation of chain-folded structures in supercooled polymer melts examined by MD simulations. *Macromolecules* 35: 1241–1252.

79 Reith, D., Meyer, H., and Müller-Plathe, F. (2001). Mapping atomistic to coarse-grained polymer models using automatic simplex optimization to fit structural properties. *Macromolecules* 34: 2335–2345.

80 Luo, C. and Sommer, J.U. (2009). Coding coarse grained polymer model for LAMMPS and its application to polymer crystallization. *Computer Physics Communications* 180: 1382–1391.

81 Paul, W., Yoon, D.Y., and Smith, G.D. (1995). An optimized united atom model for simulations of polymethylene melts. *Journal of Chemical Physics* 103: 1702–1709.

82 Hu, W.B., Frenkel, D., and Mathot, V.B.F. (2003). Free energy barrier to melting of single-chain polymer crystallite. *Journal of Chemical Physics* 118: 3455–3457.

83 Yang, J., Wang, C., Wang, K. et al. (2009). Direct formation of nanohybrid shish-kebab in the injection molded bar of polyethylene/multiwalled carbon nanotubes composite. *Macromolecules* 42: 7016–7023.

84 Nie, Y., Huang, G., Qu, L. et al. (2011). New insights into thermodynamic description of strain-induced crystallization of peroxide cross-linked natural rubber filled with clay by tube model. *Polymer* 52: 3234–3242.

85 Nie, Y., Qu, L., Huang, G. et al. Improved resistance to crack growth of natural rubber by the inclusion of nanoclay. *Polymers for Advanced Technologies* 23: 85–91.

86 Li, C.Y., Li, L., Cai, W. et al. (2005). Nano-hybrid shish-kebab: polymer decorated carbon nanotubes. *Advanced Materials* 17: 1198–1202.

87 Xu, J.Z., Chen, T., Yang, C.L. et al. (2010). Isothermal crystallization of poly(L-lactide) induced by graphene nanosheets and carbon nanotubes: a comparative study. *Macromolecules* 43: 5000–5008.

88 Yang, H., Chen, Y., Liu, Y. et al. (2007). Molecular dynamics simulation of polyethylene on single wall carbon nanotube. *Journal of Chemical Physics* 127: 094902.

89 Yang, J.S., Yang, C.L., Wang, M.S. et al. (2011). Crystallization of alkane melts induced by carbon nanotubes and graphene nanosheets: a molecular dynamics simulation study. *Physical Chemistry Chemical Physics* 13: 15476–15482.

90 Zong, G., Zhang, W., Ning, N. et al. (2013). Study of PE and iPP orientations on the surface of carbon nanotubes by using molecular dynamic simulations. *Molecular Simulation* 39: 1013–1021.

91 Ning, N., Fu, S., Zhang, W. et al. (2012). Realizing the enhancement of interfacial interaction in semicrystalline polymer/filler composites via interfacial crystallization. *Progress in Polymer Science* 37: 1425–1455.

92 Wang, L. and Duan, L. (2012). Isothermal crystallization of a single polyethylene chain induced by graphene: a molecular dynamics simulation. *Computational and Theoretical Chemistry* 1002: 59–63.

93 Patil, N., Balzano, L., Portale, G., and Rastogi, S. (2010). A study on the chain-particle interaction and aspect ratio of nanoparticles on structure development of a linear polymer. *Macromolecules* 43: 6749–6759.

94 Cui, K., Ma, Z., Tian, N. et al. (2018). Multiscale and multistep ordering of flow-induced nucleation of polymers. *Chemical Reviews* 118: 1840–1886.

95 Ma, Z., Balzano, L., and Peters, G.W.M. (2016). Dissolution and re-emergence of flow-induced shish in polyethylene with a broad molecular weight distribution. *Macromolecules* 49: 2720–2730.

96 Pérez, R.A., Córdova, M.E., López, J.V. et al. (2014). Nucleation, crystallization, self-nucleation and thermal fractionation of cyclic and linear poly(ε-caprolactone)s. *Reactive & Functional Polymers* 80: 71–82.

97 Zardalidis, G., Mars, J., Allgaier, J. et al. (2016). Influence of chain topology on polymer crystallization: poly (ethylene oxide) (PEO) rings vs. linear chains. *Soft Matter* 12: 8124–8134.

98 Iyer, K. and Muthukumar, M. (2018). Langevin dynamics simulation of crystallization of ring polymers. *Journal of Chemical Physics* 148: 244904.

99 Córdova, M.E., Lorenzo, A.T., Müller, A.J. et al. (2011). A comparative study on the crystallization behavior of analogous linear and cyclic poly (ε-caprolactones). *Macromolecules* 44: 1742–1746.

100 Samsudin, S.A., Kukureka, S.N., and Jenkins, M.J. (2017). Crystallisation kinetics of cyclic and linear poly (butylene terephthalate). *Journal of Thermal Analysis and Calorimetry* 128: 457–463.

101 Su, H.H., Chen, H.L., Díaz, A. et al. (2013). New insights on the crystallization and melting of cyclic PCL chains on the basis of a modified Thomson-Gibbs equation. *Polymer* 54: 846–859.

102 Li, L., Séréro, Y., Koch, M.H., and de Jeu, W.H. (2003). Microphase separation and crystallization in an asymmetric diblock copolymer: coupling and competition. *Macromolecules* 36: 529–532.

103 Loo, Y.L., Register, R.A., and Ryan, A.J. (2002). Modes of crystallization in block copolymer microdomains: breakout, templated, and confined. *Macromolecules* 35: 2365–2374.

104 Zha, L. and Hu, W. (2016). Molecular simulations of confined crystallization in the microdomains of diblock copolymers. *Progress in Polymer Science* 54: 232–258.

105 Nandan, B., Hsu, J.Y., and Chen, H.L. (2006). Crystallization behavior of crystalline-amorphous diblock copolymers consisting of a rubbery amorphous block. *Journal of Macromolecular Science, Part C: Polymer Reviews* 46: 143–172.

106 Wang, M., Hu, W., Ma, Y., and Ma, Y.Q. (2006). Confined crystallization of cylindrical diblock copolymers studied by dynamic Monte Carlo simulations. *The Journal of Chemical Physics* 124: 244901.

107 Wu, S., Wu, J., Huang, G., and Li, H. (2015). A shish-kebab superstructure in low-crystallinity elastomer nanocomposites: morphology regulation and load-transfer. *Macromolecular Research* 23: 537–544.

108 Wang, W., Xie, X., and Ye, X. (2010). Crystallization induced block copolymer decoration on carbon nanotubes. *Carbon* 48: 1680–1683.

109 Colson, J.P. and Eby, R.K. (1966). Melting temperatures of copolymers. *Journal of Applied Physics* 37: 3511–3514.

110 Alamo, R.G. and Mandelkern, L. (1991). Crystallization kinetics of random ethylene copolymers. *Macromolecules* 24: 6480–6493.

111 Hu, W., Mathot, V.B., and Frenkel, D. (2003). Phase transitions of bulk statistical copolymers studied by dynamic Monte Carlo simulations. *Macromolecules* 36: 2165–2175.

112 Mackay, M.E., Tuteja, A., Duxbury, P.M. et al. (2006). General strategies for nanoparticle dispersion. *Science* 311: 1740–1743.

113 Salavagione, H.J., Martínez, G., and Ellis, G. (2011). Recent advances in the covalent modification of graphene with polymers. *Macromolecular Rapid Communications* 32: 1771–1789.

114 Zhou, B., Tong, Z.Z., Huang, J. et al. (2013). Isothermal crystallization kinetics of multi-walled carbon nanotubes-graft-poly (ε-caprolactone) with high grafting degrees. *Crystengcomm* 15: 7824–7832.

115 Zhou, B., He, W.N., Jiang, X.Y. et al. (2014). Effect of molecular weight on isothermal crystallization kinetics of multi-walled carbon nanotubes-graft-poly(ε-caprolactone). *Composites Science and Technology* 93: 23–29.

116 Hao, T., Ming, Y., Zhang, S. et al. (2019). The effect of grafting density on the crystallization behaviors of polymer chains grafted onto one-dimensional nanorod. *Advances in Polymer Technology* 2019: 1–10.

117 Hao, T., Ming, Y., Zhang, S. et al. (2019). The influences of grafting density and polymer–nanoparticle interaction on crystallisation of polymer composites. *Molecular Simulation* 46: 678–688.

118 Ming, Y., Zhou, Z., Xu, D. et al. (2018). The effect of molecular weight of polymers grafted in two-dimensional filler on crystallization behaviors studied by dynamic Monte Carlo simulations. *Computational Materials Science* 155: 144–150.

119 Hao, T., Xu, D., Ming, Y. et al. (2020). Correlation between molecular weight and confined crystallization behavior of polymers grafted onto a zero-dimensional filler. *CrystEngComm* 22: 1779–1788.

120 Kim, J., Kwak, S., Hong, S.M. et al. (2010). Nonisothermal crystallization behaviors of nanocomposites prepared by in situ polymerization of high-density polyethylene on multiwalled carbon nanotubes. *Macromolecules* 43: 10545–10553.

121 Ming, Y., Zhou, Z., Hao, T. et al. (2021). Insights into the crystallization of polymer nanocomposite systems blended with grafted and free chains studied by molecular simulation. *Crystal Growth & Design* 21: 2243–2254.

122 Jana, R.N. and Cho, J.W. (2010). Thermal stability, crystallization behavior, and phase morphology of poly(ε-caprolactone) diol-grafted-multiwalled carbon nanotubes. *Journal of Applied Polymer Science* 110: 1550–1558.

123 Murariu, M. and Dubois, P. (2016). PLA composites: from production to properties. *Advanced Drug Delivery Reviews* 107: 17–46.

124 Vink, E.T.H., Rábago, K.R., Glassner, D.A., and Gruber, P.R. (2003). Applications of life cycle assessment to NatureWorks™ polylactide (PLA) production. *Polymer Degradation and Stability* 80: 403–419.

125 Chen, C.C., Chueh, J.Y., Tseng, H. et al. (2003). Preparation and characterization of biodegradable PLA polymeric blends. *Biomaterials* 24: 1167–1173.

126 Henmi, K., Sato, H., Matsuba, G. et al. (2016). Isothermal crystallization process of poly(l-lactic acid)/poly(d-lactic acid) blends after rapid cooling from the melt. *ACS Omega* 1: 476–482.

127 Ikada, Y., Jamshidi, K., Tsuji, H., and Hyon, S.H. (1987). Stereocomplex formation between enantiomeric poly (lactides). *Macromolecules* 20: 904–906.

128 Tsuji, H. and Ikada, Y. (1993). Stereocomplex formation between enantiomeric poly (lactic acids). 9. Stereocomplexation from the melt. *Macromolecules* 26: 6918–6926.

129 Zhang, P., Tian, R., Na, B. et al. (2015). Intermolecular ordering as the precursor for stereocomplex formation in the electrospun polylactide fibers. *Polymer* 60: 221–227.

130 Zhang, J., Sato, H., Tsuji, H. et al. (2005). Infrared spectroscopic study of CH_3...OC interaction during poly(L-lactide)/poly(D-lactide) stereocomplex formation. *Macromolecules* 38: 1822–1828.

131 Zhang, Z.C., Sang, Z.H., Huang, Y.F. et al. (2017). Enhanced heat deflection resistance via shear flow-induced stereocomplex crystallization of polylactide systems. *ACS Sustainable Chemistry & Engineering* 5: 1692–1703.

132 Tsuji, H., Nakano, M., Hashimoto, M. et al. (2006). Electrospinning of poly (lactic acid) stereocomplex nanofibers. *Biomacromolecules* 7: 3316–3320.

133 Tsuji, H., Hyon, S.H., and Ikada, Y. (1991). Stereocomplex formation between enantiomeric poly (lactic acid) s. 4. Differential scanning calorimetric studies on precipitates from mixed solutions of poly (D-lactic acid) and poly (L-lactic acid). *Macromolecules* 24: 5657–5662.

134 Zhou, K.Y., Li, J.B., Wang, H.X., and Ren, J. (2017). Effect of star-shaped chain architectures on the polylactide stereocomplex crystallization behaviors. *Chinese Journal of Polymer Science* 35: 974–991.

135 Pan, P., Bao, J., Han, L. et al. (2016). Stereocomplexation of high-molecular-weight enantiomeric poly (lactic acids) enhanced by miscible polymer blending with hydrogen bond interactions. *Polymer* 98: 80–87.

136 He, S., Bai, H., Bai, D. et al. (2019). A promising strategy for fabricating high-performance stereocomplex-type polylactide products via carbon nanotubes-assisted lowtemperature sintering. *Polymer* 162: 50–57.

137 Liu, H., Zhou, W., Chen, P. et al. (2020). A novel aryl hydrazide nucleator to effectively promote stereocomplex crystallization in high-molecular-weight poly(L-lactide)/poly(D-lactide) blends. *Polymer* 210: 122873.

138 Zhang, R., Zha, L.Y., and Hu, W.B. (2016). Intramolecular crystal nucleation favored by polymer crystallization: a Monte Carlo simulation evidence. *Journal of Physical Chemistry B* 120: 6754–6760.

139 Pan, P., Han, L., Bao, J. et al. (2015). Competitive stereocomplexation, homocrystallization, and polymorphic crystalline transition in poly (L-lactic acid)/poly (D-lactic acid) racemic blends: molecular weight effects. *Journal of Physical Chemistry B* 119: 6462–6470.

140 Tsuji, H. and Bouapao, L. (2012). Stereocomplex formation between poly (L-lactic acid) and poly (D-lactic acid) with disproportionately low and high molecular weights from the melt. *Polymer International* 61: 442–450.

141 Azzurri, F. and Alfonso, G.C. (2005). Lifetime of shear-induced crystal nucleation precursors. *Macromolecules* 38: 1723–1728.

142 Cavallo, D., Azzurri, F., Balzano, L. et al. (2010). Flow memory and stability of shear-induced nucleation precursors in isotactic polypropylene. *Macromolecules* 43: 9394–9400.

143 Hayashi, Y., Matsuba, G., Zhao, Y. et al. (2009). Precursor of shish-kebab in isotactic polystyrene under shear flow. *Polymer* 50: 2095–2103.

144 Kanaya, T., Polec, I.A., Fujiwara, T. et al. (2013). Precursor of shish-kebab above the melting temperature by microbeam X-ray scattering. *Macromolecules* 46: 3031–3036.

145 Seki, M., Thurman, D.W., Oberhauser, J.P., and Kornfield, J.A. (2002). Shear-mediated crystallization of isotactic polypropylene: The role of long chain-long chain overlap. *Macromolecules* 35: 2583–2594.

146 Kimata, S., Sakurai, T., Nozue, Y. et al. (2007). Molecular basis of the shish-kebab morphology in polymer crystallization. *Science* 316: 1014–1017.

147 Zhang, C., Hu, H., Wang, X. et al. (2007). Formation of cylindrite structures in shear-induced crystallization of isotactic polypropylene at low shear rate. *Polymer* 48: 1105–1115.

148 Sun, T., Chen, F., Dong, X. et al. (2009). Shear-induced orientation in the crystallization of an isotactic polypropylene nanocomposite. *Polymer* 50: 2465–2471.

149 Somani, R.H., Yang, L., Zhu, L., and Hsiao, B.S. (2005). Flow-induced shish-kebab precursor structures in entangled polymer melts. *Polymer* 46: 8587–8623.

150 Hsiao, B.S., Yang, L., Somani, R.H. et al. (2005). Unexpected shish-kebab structure in a sheared polyethylene melt. *Physical Review Letters* 94: 117802.

151 Murase, H., Ohta, Y., and Hashimoto, T. (2011). A new scenario of shish-kebab formation from homogeneous solutions of entangled polymers: Visualization of structure evolution along the fiber spinning line. *Macromolecules* 44: 7335–7350.

152 Murase, H., Ohta, Y., and Hashimoto, T. (2009). Shear-induced phase separation and crystallization in semidilute solution of ultrahigh molecular weight polyethylene: phase diagram in the parameter space of temperature and shear rate. *Polymer* 50: 4727–4736.

153 Wang, Z., Ma, Z., and Li, L. (2016). Flow-induced crystallization of polymers: Molecular and thermodynamic considerations. *Macromolecules* 49: 1505–1517.

154 Tang, H., Chen, J.B., Wang, Y. et al. (2012). Shear flow and carbon nanotubes synergistically induced nonisothermal crystallization of poly (lactic acid) and its application in injection molding. *Biomacromolecules* 13: 3858–3867.

155 Chen, Y.H., Zhong, G.J., Lei, J. et al. (2011). In situ synchrotron X-ray scattering study on isotactic polypropylene crystallization under the coexistence of shear flow and carbon nanotubes. *Macromolecules* 44: 8080–8092.

156 Zhao, Y.F., Hayasaka, K., Matsuba, G., and Ito, H. (2013). In situ observations of flow-induced precursors during shear. *Macromolecules* 46: 172–178.

157 Zhao, Y.F., Matsuba, G., Moriwaki, T. et al. (2012). Shear-induced conformational fluctuations of polystyrene probed by 2D infrared microspectroscopy. *Polymer* 53: 4855–4860.

158 Balzano, L., Kukalyekar, N., Rastogi, S. et al. (2008). Crystallization and dissolution of flow-induced precursors. *Physical Review Letters* 100: 048302.

159 Pogodina, N.V., Lavrenko, V.P., Srinivas, S., and Winter, H.H. (2001). Rheology and structure of isotactic polypropylene near the gel point: quiescent and shear-induced crystallization. *Polymer* 42: 9031–9043.

160 Pogodina, N.V., Siddiquee, S.K., van Egmond, J.W., and Winter, H.H. (1999). Correlation of rheology and light scattering in isotactic polypropylene during early stages of crystallization. *Macromolecules* 32: 1167–1174.

161 Peterlin, A. (1971). Molecular model of drawing polyethylene and polypropylene. *Journal of Materials Science* 6: 490–508.

162 Flory, P.J. and Yoon, D.Y. Molecular morphology in semi-crystalline polymers. *Nature* 272: 226–229.

163 Men, Y.F., Rieger, J., and Strobl, G. (2003). Role of the entangled amorphous network in tensile deformation of semicrystalline polymers. *Physical Review Letters* 91: 095502.

164 Hiss, R., Hobeika, S., Lynn, C., and Strobl, G. (1999). Network stretching, slip processes, and fragmentation of crystallites during uniaxial drawing of polyethylene and related copolymers. a comparative study. *Macromolecules* 32: 4390–4403.

165 Men, Y. (2020). Critical strains determine the tensile deformation mechanism in semicrystalline polymers. *Macromolecules* 53: 9155–9157.

166 Lu, Y. and Men, Y.F. (2018). Initiation, development and stabilization of cavities during tensile deformation of semicrystalline polymers. *Chinese Journal of Polymer Science* 36: 1195–1199.

167 Sheng, J., Chen, W., Cui, K., and Li, L. (2022). Polymer crystallization under external flow. *Reports on Progress in Physics* 85: 036601.

168 Nie, C., Peng, F., Cao, R. et al. (2022). Recent progress in flow-induced polymer crystallization. *Journal of Polymer Science* https://doi.org/10.1002/pol.20220330.

169 Barham, P.J. and Keller, A. (1985). High-strength polyethylene fibres from solution and gel spinning. *Journal of Materials Science* 20: 2281–2302.

170 Xu, J., Zhong, G., Hsiao, B.S. et al. (2013). Low-dimensional carbonaceous nanofiller induced polymer crystallization. *Progress in Polymer Science* 39: 555–593.

171 Li, L.Y., Li, C.Y., and Ni, C.Y. (2006). Polymer crystallization-driven, periodic patterning on carbon nanotubes. *Journal of the American Chemical Society* 128: 1692–1699.

172 Hu, X., An, H., Li, Z. et al. (2009). Origin of carbon nanotubes induced poly (L-lactide) crystallization: surface induced conformational order. *Macromolecules* 42: 3215–3218.

173 Malagù, M., Lyulin, A., Benvenuti, E., and Simone, A. (2016). A molecular-dynamics study of size and chirality effects on glass-transition temperature and ordering in carbon nanotube-polymer composites. *Macromolecular Theory and Simulations* 25: 571–581.

174 Galesk, A. (2003). Strength and toughness of crystalline polymer systems. *Progress in Polymer Science* 28: 1643–1699.

175 Keller, A. and Polym, J. (1955). Unusual orientation phenomena in polyethylene interpreted in terms of the morphology. *Science* 15: 31–49.

176 Keller, A., Macromol, J., and Machin, M. (1967). Oriented crystallization in polymers. *Science, Part B: Physics* 1: 41–91.

177 Pennings, A.J. and Kiel, A.M. (1965). Fractionation of polymers by crystallization from solution, III. On morphology of fibrillar polyethylene crystals grown in solution. *Colloid and Polymer Science* 205: 160–162.

178 De Gennes, P.G. (1974). Coil-stretch transition of dilute flexible polymers under ultrahigh velocity gradients. *Journal of Chemical Physics* 60: 5030.

179 Chen, X., Meng, L., Zhang, W. et al. (2019). Frustrating strain-induced crystallization of natural rubber with biaxial stretch. *ACS Applied Material Interfaces* 11: 47535–47544.

180 Nie, C., Peng, F., Xu, T.Y. et al. (2021). A unified thermodynamic model of flow-induced crystallization of polymer. *Chinese Journal of Polymer Science* 39: 1489–1495.

181 Yamamoto, T. (2013). Molecular dynamics in fiber formation of polyethylene and large deformation of the fiber. *Polymer* 54: 3086–3097.

12

Application, Recycling, Environmental and Safety Issues, and Future Prospects of Crystalline Polymer Composites

Busra Cetiner[1,2], Havva Baskan-Bayrak[2], and Burcu S. Okan[1,2]

[1] Sabanci University, Faculty of Engineering and Natural Sciences, Materials Science and Nano Engineering, Tuzla, Istanbul 34956, Turkey
[2] Sabanci University, Integrated Manufacturing Technologies Research and Application Center & Composite Technologies Center of Excellence, Sanayi Street, Teknopark, Istanbul 34906, Turkey

12.1 Introduction

Crystalline polymers that have high crystallinity supplied by the high regularity of polymer chains are of great interest as engineering plastics in various industries. High crystallinity results in increased density, hardness, melting, and crystallization temperature [1].

Crystalline polymers are divided into two categories as fully crystalline and semi-crystalline polymers. Semi-crystalline polymers have a smaller degree of crystallinity and crystal size than highly crystalline polymers. Crystalline polymers are preferred to be used in composite form in order to enhance the structural and performance characteristics of them, and thus the development and utilization of polymer composite structures date back to the 1940s. Moreover, crystalline polymer composites are generally reinforced with organic, inorganic, or natural-based fillers in order to meet the requirements and demands of the industry, such as high strength, light weight, flexibility, and cost efficiency. The reinforcement materials generate a synergistic effect between the polymer matrix and filler, resulting in a large interphase area and restricting the chain mobility of polymer chains. Hence, the characteristics of crystalline polymer composites are improved, which enables them to be successfully used in many application areas such as automotive, marine, military, defense, aerospace, and biomedical [2].

Since the demand and utilization of crystalline polymer composites increased in industry, the remnants have started to accumulate. The accumulation of wastes is not eco-friendly, it is toxic to human health and risk to the safety of the environment. To this end, recycling comes into prominence to diminish the accumulation of wastes and reduce the environmental and health safety risks and impacts [3].

In this chapter, the evolution, and the principles of production of crystalline polymer composite structures are described, as well as the recent progress of crystalline polymer composites, in detail.

Polymer Crystallization: Methods, Characterization, and Applications, First Edition.
Edited by Jyotishkumar Parameswaranpillai, Jenny Jacob, Senthilkumar Krishnasamy, Aswathy Jayakumar, and Nishar Hameed.
© 2023 WILEY-VCH GmbH. Published 2023 by WILEY-VCH GmbH.

In addition, the current applications of crystalline polymer composites and waste management techniques are discussed in terms of reinforcing agent type. Moreover, the environmental impacts, safety issues, and future aspects of crystalline polymer composites are concluded.

12.2 Crystalline Polymers and Composites

12.2.1 Crystalline Polymers

Crystalline polymers, which are considered as emerging class of polymers from the industrial and scientific view, offer great chemical, thermal, and mechanical properties [4]. They enable the polymeric structures to behave physically as charge transport agents or gas permeability barriers [5, 6].

The crystallinity of polymers is fall into three categories as highly crystalline (90% crystallinity), semi-crystalline (80–10% crystallinity), and amorphous (0% crystallinity) [4] according to the orientation of polymer chains [7]. The degree of crystallinity is characterized by various parameters, including the molecular weight, tacticity, chain crosslinking, density, reinforcement presence, and cooling rate of the polymer melt [8, 9]. Lopez and Wilkes examined the effect of molecular weight of poly(p-phenylene sulfide) on the crystallinity and found out that the crystallinity of polymer decreases with the increased molecular weight [10]. Molecular weight effect on the crystallinity of a poly(3-hexylthiophene) was studied by Zen et al., who obtained that low molecular weight segments of poly(3-hexylthiophene) contain highly crystalline structures [11]. It can be concluded that high molecular weight fractions result in lower degree of crystallinity. The effect of tacticity, which is the configuration of repeating side chains, creates syndiotactic, isotactic, and atactic configurations, is mainly studied together with polypropylene (PP). Choi and coworkers studied the three different tacticities of methyl groups on the side chain of PP and mentioned that due to ordered pack structures of side chains, isotactic alignments showed a higher degree of crystallinity than syndiotactic configuration, while atactic alignment did the least [12, 13]. In another study of tacticity influence on the crystallization of isotactic PP was performed by Rungswangs et al., who observed that higher tacticity of isotactic PP advocated the crystallization process by improving nucleation rate and process [14]. Hence, tacticity is an important parameter of polymers that higher tacticity of polymer results in enhanced nucleation rate and so increased degree of crystallization. Crosslinking effect on the crystallization of ultra-high molecular weight polyethylene (PE) was investigated by Ries and Pruitt and showed that free radicals are neutralized during the crosslinking, which caused decrease in crystallization level [15]. An additional study on the impact of crosslinking on the crystallization of polymers, specifically on poly(methyl methacrylate) was considered by Hussein et al. The increased crosslinking degree resulted in higher crystallinity in both aromatic and aliphatic polymer configurations [16]. The density of polymers is another parameter that alters the total crystallinity of polymers. The crystallinity of PE was examined by taking into account its different densities by Li et al. It was concluded that density increment led to obtain high degree of crystallinity [17]. In addition, grafting density of a polymer and its influence on crystallinity were simulated by Zhang et al. by mentioning that in confined systems, low grafting density caused high crystallization [18]. Hence, it can be fairly said that while high-density polymers hold high crystallinity, grafting density of restricted regions has contrary effects on the total crystallinity. Otherwise, reinforcement presence alters the mechanical and thermal properties of polymers. Li and Qui said that cellulose nanocrystal-reinforced poly(butylene succinate-co-1,2-decylene succinate)-(PBDS) had doubled tensile strength and modulus compared to its virgin counterparts [19]. Vardai et al. stated that the crystallinity of PP matrix was enhanced by reinforcing it with talc since talc acted as a nucleating agent [20]. Furthermore, Cousineau and coworkers informed that slow cooling rate of PP contributes to obtain high crystallinity degrees [21]. Since it is hard to encounter fully crystalline polymers except in single crystals [22], in order to increase the crystallinity of a polymer, the parameters given above should be regarded.

Highly crystalline polymers consist of regularly folded chains, which are known as lamella (shown in Figure 12.1). The crystalline formation is initiated when the lamella structures are folded and aligned in the same direction [24]. In the absence of a thermal gradient, the obtained crystalline formations turn into spheroidal and spherulite structures (Figure 12.1) with the effect of translation [25, 26]. As a last stage, the crystallinity of polymers can be achieved by several processes, such as cooling of the melt-phase polymer, diluting the polymer solution under melting temperature, or initiating a stretching effect on the polymer chain [23].

Strong intermolecular forces, Vander-Walls, and hydrogen bonds, regulate the polymer chains to generate crystal formation. The repelling interaction of the polymer backbone chain results in crystallization, as well. On the other hand, semi-crystalline polymers are generally in metastases state that contains both crystalline and amorphous regions, in which the crystalline components are randomly distributed and connected to each other by amorphous chains (shown in Figure 12.2).

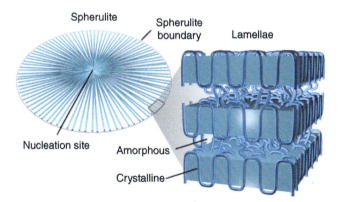

Figure 12.1 The representative image of lamella and spherulite structures [23]. Source: Reproduced with the permission of the Elsevier.

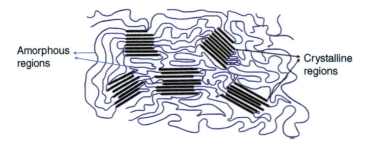

Figure 12.2 The illustration of structure of semi-crystalline polymers consisting of both crystalline and amorphous regions.

Semi-crystalline polymers have imperfect crystalline structure and non-consistency in their particle size [27]. However, they have good stiffness, strength and toughness, wear resistance, chemical resistance, and good mechanical properties [28]. They do not soften gradually when the temperature gradient rises, instead they remain stable until a particular heat is absorbed and reached to the transition temperatures. Semi-crystalline polymers have a melting temperature below that of highly crystalline polymers because of the imperfection of polymer backbone structure consisting of small and nonuniform crystals [27]. The thermal properties and the degree of crystallinity of semi-crystalline polymers allow to design and mold them according to the required demands of applications. The crystallinity degrees of several significant semi-crystalline polymers have been examined and reported in the literature. For instance, the crystallinity of polyethylene terephthalate (PET) was found to be 34.8 [29], while the crystallinity of polybutylene terephthalate (PBT) changed in the range of 24.9–44.5% [30]. In addition, the crystallinity degrees of low-density polyethylene (LDPE), Nylon 66, syndiotactic polystyrene (PS), polyoxymethylene (POM), polytetrafluoroethylene (PTFE), polyphenylene sulfide (PPS), polyether ether ketone (PEEK), and polyvinyl chloride (PVC) were calculated as 39.4% [31], 39.4% [32], 50% [33], 51.1 % [34], 53% [35], 56% [36], 20% [37], and 14.3% [38], respectively. The crystallinity values of these polymers are also summarized in Table 12.1.

12.2.2 Crystalline Polymer Composites

Polymer composites are the foremost materials used to reach and improve the demands such as high specific abrasion, high corrosion resistance, high modulus and strength, lightweight, and flexibility for many applications that could not be supplied by pure single polymers [39]. The improvements in the material performance are mainly achieved by the substantial interface that occurred between the virgin polymer chains and nano/micro-sized reinforcement heterogeneities. By this interaction, efficient stress transfer and the required space can be supplied for the crystallization of polymer chains [2, 40]. Such application of reinforcements to the polymer composites creates a nucleation effect, which initiates the crystallization of polymer chains and results in enhancement in total crystallinity of the

Table 12.1 The crystallinity values of different semi-crystalline polymers.

Type of polymer	Crsytallinity (%) of the polymer	References
Polyethylene terephthalate (PET)	34.8	[29]
Polybutylene terephthalate (PBT)	24.9–44.5	[30]
Low-density polyethylene (LDPE)	39.4	[31]
Nylon 66	39.4	[32]
Syndiotactic polystyrene (PS)	50	[33]
Polyoxymethylene (POM)	51.1	[34]
Polytetrafluoroethylene (PTFE)	53	[35]
Polyphenylene sulfide (PPS)	56	[36]
Polyether ether ketone (PEEK)	20	[37]
Polyvinyl chloride (PVC)	14.3	[38]

polymer composite [41, 42]. The loading concentration of the reinforcements has a paramount importance on the dynamics of the polymer chains. At low loading concentrations of reinforcements, the crystallinity of the final material is increased since reinforcements are replaced by primary nuclei, whereas at high loading concentrations of them, the space between each particle is reduced, and hence agglomeration of particles occurs hindering the growth of the crystalline structure of polymer matrices [43–47].

The nucleation effect and activity of fillers absolutely alter the thermal and mechanical properties of polymer composites. The conceptual idea of both advancements relies on the type of interface interactions between polymer matrix and filler. Strong, synergetic, and favorable interactions lead to increments, while inadequate and unfavorable interactions lead to reductions in the resultant polymer composite features [46].

Aside from the nucleating agent, the crystallization behavior intensely depends on homogeneous distribution and the alignment of fillers in the polymer matrix [45, 46]. The homogeneous dispersion of particles leads to perfect interfacial interactions and large interphase between the particles and polymer chains [47, 48]. Moreover, distribution of nanoparticles takes place according to the entropy of each particle, and the formed locations have significant influences on the crystal structure developments. As an example, the phase-separated, intercalated, and exfoliated dispersion behavior of layered clay particles is given in Figure 12.3.

In case of the particles having strong interactions with each other, the polymer matrix cannot interfere with them; thus, only phase-separated dispersion appears (Figure 12.3a). When the polymer chains can intercalate between the layers, the formation of perfectly-ordered multilayer structure is obtained (Figure 12.3b). If the polymer chains are completely and uniformly distributed with an average distance between the layers, exfoliated layers in polymer matrix are seen

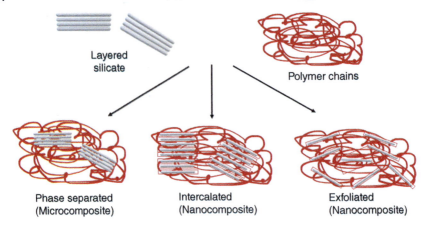

Figure 12.3 Scheme of the main dispersion behaviors of layered clay particles in polymer matrix: (a) microcomposite, (b) intercalated nanocomposite, and (c) exfoliated nanocomposite.

(Figure 12.3c). Exfoliated crystalline polymer structures possess higher surface area with maximum reinforcement but lower filler content than intercalated dispersed composites [2, 49].

Since the size and type of the filler/reinforcing agent determine the final properties of the crystalline polymer composites, it is a need to consider the crystalline polymer composites based on their reinforcement type. The crystalline polymer composites can be examined under the headings as crystalline polymer composites, including organic-based nano/micromaterials, inorganic-based nano/micromaterials, or natural nano/micromaterials as reinforcements.

12.2.2.1 Crystalline Polymer Composites with Organic Reinforcements

Organic-based reinforcement materials are mainly the nano/micro sized materials consisting of mostly carbon and oxygen atoms and small portions of other atoms. Carbon nanotubes (CNTs), carbon fibers (CFs), graphene derivatives as graphene nanosheets (GNSs), graphene nanoplates (GNPs), graphene nanoflakes (GNFs), graphene oxide (GO), fullerenes, carbon black, aramid, and Kevlar are the mostly utilized reinforcement materials to enhance the chemical/mechanical properties of the crystalline polymer composites [50]. For example, Cadek et al. showed that PVA-multiwalled carbon nanotube (MWCNT) polymer composite had an increased crystallinity of up to 27% as well as enhanced Young's modulus by the factor of 1.8 compared to the virgin PVA polymer [51]. Liang et al. mentioned that single-walled carbon nanotube (SWCNT)-reinforced PP composite had improved compressive strength, tensile strength, and flexural strength in the percentages of 133, 74, and 53, respectively. In addition, the presence of SWCNT in the polymer composite structure contributed to obtaining a great electromagnetic interference (EMI) shielding performance of 48.3 dB at 2.2 mm thickness [52]. In the other study, Liang et al. reported that by 20 wt% CF loading, 50% of boosted tensile strength, and 60% of enhanced Young's modulus of the composite could be maintained due to the contribution of CF to crystallization of polymer chains [53]. A further study was performed by Batista et al. to determine the effect of CF reinforcement on the crystallization degree of PPS matrix under different cooling rates. When the cooling was changed from rapid cooling to slow cooling and further air cooling, the crystallinity degree of the crystalline composite material was determined as 51%, 58%, and 62%, respectively [54].

Among the graphene derivatives, GNSs were incorporated into the highly crystalline polybutylene succinate structure and caused 21% of tensile strength and 24% of storage modulus increase even by the addition of 2 wt% [55]. On the other hand, its influence on the isotactic PP polymer matrix crystallization was also studied and it was observed that the half crystallization time of the isotactic PP-based GNSs-reinforced composites was diminished at least to 50% compared to that of neat PP [56]. Beyond GNSs, GNPs, GNFs also had a significant enhancement on the mechanical, thermal, and electrical features of the crystalline polymer composites. For example, GNPs reinforcement to the liquid crystalline polymer nanocomposites enhanced the modulus values 55% [57], while GNP incorporation to HDPE polymer matrix resulted in improved thermal stability [58], and GNFs addition to polyvinylidene fluoride (PVDF) matrix boosted the electrical conductivity of the polymer composite up to $1600\,S\,m^{-1}$ [59] as well as the thermal transfer to $10\,W\,mK^{-1}$ [60]. In addition, GO containing ultra-high molecular weight PE (UHMWPE) showed reduced creep resistance up to 9% with 80% of improved tensile strength [61]. A further study related to GO reinforcement was examined by Wan et al. by comparing the interfacial interactions of four different polymer matrices as polycaprolactone (PCL), poly-L-lactic acid (PLLA), PS, and high-density polyethylene (HDPE). The existence of GO showed great compatibility with polymer matrices due to its surface chemistry and high interfacial interactions, which demonstrated advancements on the Young's modulus of PLLA, PCL, PS, and HDPE as 60%, 109%, 95%, and 35%, respectively [62].

Furthermore, a further study was conducted to search for the changes in the electrical properties of crystalline polymer composites. To this end, Burmistrov et al. prepared carbon black-containing ethylene-octene copolymer and stated that the incorporation of carbon black into the polymer structure improved the electrical conductivity of the composite by 5 times magnitude compared to the pure copolymer [63].

Last but not the least, short Kevlar fibers were utilized in PP composites and waste PE composites. Addition of Kevlar fibers improved the tensile strength of PP composite by 53%, increased the crystallization temperature, and the degree of crystallinity, and improved thermal stability, as well [64]. Additionally, the presence of Kevlar fibers in the waste PE composite had a great contribution (151%) on the toughness and 33% of enhancement on the tensile strength of final material [65].

12.2.2.2 Crystalline Polymer Composites with Inorganic Reinforcements

The inorganic-based reinforcers generally exist in the ground generated by minerals, and they boost the mechanical performance of the composites as well as their crystallinity. The fundamental inorganic-based fillers are boron nitride (BN), clay-like and ceramic-based fillers materials such as silica, montmorillonite (MMT), talc, calcium carbonate ($CaCO_3$), and glass fibers [66]. In a study, conducted by Chen et al., hexagonal boron nitride (h-BN) and alumina particles were incorporated into PTFE structure, and it was mentioned that the friction coefficient of crystalline composites enhanced 21.32% by lessening the wear rate by 48.09% [67]. Moreover, Cui et al. underlined a synergistic effect that occurred between BN and PS, and BN and PA6. Introduction to BN to PS and PA6 polymer matrices contributed to the improvements in the thermal conductivities of polymer composites in the percentages of 38, and 34, respectively [68]. In one of the works, the effect of organically montmorillonite (OMMT) incorporation on the thermal and mechanical features of polylactic acid (PLA) and poly(butylene adipate-co-terephthalate) (PBAT) blend was examined, and it was observed that 48.26 MPa tensile strength with improved glass transition and crystallization temperature could be achieved even by the addition of 5 wt % OMMT [69]. On the other hand, Liu et al. reinforced UHMWPE with MMT and stated that the integration of MMT into UHMWPE decreased the crystallization rate constant and enlarged the range of crystal growth temperature [70]. In an another study, Makhlouf et al. examined the crystallinity of PP/submicronic-talc composites and found that the crystallinity of PP polymer was increased up to 95% by addition of talc filler [71]. Jain et al. also investigated the influence of talc on thermal, mechanical, and rheological behavior of PLA polymer and informed that the crystallinity of the PLA was advanced from 4.3% to 35%, while tensile and storage modulus were enhanced by 85% and 37%, respectively [72]. Moreover, the significant improvement in the tensile modulus of the crystalline polymer composites was ensured by Lapcik et al. by reinforcing the HDPE composites with spherical hollow sphere $CaCO_3$ particles [73].

12.2.2.3 Crystalline Polymer Composites with Natural Reinforcements

Except from the organic and inorganic-based reinforcements, natural fibers are also appropriate reinforcement materials used to support the crystallization and mechanical performance of the polymer composites. They are easily found in nature in plant forms such as cellulose-based fibers and crystals, cotton kenaf, flax, sisal, hemp, jute, ramie, and kraft pulp fibers [74]. It was observed that the cellulose fiber-reinforced PU polymer matrix gained 400% of increased modulus values whereas the thermal features of them were slightly tuned [75]. The Young's modulus of the PLA composites was also raised by 50% in case of nanocrystalline cellulose reinforcement [76]. In addition, Debeli et al. mentioned that the addition of ramie fibers into PLA composites led to 25% of enhancement in the degree of crystallinity, 50% increase in tensile strength, and 35% of tensile modulus improvements, as well [77]. In an another study performed by Bayart et al. it was observed that inclusion of titanium dioxide-coated flax fibers into PLA matrix contributed to obtain 49% of boosted crystallinity degree, 15% of enhanced interfacial strength, and 40% of improved impact strength compared to that of non-reinforced PLA matrix [78]. On the other hand, Selvakumar et al. emphasized that the tensile strength of the jute and human hair-reinforced polymer composites was increased from 16 to 23.5 MPa [79]. In one of the works, loading 2% kenaf fibers onto the starch-based film resulted in augmented the water solubility, water absorption, and crystallinity index in the amounts of 55%, 170%, and 40%, respectively, as well as developing the tensile features of the composite [80].

12.3 Applications of Crystalline Polymer Composites

Crystalline commodity, engineering, and high-performance plastics are widely used in several industries due to their excellent processability and moldability. However, the properties of a single polymer could not meet the demands of industry; thus, polymer composites are great candidates to improve the characteristics, in particular the mechanical properties of the final products used in many applications [81].

12.3.1 Automotive Applications of Crystalline Polymer Composites

Automotive industry is one of the leading industries using polymer composites since lightweight, processability, high stiffness, and impact strength are required in automotive applications. The most abundant plastic polymer in a car is PP (43 wt%), followed by PU (14 wt%) and engineering thermoplastics like PA (12 wt%) [81, 82]. PP has been utilized in several parts in an automotive, such as bumpers, seats, wheel covers, cable insulation, carpet fibers, and chemical tanks, due to its strong mechanical properties, easy processability, and low density. For example, Chevrolet applied glass fiber-reinforced composites in body panels. Toyota commercialized timing belt covers constituted of clay nanoparticles-reinforced Nylon 6 polymer composites. General Motors Company commercialized new products consisting of PP/clay nanoparticles for the step-assist, sail panel, center bridge, box-rail protector, and door structure of cars. In addition, Exatec LLC commercialized polysiloxane-based nanocomposite coating in panels, which is used in car panels while Pirelli Tyre commercialized a composite structure of Kevlar, clay, and rubber for their branded tires. Moreover, i3 and i8 models of BMW have carbon or glass fiber-reinforced polymer composite structures in passenger seats, roof cover, and trunk lid [83].

On the other hand, Batista et al. incorporated cellulose hybrid fibers, a combination of long glass fiber, short glass fiber, or talc, into PP matrix for automotive applications and studied the mechanical properties of PP matrix. As a result, 86% of tensile stress, 252% of Young's modulus, and 23% of impact strength improvements were determined [84]. In another study, Ajorloo et al. integrated feasible recycled PP and fly ash into PP matrix for the replacement of elastomeric-based auto parts with an environment-friendly and cost-effective alternative. The integration of recycled PP content developed the mechanical properties by decreasing the ductility and impact strength of recycled PP and fly ash incorporated composite [85]. The other studies on the usage of crystalline polymer composites in automotive applications in terms of the polymer matrix, reinforcing agent, the type of reinforcing agent, application part, and the key points for the application are summarized in Table 12.2 shown below.

To sum up, it could be seen that crystalline polymer composites, which are mainly composed of a polymer matrix and natural, organic, or inorganic reinforcing agents,

Table 12.2 Literature studies on the usage of crystalline polymer composites in automotive applications in terms of the polymer matrix, reinforcing agent, the type of reinforcing agent, application part, and the key points for the application.

Polymer matrix	Reinforcing agent	Source of reinforcing agent	Application	Key points and aim	References
Polypropylene	Talc	Inorganic	Exterior trim part of automotive industry	Lightweight Improvements in mechanical and thermal properties	[86]
Polypropylene	Palm, Flax fiber	Natural	Interior parts in automotive industry	Lightweight Reduced CO_2 emission	[87]
Polypropylene	Glass/carbon fiber	Organic and inorganic	Bumper beam of automotives	Lightweight bumper beam Improved impact performance	[88]
Polypropylene	Multi-layered flat-bed weft-knitted glass fiber and glass fiber	Inorganic	Structural parts of automotives	Benchmarking	[89]
Polypropylene	Coconut/coir fiber	Natural	Automotive interior applications such as door trim and cabin linings	Lightweight and cost-effective alternative for glass fiber	[90]
Polypropylene	Long carbon fiber	Organic	High-volume automotive parts	Improved stiffness Lightweight	[91]
Polyamide 11	Wood fibers	Organic	Door car handle	Alternative to commercial and commodity products	[92]
Polyamide 6,6	Long glass fibers	Inorganic	Substitutes for metallic materials in automotive applications	Improvement in mechanical properties	[93]
Polypropylene-Polyamide 6	Carbon fiber/MWCNT	Organic	Automotive parts	Lightweight Improvement in mechanical properties	[94]

Table 12.2 (Continued)

Polymer matrix	Reinforcing agent	Source of reinforcing agent	Application	Key points and aim	References
Polyamide 66	Recycled carbon fiber	Organic	Automotive parts	Cost-reduction	[95]
Polyamide 1010/Poly-thioamide	Graphene nanoplatelets	Organic	High-temperature-resistant automotive parts	Candidate on utilization for automotive parts	[96]
Recycled polyamide 6/recycled polyamide 610	Cellulose fibers	Organic	Automotive parts	Cost-effectiveness Lightweight Improved mechanical properties	[97]
Biobased polyphthala-mides	Glass fiber	Inorganic	Automotive fuel supplying system	Improved barrier properties Similar mechanical properties of fuel tank	[98]
Polylactic acid	Carbon fiber	Organic	3D printing of automotive components	Lightweight Decrease in print duration Improved mechanical properties	[99]
Thermoplastic polyurethane	Sugar palm/glass fibers	Organic/inorganic	Automotive parts	Improved thermal and mechanical properties	[100]
Polyvinyl chloride	Glass beads	Inorganic	Wire and cable insulation	Improved mechanical properties and resistance	[101]
Acrylonitrile-butadiene/ethylene-propylene-diene (NBR/EPDM) rubber	Carbon black	Inorganic	Brake hoses, motor mounts, and belts	Improved rheological and mechanical properties	[102]
PTP vegetable-based thermoset resins	Hemp fibers	Natural	Exterior components of bus body	Recyclable, economical, and improved mechanical properties	[103]

have been significantly applied in automotive industry from exterior parts to interior and fuel tank structures due to their admirable characteristics.

12.3.2 Biomedical Applications of Crystalline Polymer Composites

Crystalline polymer composites are of great interest in biomedical applications such as tissue engineering, wound healing, smart drug delivery systems, regenerative medicine, and dentistry due to their high resistance, high strength, and lightweight. The most commonly utilized crystalline polymers in biomedical field are PLA, polyacrylic acid (PAA), polyvinyl alcohol (PVA), PCL, and polyglycolide (PGA) [104]. The fillers utilized in biomedical applications are mostly natural-based ones since biocompatibility and biodegradability are the major considerations. Although, there are some challenges in the production and development of composites for biomedical applications due to the difficulties in obtaining feasible, durable, effective, and multipurpose integrated advances, many studies have been reported on the fabrication of crystalline polymer composites for specific biomedical applications. For instance, Khan et al. prepared a microsphere-based scaffold structure of crystalline poly(lactide-co-glycolide) polymer-based calcium phosphate-doped composite structure as a bone implant tissue Incubation of the crystalline polymer composite took eight weeks and resulted in 11–14% decrease in molecular weight, showing that the calcium phosphate reprecipitated and turned into free calcium ions, which boost the integration of implant into bone structure [105]. In another study, Ahmed et al. studied the examination of the mechanical and biological properties of the calcium phosphate-doped glass fiber integrated PCL composite structure for bone tissue engineering applications. The binary calcium phosphate glass fiber-reinforced PCL composite had 30 MPa of flexural strength and 2.5 GPa of flexural modulus, while 20% of mass loss occurred in bone fracture surface, ensuring the improved mechanical and bio-based features of the obtained fiber composites [106]. On the other hand, since it is challenging to produce blood vessels with the required features in tissue engineering applications, an artificial blood vessel was created by Tang et al. with bacterial nanocellulose (BNC) tubes incorporated PVA polymer. PVA-based BNC-reinforced composites had porous structure outside, allowing the transfer of nutrients and helping regeneration; therefore, the water permeability and water uptake as well as tensile properties were improved in the resultant composites [107]. Additionally, macro-porous blood vessels made from chitosan-based gelatin-reinforced composite structures were manufactured by Badhe et al. The gelatin-filled composite having a large surface area and 82% porosity as well as high flexibility, and elasticity was also found to be biodegradable even after two weeks. Although 50% of composite degraded in in vitro experiments, it contributed to cell proliferation and adhesion for 20 days after the proliferation of the fibroblasts [108]. The regeneration and proliferation of human keratinocytes, fibroblasts, and mesenchymal stem cells were studied by Bhowmick et al. by using the composite structure of modified hydroxyapatite and chondroitin sulfate. The successful proliferation of the three cell types was observed on the composite surfaces [109]. Pereira et al. fabricated the aloe vera-incorporated alginate hydrogel films with 100% of improved water absorption, and 78% of light transmission to enhance the thermal stability in skin treatment applications [110]. Furthermore, in Table 12.3, more detailed specific relevant studies about the biomedical applications are shown.

As seen in Table 12.2, biomedical applications of crystalline polymer composites have been performed in many specific targeted areas and also will be popular in the future due to their novel characteristics.

12.3.3 Defense and Aerospace Applications of Crystalline Polymer Composites

Crystalline polymer composites are also popular in the defense industry and used in many specific applications such as aircraft, land vehicles, body armor, arms, and ammunition since World War II led developments in aerospace industry, as well [129]. In defense and aerospace industries, the most widely preferred crystalline polymers are polyamides (PA-6, PA-12), PP, PPS, PTFE, polyamide-imide (PAI), PEEK and polyether ketone ketone (PEKK). In addition, numerous fibers, tubes, and platelets, including Kevlar, glass and CF, CNT, graphene, and GNPs, have been utilized as reinforcement agents in the production of crystalline polymer composites. For example, Airbus used crystalline thermoplastic composites reinforced with CF in the aircraft design from the A310 to A380 family [130, 131]. Boeing also applied fiber-reinforced composite in tail cone of Boeing 777 [132]. Maia et al. focused on lightweight solutions for defense industry. In the study, virgin carbon fiber was replaced with the recycled CFs and integrated into PP/PA12 polymer blend. The recycled CF incorporated PP/PA12 blend improved the mechanical strength, and impact energy in the amounts of 17% and 16%, respectively [133]. In an another study, Raja et al. investigated the effect of using hybrid bamboo/glass fiber integrated PP on aerospace industry and mentioned that the hybrid bamboo/glass fiber showed an 18% increase in tensile strength and advanced the fatigue life of the products [134]. Moreover, Lee et al. utilized PP-based recycled CF composite and applied surface treatment to the composite in order to improve the adhesion between the nanofiller and polymer matrix. At the end, it was suggested that the surface treatment had a considerable impact on the enhancement of the adhesion and thus surface activity of recycled CFs, by eliminating the drawbacks of the recycled CFs [135]. In addition, King et al. examined three different carbon nanofillers, which were carbon black, CF, and GNPs, and their integration into PEEK polymer matrix to observe the electrical and thermal conductivities as well as mechanical properties of the resultant materials. It was concluded that the hybrid CF/carbon black composite raised the mechanical properties as well as the thermal and electrical conductivity higher than the other filler formulations [136]. In an additional study, Fischer et al. developed a PEKK-based CF-reinforced composite for laser sintering and interior cabin ventilation system in aerospace. A high-performance composite structure with 60% of refreshing rates of ventilation air was successfully obtained [137]. In one of the works, Kevlar, one of the strongest fibers, and powdered activated carbon were incorporated into the crystalline

Table 12.3 Literature studies on the specific relevant examples of crystalline polymer composites in biomedical applications.

Polymer	Reinforcing agent	Application area	Specific application	Key points	References
PLA	Bio-glass fibers	Tissue engineering	Weight-bearing long bone fractures	Improvements on 31% of tensile and 13.5% of flexural strength. A bone-like calcium layer was deposited onto the surface of implant showing bone healing	[111]
PLA	Magnesium fluoride (MgF_2) coated Mg rods	Tissue engineering	Bone fracture	107% of improvement on the bending strength. After immersion of composites into body fluid, the composite had tensile strength of 80 MPa for 8 wk	[112]
Alginate (AL)	Hydroxyapatite (HA) – Silk fibroin (SF)	Tissue engineering	Bone regeneration	Successful bone formation. Enhanced metabolic activity and growth of cells	[113]
PLA	Ethyl cellulose (EC) – Hydroxyapatite (HA)	Tissue engineering	Bone substitute for regeneration	Scaffolds had porous structure, kept its dimensions through degradation, and possessed optimum mechanical properties	[114]
Polypropylene carbonate (PPC)/poly (D-lactic acid) (PDLA)	Tricalcium phosphate (TCP)	Tissue engineering	Bone defect repair	Biocompatible, flexible, biodegradable, and osteoconductive composite	[115]
PLA	Chitosan fibers	Tissue engineering	Constructing cardiac tissue	Mimicking the extracellular matrix, promoting the adhesion, growth, and viability of cardiomyocyte. Composites with aligned fibers had higher mechanical strength and biocompatibility than those with random fibers	[116]

PLC/Chitosan	Gelatin	Tissue engineering	Cardiac tissue engineering	Hydrogel scaffold had porous structure, similar and adequate compression modulus to native tissue	[117]
PCL	Gelatin	Tissue engineering	Blood vessel endothelial layer scaffold	Scaffolds had tailorable biophysical as well as comparable mechanical characteristics to human coronary arteries with powerful interactions of composite and cells	[118]
PCL	Gelatin	Tissue engineering	Wound healing and layered dermal reconstruction	Enabling the cells to adhere, proliferate, and grow Suitable 3D scaffold for autogenous fibroblast cells	[119]
PCL	Gelatin	Tissue engineering	Nerve tissue engineering	Improved the nerve specialization and proliferation.	[120]
PCL	Gelatin	Tissue engineering	Muscle tissue engineering	Proliferation of myoblasts was higher in crosslinked PCL-gelatin composite than virgin PCL nanofibers. Regulating proliferation and specialization	[121]
PCL	Chitosan (CH)-Caffeic acid (CA)	Skin tissue engineering	Wound dressing	Human dermal fibroblast adhesion and proliferation as well as antimicrobial properties	[122]
PU	Kaolin	Wound dressing	Acute wounds hemostatic dressing	Extreme release of ciprofloxacin over 2 h and outstanding hemostatic role and absorptive properties	[123]
Chitosan	Titanium dioxide (TiO$_2$)	Wound dressing	Wound dressing and skin restoration	Composites supplied the cell proliferation and prevented apoptosis as well as showed antibacterial effect	[124]

(continued)

Table 12.3 (Continued)

Polymer	Reinforcing agent	Application area	Specific application	Key points	References
Chitosan	Banana peel powder	Wound dressing	Wound infection treatments	Composite structure had antimicrobial properties, decreased water absorption so less swelling, and lower dielectric properties	[125]
Polycaprolactone	Silver (Ag) – Cobalt (Co) doped bioglass nanoparticles	Wound dressing	Wound healing	50% of Young's modulus improvement, porous surface, high surface area with antibacterial and angiogenic properties	[126]
Silk fibroin (SF)	Graphene oxide (GO)	Wound dressing	Wound dressing	Improved biocompatibility, so cell adhesion and proliferation as well as 45% of antibacterial activity	[127]
Poly co-glycolide copolymer (PLGA)	Polyglycolic acid (PGA) fiber mesh	Dental resin	Regeneration of mammalian dental tissues	Bio-engineered tooth tissues Dental epithelial and mesenchymal stem cells were obtained in vitro for 6 d.	[128]

unsaturated polyester polymer matrix by Xia et al. to observe the effect of hybrid composite form on the improvements of EMI shielding. The EMI shielding and absorption of composites showed an increase up to 93% and 64%, respectively [138]. Tian et al. proposed a new feasible fabrication technique for 3D printing of continuous CF-reinforced PLA matrix for aviation and aerospace industry. The mechanical test results showed that 27% of fiber loading exhibited high performance as 335 MPa of flexural strength and 30 GPa of flexural modulus [139].

To sum up, specifically, carbon/glass fiber-reinforced crystalline polymer composites are the leading materials in aerospace and defense applications due to their availability, low cost, lightweight, and improved mechanical properties that are advantageous to supply the industrial demands effectively.

12.3.4 Other Applications of Crystalline Polymer Composites

Overall, not only the applications discussed above but also many other areas use crystalline polymer composites in their applications. In the past three decades, crystalline polymer composites have been widely utilized in civil engineering applications. For instance, magnesium phosphate cement-based glass fiber-reinforced composite was studied by Fang et al. for the achievement of fire-retardancy. The obtained cement composite had a high spread fluidity of 200 mm and bonding strength of 0.6 MPa. Additionally, incorporated glass fibers had a vital role in inhibiting cracks in cement structures [137]. Gopinath et al. reported the performance of vinyl ester-based glass and jute fibers-reinforced hybrid composite structure for deck panels. It was told that the composite structure produced by the addition of stiffeners obeyed the design obligations of ODOT, USA [140]. Moreover, Amiri et al. manufactured a bicycle frame, made of hybrid flax and CF composite, had 2.1 kg weight in total and demonstrated superb damping properties over steel [141]. In marine applications, mechanical properties of plastics and compartment materials that are used in the body of vehicles are weakened due to the seawater aging. For this reason, Jethsi et al. attempted to improve the mechanical properties of the fabric by creating a hybrid composite structure of glass and CFs embedded in Bisphenol A polymer. Finally, 14% of tensile strength, 43% of flexural strength, and 64% of modulus enhancements of composites could be obtained. In addition, Kumar et al. designed a CF composite for marine propellers. This newly designed composite structure exhibited high strength-to-weight ratio, elevated efficiency, and hydro-elasticity based on the response characteristic [142].

Technological developments lead to different applications of crystalline polymer composites. In this sense, crystalline polymer composites are used in electronics to supply the thermal conductivity requirements for encapsulation. Zhou et al. proposed to incorporate h-BN into polyimide (PI)-modified aluminum nitride (AlN) polymer to reach $2.03\,W\,mK^{-1}$ of thermal conductivity, which was higher than the virgin polymer [143]. In a different study, Zhao et al. designed a poplar biofiber filler incorporated PLA composite with approximately 60% of developed tensile strength for 3D printing [144]. Natural fibers are also utilized in sports and leisure goods, such as an Australian surfboard production company that produced environmentally friendly and sustainable surfboards made of commingled flax/PP, commingled flax/PLA, and flax fibers [145]. As a summary, it can be clearly said that the crystalline polymer composites can provide the requirements and create synergetic effects on the properties of polymers.

12.4 Recycling, Environmental, and Safety Issues of Crystalline Polymer Composites

Plastic wastes have become a serious concern with the increasing demand for plastic production and due to increasing population. Plastic scraps and products must be accordingly regained and recovered according to the recent prospects of waste management and regulations of environmental issues. To this end, there are number of attempts to convert the plastic wastes into high-value-added products since energy content of wastes is great. One of the main sources of plastic wastes are thermoplastics, which can be easily melted and solidified by cooling and reheating, which enables re-melting and reprocessing. Thermosets are the other type of plastic waste sources, which can be melted and molded, but reheating and reprocessing them are impossible due to the degradation of polymer chains by heat treatment. Increased accumulation of wastes cannot be thought of only as polymer matrices but also as polymer composites [146].

Recycling is a prominent method to deplete the accumulation of waste plastics and reduce their detrimental environmental impacts [147]. Recycling techniques can be summarized into three main categories: thermal, chemical, and mechanical processes. Mechanical recycling is the process of reducing the size of the composite into its components by mechanical crushing techniques [148]. Chemical recycling covers the conversion of hydrocarbon-based waste into hydrocarbon-based oil/product through the treatment with appropriate chemicals [149], while in the thermal recycling process, the plastic scrap is turned into energy and valuable materials via microwave pyrolysis, combustion pyrolysis, or fluidized-bed pyrolysis. Compared to mechanical or chemical recycling, thermal recycling provides the cleanest product with maximum efficiency [150].

From production to end-of-life of a product and its recycling methods are represented below, in Figure 12.4.

In this chapter, the recycling of crystalline polymer composites is given in terms of reinforcement types of the composites.

12.4.1 Recycling of Glass Fiber-Reinforced Crystalline Polymer Composites

Recycling process of GF-reinforced polymer composite has three strategies, regardless of the crystalline structure of polymer matrix. Successful recycling of glass fiber composites can be easily processed by mechanical, chemical, and thermal recycling techniques and several studies on the recycling of glass fiber-filled crystalline composites have been reported in the literature [151]. Kiss et al. mentioned that

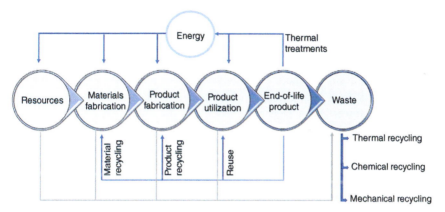

Figure 12.4 Schematic diagram of a product from the production to end-of-life and recycling methods.

while the recycled PP/GF and PA6/GF composites had weak features due to having shorter fiber formations and random distribution of them, recycled laminates had comparable results to virgin composites [152]. Cousins et al. utilized wind turbines, which are composed of glass fiber-embedded crystalline Elium thermoplastic composite and examined the feasibility, and recyclability of the turbines. Recyclability studies were performed through chemical recycling of polymer dissolution, where polymer matrix was regained, and glass rovings were recovered [153]. Furthermore, Pietroluongo et al. studied the mechanical recycling impact on the rheological and mechanical properties of a crystalline PA66-based glass fiber-filled composite. However, the thermal and mechanical features of the recycled composites were not as satisfactory as the virgin composites [154]. Despite the fact that literature possesses different strategies on glass fiber composites recycling, the route of waste glass fiber-reinforced crystalline polymers has not been established well enough, so misguidance of those materials can occur [155]. For this reason, there is a legislation of European Union (EU) 2008/98/EC directive, and this legislates taxes and waste management routes of recycling to decrease the environmental impact [156]. Hence, countries such as Germany and United Kingdom have obeyed the advice of this directive and have made new legislations of glass fiber-reinforced polymer composites to ban the landfilling of waste disposals to conserve the soil [157, 158].

12.4.2 Recycling of Carbon Fiber-Reinforced Crystalline Polymer Composites

CF-reinforced composite structures have been an accelerating topic of recycling in the past two decades since carbon-based composites have shown an incredible growth due to their synergetic properties such as durability and high strength to low weight ratio [151]. However, preserving the high modulus of CFs is the main challenge of recycling CF-based composites in the commercial site. Although mechanical recycling is relatively cheaper and simpler compared to thermal or chemical recycling, it provides shorter fibers, which results in lower mechanical properties [159]. However, the high modulus of the composites can be preserved through thermal recycling since it supplies longer fiber structures. For instance, recycling of CF compositesq at 823 K under oxygen through fluidized-bed pyrolysis offers clean and pure fibers without resin, allowing to obtain efficient energy recovery from heat energy, while this is not the case for the pyrolysis at higher temperatures [160]. Tian et al. examined the utilization of recycled PLA based-CF-reinforced composites in 3D printing. The tensile properties of recycled composites revealed 25% higher bending strength and improved tensile forces [161]. Colucci et al. used both virgin and recycled PA66 polymer matrix in automotive parts. PA66 polymer matrices were reinforced with CFs, and the mechanical properties of obtained composites were compared. The final mechanical properties of recycled composite could compete with those of virgin PA66 CF-filled composite; hence, recycled composite was an appropriate candidate for automotive applications [162]. In another study, Tapper et al. investigated the effects of closed-loop chemical recycling, a new method for recycling, on the mechanical features of CF-based PP composite. It was concluded that 26% improvement in ultimate tensile strength and 43% improvement in ultimate strain could be achieved with short but re-aligned fibers by this new recycling method [163]. The same methodology of closed-loop chemical recycling was also applied by Tapper et al. on the examination of mechanical features of CF-reinforced PA6 polymer composites after recycling. However, 40% of decrease in tensile strength and modulus was observed due to the agglomeration of CFs during the recycling [164].

12.4.3 Recycling of Carbon Nanotubes-Reinforced Crystalline Polymer Composites

Although recycling of carbon-based composites has been strategized well, there are still some challenges, which are agglomeration caused by the distribution of the fillers in the polymer matrix and health issues like high carcinogenic risks. Zhang et al. recycled the CNTs-reinforced PP composite by mechanical processes and revealed that recycling of CNT-reinforced composites had no dramatic effect on the mechanical features of the recyclates, which indicated that recycled CNT-based PP composite had a potential of being reused. However, grinding of carbon-based nanomaterials has a severe effect on human health and is an environmental issue since high exposure to carbon is dangerous [165]. Salas et al. aimed to produce CNTs on the recycled carbon surface via pyrolysis at high temperatures and revealed that the utilization of impregnations besides high temperature boosted the growth of CNTs without leaving either resin or damage [166]. In another study, Stan et al. assessed LDPE-based MWCNT-reinforced composites in terms of their rheological, electrical, and mechanical properties. Recycled MWCNT-reinforced composites had comparable results with virgin composite and showed higher strain at break values [167]. Moreover, Stan et al. stated that the loading amount of recycled MWCNT had an indirect relationship with the Young's modulus, tensile strength, and electrical conductivity [168].

12.4.4 Recycling of Natural Fiber-Reinforced Crystalline Polymer Composites

Recycling of natural fiber-reinforced crystalline polymer composites can be easily carried out by mechanical, thermal, and chemical recycling techniques; however, when the reinforced composite is at the end of its life, acquiring natural fibers from composite is neither feasible nor cost-effective [169]. El Abbasi et al. utilized mechanically recycled short alfa fibers in PP composites and informed that storage modulus of recycled composites decreased due to the small size of the alfa fibers [170]. Bourmaud et al. studied the recyclability of natural flax fiber-reinforced PLA-PBS polymer composites as well as the impacts of recycling on mechanical, thermal, and rheological properties of the resultant material. It was summarized that while the viscosity of the composites increased and tensile strength decreased, Young's modulus of the recycled composite was not affected by the treatment [171]. In addition, flax fiber-reinforced PA11-PP composites were recycled by Gourier et al. who concluded that tensile modulus and strength of recycled composites showed a nice stability where elongation at break increased up to 14.7% in recycled PP. Similarly, recycled PA11 composite had compatible thermal, mechanical, and rheological properties with its neat counterpart. However, the fiber length decrease was higher in PA11 composites than PP composites [172]. In one of the works, Ngaowthong et al. discussed the impact of the recycling of sisal fiber incorporated PP-PLA composites on the mechanical properties, water absorption, and crystallinity. As the crystallinity of PLA composites raised after recycling, water absorption decreased in both composite forms as well as mechanical performance [173].

12.4.5 Environmental Impact and Safety Issues of Crystalline Polymer Composites

Beginning from the production processes of crystalline polymer composites to their recycling methods, greenhouse gases are released. Hence, to reduce the caused hazardous environmental impacts, specifically CO_2 emissions, the change in the processes or recycling processes needs to be considered. In the past years, innovations in the fabrication of green products have been encouraged since they bring in sustainability factors involving utilization of renewable resources, diminishing wastes, contributing biodegradability, efficient energy use, and cost-effective solutions. Due to being environmentally friendly, biodegradable, and cost-effective, biocomposites produced from polymer matrix and natural-based reinforcers are trendy in composite manufacturing field [3]. In this sense, Sahmaran et al. produced self-healing cementitious composites to reduce the CO_2 emission during processing of cement, which generates 85% of CO_2 [174]. Ashori et al. investigated the possibility of usage of wood–plastic composites as green alternatives to conventional composites for automotive applications. Wood–plastic composites were found to be proper candidates since they were promising, sustainable, and biodegradable green materials to achieve durability without using toxic chemicals and reduce the CO_2 emission [175]. Additional studies are still ongoing to assess the safety issues and find alternative solutions. For instance, fiber glass composite manufacturing and application have increased its popularity due to its low toxicity since it has been proven that, specifically, being direct exposure to high concentrations of carbon-based nanomaterials treats human health adversely [176].

On the other hand, as being the most effective solution to reduce the greenhouse gas emissions, recycling techniques and upcycling of waste plastics into high-value-added products have become prominent. Moreover, by using life cycle assessment technique, the environmental impacts of the processes or products can be easily controlled. It is effectively used to examine the environmental consequences and sustainability of a certain manufacturing process or the overall product. Wicik et al. conducted LCA technique to evaluate the environmental benefits of the recycling of carbon fiber-reinforced plastic (CFRP) waste and to assess the potential benefits of material reuse. It was concluded that although ecological limitations of recycling the CFRPs were still present, it was exactly environmentally suitable to decrease the CFRP waste in specific cases [177]. As a conclusion, in order to have a pleasant habitat around and in the future, the accumulated wastes must be regained, reused, and recycled/upcycled regarding the environmental impact and safety issues.

12.5 Future Prospects of Crystalline Polymer Composites

Crystalline polymer composites have been a trending topic both in industry and academy due to their ability to provide improved performance, outstanding properties, and high efficiency through integration of reinforcements into polymer matrices. Up to now, there are several applications of crystalline polymer-based composites, including transportation, marine, civil construction, military, biomedical, sport and leisure, electronics, and food packaging. Future trends in crystalline polymer composites depend on the utilization of reinforced nanomaterials with elevated features [178]. Specifically, fiber-reinforced crystalline polymer composites have a great potential due to excellent compatibility of fibers with crystalline polymers, resulting in lightweight and enhanced mechanical properties in different industries. For example, it is expected that synthetic fiber utilization will surpass US$ 9 billion by the end of 2027 [179]. In particular, the future of CF-reinforced crystalline polymers is limitless. Future applications of CFs are thought to be as alternative materials in wind turbines, energy storage, transportation, and fuel-efficient automobiles in large production series cars. The market of CF industry is expected to rise to US$ 64 billion by recycling and standardization progresses by the end of 2030 [180]. Furthermore, CNTs-reinforced crystalline composites are believed to receive more interest in the future since CNTs are highly electrically conductive, low cost, and have the features to replace the metal wires. The semiconducting characteristics of CNTs enable them to be applied in computer chips, effectively. Moreover, the futuristic applications of CNTs can replace the CF requirement in applications where lightweight, flame retardancy, and ecological and biological features are desired. For instance, in one of the works, CNTs were injected into a kidney tumor of a mouse to kill the tumor cells by the vibration of the nanotubes [181].

Optimistically, by the directions and motivations described above, CNTs and fibers will be the future applications of crystalline polymer composites. Besides, based on a report published in 2020, more than 400 million tons of plastic are produced per year. Therefore, several techniques are ongoing to recycle, reuse, and reduce production of waste. For this purpose, prospects of crystalline polymer composites allow to conduct new recycling techniques, which require less energy consumption, low cost, and less time.

Green nanotechnology is the route to green manufacturing for environmentally friendly composite production by diminishing the usage of hazardous volatile organic compounds and solvents such as toluene, dimethyl formaldehyde, and methylene chloride that treat human health and biodiversity. Nowadays, instead of using toxic solvents, the utilization of supercritical carbon dioxide liquid is offered as a new material to dissolve polymer matrices [182]. Likewise, the next generation of crystalline polymer composites should have multifunctionality to generate smart composites. The multifunctionality of composites possesses different technological residence levels (TRL), including improved self-healing, energy harvesting and storage, embedded sensors and actuators, data and power transmission, and electrical and thermal conductivity. The market of total multifunctional and smart composites is estimated to surpass 5 kilotons by 2029 [179].

12.6 Conclusions

This chapter aimed to give a detailed information on the crystalline polymer composites according to the reinforcement types, such as organic, inorganic, and natural kinds. Furthermore, the automotive, biomedical, defense, and aerospace applications of crystalline polymer composites are explained in detail as well as their other applications. Since recycling has a paramount importance to preserve the environment and overcome the problems of waste disposal, a comprehensive understanding is given of the recycling, environmental, and safety issues of crystalline polymer composites. Consequently, the future prospects of crystalline polymer composites are addressed to contribute to the circular economy, sustainability, technology, and science.

References

1 Cheng, S.Z.D. and Jin, S. (2002). Crystallization and melting of metastable crystalline polymers. In: *Handbook of Thermal Analysis and Calorimetry*, vol. 3 (ed. S.Z.D. Cheng), 167–195. Elsevier Science B.V.
2 de Dantas, Oliveira, A. and Augusto Gonçalves Beatrice, C. (2019). Polymer nanocomposites with different types of nanofiller. In: *Nanocomposites – Recent Evolutions* (ed. S. Sivasankaran), 103–127. InTechOpen.

3 Yu, L., Dean, K., and Li, L. (2006). Polymer blends and composites from renewable resources. *Progress in Polymer Science (Oxford)* 31 (6): 576–602.

4 Balani, K., Verma, V., Agarwal, A., and Narayan, R. (2015). Physical, thermal, and mechanical properties of polymers. In: *Biosurfaces: A Materials Science and Engineering Perspective* (ed. K. Balani, V. Verma, A. Agarwal, and R. Narayan), 329–344. John Wiley & Sons, Inc.

5 Gu, K., Onorato, J.W., Luscombe, C.K., and Loo, Y.L. (2020). The role of tie chains on the mechano-electrical properties of semiconducting polymer films. *Advanced Electronic Materials* 6 (4): 1901070.

6 Wang, H., Keum, J.K., Hiltner, A. et al. (2009). Confined crystallization of polyethylene oxide in nanolayer assemblies. *Science* 323 (5915): 757–760.

7 Keller, A. (1952). Morphology of crystallizing polymers. *Nature* 169 (4309): 913–914.

8 Zachmann, H.G. (1974). Theory of nucleation and crystal growth of polymers in concentrated solutions. *Pure and Applied Chemistry* 38 (1, 2): 79–96.

9 Arif, P.M., Kalarikkal, N., and Thomas, S. (2018). Introduction on crystallization in multiphase polymer systems. In: *Crystallization in Multiphase Polymer Systems* (ed. S. Thomas, P.M. Arif, E. Bhoje Gowd, and N. Kalarikkal), 1–16. Elsevier.

10 López, L.C. and Wilkes, G.L. (1988). Crystallization kinetics of poly (p-phenylene sulphide): effect of molecular weight. *Polymer* 29 (1): 106–113.

11 Zen, A., Saphiannikova, M., Neher, D. et al. (2006). Effect of molecular weight on the structure and crystallinity of poly(3-hexylthiophene). *Macromolecules* 39 (6): 2162–2171.

12 Quirk, R.P. (1981). Stereochemistry and macromolecules: principles and applications. *Journal of Chemical Education* 58 (7): 540.

13 Choi, D. and White, J.L. (2002). Crystallization and orientation development in melt spinning isotactic PP of varying tacticity. *International Polymer Processing* 17 (3): 233–243.

14 Rungswang, W., Jarumaneeroj, C., Patthamasang, S. et al. (2019). Influences of tacticity and molecular weight on crystallization kinetic and crystal morphology under isothermal crystallization: Evidence of tapering in lamellar width. *Polymer* 172: 41–51.

15 Ries, M.D. and Pruitt, L. (2005). Effect of cross-linking on the microstructure and mechanical properties of ultra-high molecular weight polyethylene. *Clinical Orthopaedics and Related Research* 440: 149–156.

16 Hussein, M.A., El-Shishtawy, R.M., Abu-Zied, B.M., and Asiri, A.M. (2016). The impact of cross-linking degree on the thermal and texture behavior of poly(methyl methacrylate). *Journal of Thermal Analysis and Calorimetry* 124 (2): 709–717.

17 Li, D., Zhou, L., Wang, X. et al. (2019). Effect of crystallinity of polyethylene with different densities on breakdown strength and conductance property. *Materials* 12 (11): 1746.

18 Zhang, S., Ming, Y., Wei, Y. et al. (2021). The effect of grafting density on the crystallization behavior of one-dimensional confined polymers. *Journal of Applied Polymer Science* 138 (12): 50064.

19 Li, J. and Qui, Z. (2022). Fully biodegradable poly(butylene succinate-co-1,2-decylene succinate)/Cellulose nanocrystals composites with significantly enhanced crystallization and mechanical property. *Polymer* 252: 124946.

20 Várdai, R., Schäffer, F., and M., Lummerstorfer, T., Jerabek, M., Gahleitner, M., Faludi, G., Móczó, J., and Pukánszky, B. (2022). Crystalline structure and reinforcement in hybrid PP composites. *Journal of Thermal Analysis and Calorimetry* 147 (1): 145–154.

21 Chen, S.C., Lai, S.A., and Cousineau, J.S. (2013). Effect of cooling rate and mold counter pressure on the crystallinity and foaming control in microcellular injection molded polypropylene parts. In: *Annual Technical Conference – ANTEC, Conference Proceedings, Society of Plastics Engineers*, vol. 2. Ohio, USA.

22 Dyamenahalli, K., Famili, A., and Shandas, R. (2015). Characterization of shape-memory polymers for biomedical applications. In: *Shape Memory Polymers for Biomedical Applications* (ed. L. Yahia), 35–63. Elsevier.

23 Ramanathan, M. and Darling, S.B. (2011). Mesoscale morphologies in polymer thin films. *Progress in Polymer Science (Oxford)* 36 (6): 793–812.

24 Carraher Jr., C.E. (2021) Seymour/Carraher's Polymer Chemistry, 1–776.

25 Popelka, A., Zavahir, S., and Habib, S. (2020). Morphology analysis. In: *Polymer Science and Innovative Applications* (ed. A.M. MAA, D. Ponnamma, and M.A. Carignano), 21–68. Elsevier.

26 Goldberg, M.W. and Allen, T.D. (1992). High resolution scanning electron microscopy of the nuclear envelope: Demonstration of a new, regular, fibrous lattice attached to the baskets of the nucleoplasmic face of the nuclear pores. *Journal of Cell Biology* 119 (6): 1429–1440.

27 Mathot, V.B.F. (2002). Characterisation of polymers by thermal analysis. *Thermochimica Acta* 389 (1, 2): 213.

28 Sastri, V.R. (2022). High-temperature engineering thermoplastics: polysulfones, polyimides, polysulfides, polyketones, liquid crystalline polymers, fluoropolymers, and polyarylamides. In: *Plastics in Medical Devices* (ed. V.R. Sastri), 233–286. Elsevier.

29 Chen, S., Xie, S., Guang, S. et al. (2020). Crystallization and thermal behaviors of poly(ethylene terephthalate)/bisphenols complexes through melt post-polycondensation. *Polymers* 12 (12): 3053.

30 Danek, M., Lutomski, M., Maniukiewicz, W., and Kozanecki, M. (2021). The crystallinity of poly(butylene terephthalate) in mass-scale extrusion products as seen by differential scanning calorimetry. *Polymers for Advanced Technologies* 32 (3): 1272–1287.

31 Abdelmouleh, M. and Jedidi, I. (2021). Development of LDPE crystallinity in LDPE/Cu composites. In: *Plastic Deformation in Materials [Working Title]* (ed. S. Kumar), 1–17. InTechOpen.

32 Navarro-Pardo, F., Martínez-Barrera, G., Martínez-Hernández, A.L. et al. (2013). Effects on the thermo-mechanical and crystallinity properties of nylon 6,6 electrospun fibres reinforced with one dimensional (1D) and two dimensional (2D) carbon. *Materials* 6 (8): 3494–3513.

33 Pasztor, A.J., Landes, B.G., and Karjala, P.J. (1991). Thermal properties of syndiotactic polystyrene. *Thermochimica Acta* 177 (C): 187–195.

34 Jiao, Q., Chen, Q., Wang, L. et al. (2019). Investigation on the crystallization behaviors of polyoxymethylene with a small amount of ionic liquid. *Nanomaterials* 9 (2): 206.

35 Wang, X.Q., Chen, D.R., Han, J.C., and Du, S.Y. (2002). Crystallization behavior of polytetrafluoroethylene (PTFE). *Journal of Applied Polymer Science* 83 (5): 990–996.

36 Nohara, L.B., Nohara, E.L., Moura, A. et al. (2006). Study of crystallization behavior of poly(phenylene sulfide). *Polimeros* 16 (2): 104–110.

37 Martin, A., Addiego, F., Mertz, G. et al. (2016). Pitch-based carbon fibre-reinforced peek composites: optimization of interphase properties by water-based treatments and self-assembly. *Journal of Material Science & Engineering* 05 (06): 1000308.

38 Liu, Y. and Zhang, C. (2007). The influence of additives on crystallization of polyvinyl chloride. *Journal Wuhan University of Technology, Materials Science Edition* 22 (2): 271–275.

39 Sperling, L.H. (2000). History and development of polymer blends and IPNS. In: *Applied Polymer Science: 21st Century* (ed. C.D. Craver and C.E. Carraher Jr..), 343–354. Elsevier.

40 Zaïri, F., Gloaguen, J.M., Naït-Abdelaziz, M. et al. (2011). Study of the effect of size and clay structural parameters on the yield and post-yield response of polymer/clay nanocomposites via a multiscale micromechanical modelling. *Acta Materialia* 59 (10): 3851–3863.

41 Saeidlou, S., Huneault, M.A., and Park, C.B. (2012). Poly(lactic acid) crystallization. *Progress in Polymer Science* 37: 1657–1677.

42 Liu, G., Zhang, X., and Wang, D. (2014). Tailoring crystallization: towards high-performance poly (lactic acid). *Advanced Materials* 26 (40): 6905–6911.

43 Choi, J., Hore, M.J.A., Meth, J.S. et al. (2013). Universal scaling of polymer diffusion in nanocomposites. *ACS Macro Letters* 2 (6): 485–490.

44 Wei, Z. and Song, P. (2018). Crystallization behavior of semicrystalline polymers in the presence of nucleation agent. In: *Crystallization in Multiphase Polymer Systems* (ed. S. Thomas, P.M. Arif, E.B. Gowd, and N. Kalarikkal), 433–469. Elsevier.

45 Das, S., Samal, S.K., Mohanty, S., and Nayak, S.K. (2018). Crystallization of polymer blend nanocomposites. In: *Crystallization in Multiphase Polymer Systems* (ed. S. Thomas, P.M. Arif, E.B. Gowd, and N. Kalarikkal), 313–339. Elsevier.

46 Pielichowski, K. and Pielichowska, K. (2018). Polymer nanocomposite. In: *Handbook of Thermal Analysis and Calorimetry* (ed. S. Vyazovkin, N. Koga, and C. Schick), 431–485. Elsevier.

47 Zhao, W., Su, Y., Gao, X. et al. (2016). Interfacial effect on confined crystallization of poly(ethylene oxide)/silica composites. *Journal of Polymer Science, Part B: Polymer Physics* 54 (3): 414–423.

48 Lin, C.C., Gam, S., Meth, J.S. et al. (2013). Do attractive polymer-nanoparticle interactions retard polymer diffusion in nanocomposites? *Macromolecules* 46 (11): 4502–4509.

49 Müller, K., Bugnicourt, E., Latorre, M. et al. (2017). Review on the processing and properties of polymer nanocomposites and nanocoatings and their applications in the packaging, automotive and solar energy fields. *Nanomaterials* 7 (4): 74.

50 S, P., KM, S., K, N., and S, S. (2017). Fiber reinforced composites - a review. *Journal of Material Science & Engineering* 06 (03): 1000341.

51 Cadek, M., Coleman, J.N., Barron, V. et al. (2002). Morphological and mechanical properties of carbon-nanotube-reinforced semicrystalline and amorphous polymer composites. *Applied Physics Letters* 81 (27): 5123.

52 Wu, H.Y., Jia, L.C., Yan, D.X. et al. (2018). Simultaneously improved electromagnetic interference shielding and mechanical performance of segregated carbon nanotube/polypropylene composite via solid phase molding. *Composites Science and Technology* 156: 87–94.

53 Liang, J., Xu, Y., Wei, Z. et al. (2014). Mechanical properties, crystallization and melting behaviors of carbon fiber-reinforced PA6 composites. *Journal of Thermal Analysis and Calorimetry* 115 (1): 209–218.

54 Batista, N.L., Olivier, P., Bernhart, G. et al. (2016). Correlation between degree of crystallinity, morphology and mechanical properties of PPS/carbon fiber laminates. *Materials Research* 19 (1): 195–201.

55 Wang, X., Yang, H., Song, L. et al. (2011). Morphology, mechanical and thermal properties of graphene-reinforced poly(butylene succinate) nanocomposites. *Composites Science and Technology* 72 (1): 1–6.

56 Xu, J.Z., Chen, C., Wang, Y. et al. (2011). Graphene nanosheets and shear flow induced crystallization in isotactic polypropylene nanocomposites. *Macromolecules* 44 (8): 2808–2818.

57 Biswas, S., Fukushima, H., and Drzal, L.T. (2011). Mechanical and electrical property enhancement in exfoliated graphene nanoplatelet/liquid crystalline polymer nanocomposites. *Composites Part A: Applied Science and Manufacturing* 42 (4): 371–375.

58 Jiang, X. and Drzal, L.T. (2012). Multifunctional high-density polyethylene nanocomposites produced by incorporation of exfoliated graphene nanoplatelets 2: crystallization, thermal and electrical properties. *Polymer Composites* 33 (4): 636–642.

59 Tarhini, A.A. and Tehrani-Bagha, A.R. (2019). Graphene-based polymer composite films with enhanced mechanical properties and ultra-high in-plane thermal conductivity. *Composites Science and Technology* 184: 107797.

60 Jung, H., Yu, S., Bae, N.S. et al. (2015). High through-plane thermal conduction of graphene nanoflake filled polymer composites melt-processed in an L-shape kinked tube. *ACS Applied Materials and Interfaces* 7 (28): 15256–15262.

61 Bhattacharyya, A., Chen, S., and Zhu, M. (2014). Graphene reinforced ultra high molecular weight polyethylene with improved tensile strength and creep resistance properties. *Express Polymer Letters* 8 (2): 74–84.

62 Wan, C. and Chen, B. (2012). Reinforcement and interphase of polymer/graphene oxide nanocomposites. *Journal of Materials Chemistry* 22 (8): 3637–3646.

63 Burmistrov, I., Gorshkov, N., Ilinykh, I. et al. (2016). Improvement of carbon black based polymer composite electrical conductivity with additions of MWCNT. *Composites Science and Technology* 129: 79–85.

64 Fu, S., Yu, B., Tang, W. et al. (2018). Mechanical properties of polypropylene composites reinforced by hydrolyzed and microfibrillated Kevlar fibers. *Composites Science and Technology* 163: 141–150.

65 Lu, W., Yu, W., Zhang, B. et al. (2021). Kevlar fibers reinforced straw wastes-polyethylene composites: combining toughness, strength and self-extinguishing capabilities. *Composites Part B: Engineering* 223: 109117.

66 Abdul Khalil, H.P.S., Chong, E.W.N., Owolabi, F.A.T. et al. (2019). Enhancement of basic properties of polysaccharide-based composites with organic and inorganic fillers: a review. *Journal of Applied Polymer Science* 136 (12): 47251.

67 Chen, H., Li, C., Yao, Q. et al. (2022). Enhanced thermal conductivity and wear resistance of polytetrafluoroethylene via incorporating hexagonal boron nitride and alumina particles. *Journal of Applied Polymer Science* 139 (3): 51497.

68 Cui, X., Ding, P., Zhuang, N. et al. (2015). Thermal conductive and mechanical properties of polymeric composites based on solution-exfoliated boron nitride and graphene nanosheets: a morphology-promoted synergistic effect. *ACS Applied Materials and Interfaces* 7 (34): 19068–19075.

69 He, H., Liu, B., Xue, B., and Zhang, H. (2022). Study on structure and properties of biodegradable PLA/PBAT/organic-modified MMT nanocomposites. *Journal of Thermoplastic Composite Materials* 35 (4): 503–520.

70 Liu, C., Qiu, H.T., Liu, C.J., and Zhang, J. (2012). Study on crystal process and isothermal crystallization kinetics of UHMWPE/CA-MMT composites. *Polymer Composites* 33 (11): 1987–1992.

71 Makhlouf, A., Satha, H., Frihi, D. et al. (2016). Optimization of the crystallinity of polypropylene/submicronic-talc composites: The role of filler ratio and cooling rate. *Express Polymer Letters* 10 (3): 237–247.

72 Jain, S., Misra, M., Mohanty, A.K., and Ghosh, A.K. (2012). Thermal, mechanical and rheological behavior of poly(lactic acid)/talc composites. *Journal of Polymers and the Environment* 20 (4): 1027–1037.

73 Lapčík, L., Maňas, D., Vašina, M. et al. (2017). High density poly(ethylene)/$CaCO_3$ hollow spheres composites for technical applications. *Composites Part B: Engineering* 113: 218–224.

74 Faruk, O., Bledzki, A.K., Fink, H.P., and Sain, M. (2014). Progress report on natural fiber reinforced composites. *Macromolecular Materials and Engineering* 299 (1): 9–26.

75 Hadjadj, A., Jbara, O., Tara, A. et al. (2016). Effects of cellulose fiber content on physical properties of polyurethane based composites. *Composite Structures* 135: 217–223.

76 Fortunati, E., Armentano, I., Zhou, Q. et al. (2012). Microstructure and non-isothermal cold crystallization of PLA composites based on silver nanoparticles and nanocrystalline cellulose. *Polymer Degradation and Stability* 97 (10): 2027–2203.

77 Debeli, D.K., Tebyetekerwa, M., Hao, J. et al. (2018). Improved thermal and mechanical performance of ramie fibers reinforced poly(lactic acid) biocomposites via fiber surface modifications and composites thermal annealing. *Polymer Composites* 39: E1867–E1879.

78 Bayart, M., Foruzanmehr, M.R., Vuillaume, P.Y. et al. (2022). Poly(lactic acid)/flax composites: effect of surface modification and thermal treatment on interfacial adhesion, crystallization, microstructure, and mechanical properties. *Composite Interfaces* 29 (1): 17–36.

79 Selvakumar, K. and Meenakshisundaram, O. (2019). Mechanical and dynamic mechanical analysis of jute and human hair-reinforced polymer composites. *Polymer Composites* 40 (3): 1132–1141.

80 Hazrol, M.D., Sapuan, S.M., Zainudin, E.S. et al. (2022). Effect of Kenaf fibre as reinforcing fillers in corn starch-based biocomposite film. *Polymers* 14 (8): 1590.

81 Gasses, G.H. (2020). *EuRIC call for Recycled Plastic Content in Cars*, 1–4. European Recycling Industries Confederation.

82 Rusu, D., Boyer, S.A.E., Lacrampe, M.F., and Krawczak, P. (2011). Bioplastics and vegetal fiber reinforced bioplastics for automotive applications. In: *Handbook of Bioplastics and Biocomposites Engineering Applications* (ed. S. Pilla), 397–449. Wiley.

83 Jacques, K., Bax, L., Vasiliadis, H. et al. (2015). Polymer composites for automotive sustainability. *European Technology Platform for Sustainable Chemistry* 56: 1–56.

84 Reale Batista, M.D., Drzal, L.T., Kiziltas, A., and Mielewski, D. (2020). Hybrid cellulose-inorganic reinforcement polypropylene composites: lightweight materials for automotive applications. *Polymer Composites* 41 (3): 1074–1089.

85 Ajorloo, M., Ghodrat, M., and Kang, W.H. (2021). Incorporation of recycled polypropylene and fly ash in polypropylene-based composites for automotive applications. *Journal of Polymers and the Environment* 29 (4): 1298–1309.

86 Nofar, M., Ozgen, E., and Girginer, B. (2020). Injection-molded PP composites reinforced with talc and nanoclay for automotive applications. *Journal of Thermoplastic Composite Materials* 33 (11): 1478–1498.

87 Al-Oqla, F.M., Sapuan, S.M., Ishak, M.R., and Nuraini, A.A. (2016). A decision-making model for selecting the most appropriate natural fiber - polypropylene-based composites for automotive applications. *Journal of Composite Materials* 50 (4): 543–556.

88 Kim, D.H., Kim, H.G., and Kim, H.S. (2015). Design optimization and manufacture of hybrid glass/carbon fiber reinforced composite bumper beam for automobile vehicle. *Composite Structures* 131: 742–752.

89 Hufenbach, W., Böhm, R., Thieme, M. et al. (2011). Polypropylene/glass fibre 3D-textile reinforced composites for automotive applications. *Materials and Design* 32 (3): 1468–1476.

90 Ayrilmis, N., Jarusombuti, S., Fueangvivat, V. et al. (2011). Coir fiber reinforced polypropylene composite panel for automotive interior applications. *Fibers and Polymers* 12 (7): 919–926.

91 Harper, L.T., Burn, D.T., Johnson, M.S., and Warrior, N.A. (2017). Long discontinuous carbon fibre/polypropylene composites for high volume structural applications. *Journal of Composite Materials* 52 (9): 1155–1170.

92 Oliver-Ortega, H., Julian, F., Espinach, F.X. et al. (2019). Research on the use of lignocellulosic fibers reinforced bio-polyamide 11 with composites for automotive parts: car door handle case study. *Journal of Cleaner Production* 226: 64–73.

93 Teixeira, D., Giovanela, M., Gonella, L.B., and Crespo, J.S. (2013). Influence of flow restriction on the microstructure and mechanical properties of long glass fiber-reinforced polyamide 6.6 composites for automotive applications. *Materials and Design* 47: 287–294.

94 Nguyen-Tran, H.D., Hoang, V.T., Do, V.T. et al. (2018). Effect of multiwalled carbon nanotubes on the mechanical properties of carbon fiber-reinforced polyamide-6/polypropylene composites for lightweight automotive parts. *Materials* 11 (3): 429.

95 Caltagirone, P.E., Ginder, R.S., Ozcan, S. et al. (2021). Substitution of virgin carbon fiber with low-cost recycled fiber in automotive grade injection molding polyamide 66 for equivalent composite mechanical performance with improved sustainability. *Composites Part B: Engineering* 221: 109007.

96 Kausar, A. (2017). Polyamide 1010/polythioamide blend reinforced with graphene nanoplatelet for automotive part application. *Advances in Materials Science* 17 (3): 24–36.

97 Birch, A., Dal Castel, C., Kiziltas, A., Mielewski, D., and Simon, L. (2015). Development of cost effective and sustainable polyamide blends for automotive applications. In: *SPE Automotive Composites Conference & Exhibition* (November), pp. 1–10. Troy, MI, USA.

98 Wei, X.F., Kallio, K.J., Bruder, S. et al. (2020). High-performance glass-fibre reinforced biobased aromatic polyamide in automotive biofuel supply systems. *Journal of Cleaner Production* 263: 121453.

99 Yao, X., Luan, C., Zhang, D. et al. (2017). Evaluation of carbon fiber-embedded 3D printed structures for strengthening and structural-health monitoring. *Materials and Design* 114: 424–432.

100 Atiqah, A., Jawaid, M., Sapuan, S.M. et al. (2018). Thermal properties of sugar palm/glass fiber reinforced thermoplastic polyurethane hybrid composites. *Composite Structures* 202: 954–958.

101 Ermis, K., Unal, H., and Gunay, M. (2021). Glass bead effects on tribological and mechanical properties of plasticized polyvinyl chloride cable used in vehicles as a filler. *Proceedings of the Institution of Mechanical Engineers, Part J: Journal of Engineering Tribology* 235 (11): 2432–2439.

102 Jovanović, V., Samaržija-Jovanović, S., Budinski-Simendić, J. et al. (2013). Composites based on carbon black reinforced NBR/EPDM rubber blends. *Composites Part B: Engineering* 45 (1): 333–340.
103 Müssig, J., Schmehl, M., von Buttlar, H.B. et al. (2006). Exterior components based on renewable resources produced with SMC technology-Considering a bus component as example. *Industrial Crops and Products* 24 (2): 132–145.
104 Zagho, M.M., Hussein, E.A., and Elzatahry, A.A. (2018). Recent overviews in functional polymer composites for biomedical applications. *Polymers* 10 (7): 739.
105 Khan, Y.M., Cushnie, E.K., Kelleher, J.K., and Laurencin, C.T. (2007). In situ synthesized ceramic-polymer composites for bone tissue engineering: Bioactivity and degradation studies. *Journal of Materials Science* 42 (12): 4183–4190.
106 Ahmed, I., Parsons, A.J., Palmer, G. et al. (2008). Weight loss, ion release and initial mechanical properties of a binary calcium phosphate glass fibre/PCL composite. *Acta Biomaterialia* 4 (5): 1307–1314.
107 Tang, J., Bao, L., Li, X. et al. (2015). Potential of PVA-doped bacterial nanocellulose tubular composites for artificial blood vessels. *Journal of Materials Chemistry B* 3 (43): 8537–8547.
108 Badhe, R.V., Bijukumar, D., Chejara, D.R. et al. (2017). A composite chitosan-gelatin bi-layered, biomimetic macroporous scaffold for blood vessel tissue engineering. *Carbohydrate Polymers* 157: 1215–1225.
109 Bhowmick, S., Rother, S., Zimmermann, H. et al. (2017). Biomimetic electrospun scaffolds from main extracellular matrix components for skin tissue engineering application – The role of chondroitin sulfate and sulfated hyaluronan. *Materials Science and Engineering C* 79: 15–22.
110 Pereira, R., Carvalho, A., Vaz, D.C. et al. (2013). Development of novel alginate based hydrogel films for wound healing applications. *International Journal of Biological Macromolecules* 52 (1): 221–230.
111 Mehboob, H., Bae, J.H., Han, M.G., and Chang, S.H. (2016). Effect of air plasma treatment on mechanical properties of bioactive composites for medical application: Composite preparation and characterization. *Composite Structures* 143: 23–32.
112 Butt, M.S., Bai, J., Wan, X. et al. (2017). Mg alloy rod reinforced biodegradable poly-lactic acid composite for load bearing bone replacement. *Surface and Coatings Technology* 309: 471–479.
113 Jo, Y.Y., Kim, S.G., Kwon, K.J. et al. (2017). Silk fibroin-alginate-hydroxyapatite composite particles in bone tissue engineering applications in vivo. *International Journal of Molecular Sciences* 18 (4): 858.
114 Mao, D., Li, Q., Bai, N. et al. (2018). Porous stable poly(lactic acid)/ethyl cellulose/hydroxyapatite composite scaffolds prepared by a combined method for bone regeneration. *Carbohydrate Polymers* 180: 104–111.
115 Chang, G.W., Tseng, C.L., Tzeng, Y.S. et al. (2017). An in vivo evaluation of a novel malleable composite scaffold (polypropylene carbonate/poly(D-lactic acid)/tricalcium phosphate elastic composites) for bone defect repair. *Journal of the Taiwan Institute of Chemical Engineers* 80: 813–819.

116 Liu, Y., Wang, S., and Zhang, R. (2017). Composite poly(lactic acid)/chitosan nanofibrous scaffolds for cardiac tissue engineering. *International Journal of Biological Macromolecules* 103: 1130–1137.

117 Pok, S., Myers, J.D., Madihally, S.V., and Jacot, J.G. (2013). A multilayered scaffold of a chitosan and gelatin hydrogel supported by a PCL core for cardiac tissue engineering. *Acta Biomaterialia* 9 (3): 5630–5642.

118 Jiang, Y.C., Jiang, L., Huang, A. et al. (2017). Electrospun polycaprolactone/gelatin composites with enhanced cell–matrix interactions as blood vessel endothelial layer scaffolds. *Materials Science and Engineering C* 71: 901–908.

119 Chong, E.J., Phan, T.T., Lim, I.J. et al. (2007). Evaluation of electrospun PCL/gelatin nanofibrous scaffold for wound healing and layered dermal reconstitution. *Acta Biomaterialia* 3 (3 SPEC. ISS): 321–330.

120 Mobarakeh, L., Prabhakaran, M.P., Morshed, M. et al. (2008). Electrospun poly(ε-caprolactone)/gelatin nanofibrous scaffolds for nerve tissue engineering. *Biomaterials* 29 (34): 4532–4539.

121 Kim, M.S., Jun, I., Shin, Y.M. et al. (2010). The development of genipin-crosslinked poly(caprolactone) (PCL)/gelatin nanofibers for tissue engineering applications. *Macromolecular Bioscience* 10 (1): 91–100.

122 Oh, G.W., Ko, S.C., Je, J.Y. et al. (2016). Fabrication, characterization and determination of biological activities of poly(ε-caprolactone)/chitosan-caffeic acid composite fibrous mat for wound dressing application. *International Journal of Biological Macromolecules* 93: 1549–1558.

123 Lundin, J.G., McGann, C.L., Daniels, G.C. et al. (2017). Hemostatic kaolin-polyurethane foam composites for multifunctional wound dressing applications. *Materials Science and Engineering C* 79: 702–709.

124 Behera, S.S., Das, U., Kumar, A. et al. (2017). Chitosan/TiO$_2$ composite membrane improves proliferation and survival of L929 fibroblast cells: application in wound dressing and skin regeneration. *International Journal of Biological Macromolecules* 98: 329–340.

125 Kamel, N.A., Abd El-messieh, S.L., and Saleh, N.M. (2017). Chitosan/banana peel powder nanocomposites for wound dressing application: preparation and characterization. *Materials Science and Engineering C* 72: 543–550.

126 Moura, D., Souza, M.T., Liverani, L. et al. (2017). Development of a bioactive glass-polymer composite for wound healing applications. *Materials Science and Engineering C* 76: 224–232.

127 Wang, S.D., Ma, Q., Wang, K., and Chen, H.W. (2017). Improving antibacterial activity and biocompatibility of bioinspired electrospinning silk fibroin nanofibers modified by graphene oxide. *ACS Omega* 3 (1): 406–413.

128 Duailibi, M.T., Duailibi, S.E., Young, C.S. et al. (2004). Bioengineered teeth from cultured rat tooth bud cells. *Journal of Dental Research* 83 (7): 523–528.

129 Palucka, T. and Bensaude-Vincent, B. (2002). Composites overview. History of Recent Science and Technology (accessed on 17 March 2023).

130 Bhat, A., Budholiya, S., Raj, S.A. et al. (2021). Review on nanocomposites based on aerospace applications. *Nanotechnology Reviews* 10 (1): 237–253.

131 Mrezova, M. (2013). Advanced composite materials of the future in aerospace industry. *Incas Bulletin* 5 (3): 139–150.

132 Seydibeyoğlu, M.Ö., Doğru, A., Kandemir, M.B., and Aksoy, Ö. (2020). Lightweight composite materials in transport structures. In: *Lightweight Polymer Composite Structures* (ed. S.M. Rangappa, J. Parameswaranpillai, S. Siengchin, and L. Kroll), 103–130. CRC Press.

133 Maia, B.S., Behravesh, A.H., Tjong, J., and Sain, M. (2022). Mechanical performance of modified polypropylene/polyamide matrix reinforced with treated recycled carbon fibers for lightweight applications. *Journal of Polymer Research* 29 (4): 1–18.

134 Raja, D.B.P. and Retnam, B.S.J. (2019). Effect of short fibre orientation on the mechanical characterization of a composite material-hybrid fibre reinforced polymer matrix. *Bulletin of Materials Science* 42 (3): 1–7.

135 Lee, H., Ohsawa, I., and Takahashi, J. (2015). Effect of plasma surface treatment of recycled carbon fiber on carbon fiber-reinforced plastics (CFRP) interfacial properties. *Applied Surface Science* 328: 241–246.

136 King, J.A., Tomasi, J.M., Klimek-McDonald, D.R. et al. (2018). Effects of carbon fillers on the conductivity and tensile properties of polyetheretherketone composites. *Polymer Composites* 39: E807–E816.

137 Fischer, S., Pfister, A., Galitz, V. et al. (2016). A high-performance material for aerospace applications: development of carbon fiber filled PEKK for laser sintering. In: *Solid Freeform Fabrication 2016: Proceedings of the 27th Annual International Solid Freeform Fabrication Symposium - An Additive Manufacturing Conference, SFF 2016*, pp. 808–813. The Minerals, Metals & Materials Society, Texas, USA.

138 Xia, C., Zhang, S., Ren, H. et al. (2016). Scalable fabrication of natural-fiber reinforced composites with electromagnetic interference shielding properties by incorporating powdered activated carbon. *Materials* 9 (1): 10.

139 Tian, X., Liu, T., Yang, C. et al. (2016). Interface and performance of 3D printed continuous carbon fiber reinforced PLA composites. *Composites Part A: Applied Science and Manufacturing* 88: 198–205.

140 Gopinath, R., Poopathi, R., and Saravanakumar, S.S. (2019). Characterization and structural performance of hybrid fiber-reinforced composite deck panels. *Advanced Composites and Hybrid Materials* 2 (1): 115–124.

141 Amiri, A., Krosbakken, T., Schoen, W. et al. (2018). Design and manufacturing of a hybrid flax/carbon fiber composite bicycle frame. *Proceedings of the Institution of Mechanical Engineers, Part P: Journal of Sports Engineering and Technology* 232 (1): 28–38.

142 Kumar, A., Lal Krishna, G., and Anantha Subramanian, V. (2019). Design and analysis of a carbon composite propeller for podded propulsion. In: *Lecture Notes in Civil Engineering*, vol. 22 (ed. K. Murali, V. Sriram, A. Samad, and N. Saha), 350. Singapore: Springer.

143 Zhou, Y., Yao, Y., Chen, C.Y. et al. (2014). The use of polyimide-modified aluminum nitride fillers in AlN@PI/Epoxy composites with enhanced thermal conductivity for electronic encapsulation. *Scientific Reports* 4: 1–6.

144 Zhao, X., Tekinalp, H., Meng, X. et al. (2019). Poplar as biofiber reinforcement in composites for large-scale 3D printing. *ACS Applied Bio Materials* 2 (10): 4557–4570.

145 Oladele, I.O., Omotosho, T.F., and Adediran, A.A. (2020). Polymer-based composites: an indispensable material for present and future applications. *International Journal of Polymer Science* 2020: 1–12.

146 Utekar, S. and V K, S., More, N., and Rao, A. (2021). Comprehensive study of recycling of thermosetting polymer composites – Driving force, challenges and methods. *Composites Part B: Engineering* 207: 108596.

147 Francis, R. (2016). *Recycling of Polymers: Methods, Characterization and Applications*, 1–288. Wiley.

148 Singh, N., Hui, D., Singh, R. et al. (2017). Recycling of plastic solid waste: A state of art review and future applications. *Composites Part B: Engineering* 115: 409–422.

149 Kumar, S., Panda, A.K., and Singh, R.K. (2011). A review on tertiary recycling of high-density polyethylene to fuel. *Resources, Conservation and Recycling* 55 (11): 893–910.

150 Bhadra, J., Al-Thani, N., and Abdulkareem, A. (2017). Recycling of polymer-polymer composites. In: *Micro and Nano Fibrillar Composites (MFCs and NFCs) from Polymer Blends* (ed. S. Thomas, R. Mishra, and N. Kalarikkal), 263–277. Elsevier.

151 Scaffaro, R., Di Bartolo, A., and Dintcheva, N.T. (2021). Matrix and filler recycling of carbon and glass fiber-reinforced polymer composites: a review. *Polymers* 13 (21): 3817.

152 Kiss, P., Stadlbauer, W., Burgstaller, C. et al. (2020). In-house recycling of carbon- and glass fibre-reinforced thermoplastic composite laminate waste into high-performance sheet materials. *Composites Part A: Applied Science and Manufacturing* 139: 106110.

153 Cousins, D.S., Suzuki, Y., Murray, R.E. et al. (2019). Recycling glass fiber thermoplastic composites from wind turbine blades. *Journal of Cleaner Production* 209: 1252–1263.

154 Pietroluongo, M., Padovano, E., Frache, A., and Badini, C. (2020). Mechanical recycling of an end-of-life automotive composite component. *Sustainable Materials and Technologies* 23: e00143.

155 Chen, C.H., Huang, R., Wu, J.K., and Yang, C.C. (2006). Waste E-glass particles used in cementitious mixtures. *Cement and Concrete Research* 36 (3): 449–456.

156 Directive (2008). Directive 2008/98/EC of the European Parliament and of the Council on Waste and Repealing Certain Directives. Official Journal of the European Union, L312, pp. 3–30. Official Journal of the European Union.

157 Job, S., Leeke, G., Mativenga, P.T. et al. (2016). Composites Recycling – Where are we now? Composites UK, 2016, p. 20 (accessed on 17 March 2023).

158 Larsen, K. (2009). Recycling wind turbine blades. *Renewable Energy Focus* 9 (7): 70–73.

159 Ogi, K., Shinoda, T., and Mizui, M. (2005). Strength in concrete reinforced with recycled CFRP pieces. *Composites Part A: Applied Science and Manufacturing* 36 (7): 893–902.

160 Yip, H.L.H., Pickering, S.J., and Rudd, C.D. (2002). Characterisation of carbon fibres recycled from scrap composites using fluidised bed process. *Plastics, Rubber and Composites* 31 (6): 278–282.

161 Tian, X., Liu, T., Wang, Q. et al. (2017). Recycling and remanufacturing of 3D printed continuous carbon fiber reinforced PLA composites. *Journal of Cleaner Production* 142: 1609–1618.

162 Colucci, G., Ostrovskaya, O., Frache, A. et al. (2015). The effect of mechanical recycling on the microstructure and properties of PA66 composites reinforced with carbon fibers. *Journal of Applied Polymer Science* 132 (29): 1–9.

163 Tapper, R.J., Longana, M.L., Yu, H. et al. (2018). Development of a closed-loop recycling process for discontinuous carbon fibre polypropylene composites. *Composites Part B: Engineering* 146: 222–231.

164 Tapper, R.J., Longana, M.L., Hamerton, I., and Potter, K.D. (2019). A closed-loop recycling process for discontinuous carbon fibre polyamide 6 composites. *Composites Part B: Engineering* 179: 107418.

165 Zhang, J., Panwar, A., Bello, D. et al. (2016). The effects of recycling on the properties of carbon nanotube-filled polypropylene composites and worker exposures. *Environmental Science: Nano* 3 (2): 409–417.

166 Salas, A., Medina, C., Vial, J.T. et al. (2021). Ultrafast carbon nanotubes growth on recycled carbon fibers and their evaluation on interfacial shear strength in reinforced composites. *Scientific Reports* 11 (1): 1–11.

167 Stan, F., Stanciu, N.-V., Fetecau, C., and Sandu, I.-L. (2019). Mechanical recycling of low-density polyethylene/carbon nanotube composites and its effect on material properties. *Journal of Manufacturing Science and Engineering* 141 (9): 091004.

168 Stan, F., Sandu, I.-L., and Constantinescu (Turcanu), A.-M., Stanciu, N.-V., and Fetecau, C. (2022). The influence of carbon nanotubes and reprocessing on the morphology and properties of high-density polyethylene/carbon nanotube composites. *Journal of Manufacturing Science and Engineering* 144 (4): 041011.

169 Parparita, E., Uddin, M.A., Watanabe, T. et al. (2015). Gas production by steam gasification of polypropylene/biomass waste composites in a dual-bed reactor. *Journal of Material Cycles and Waste Management* 17 (4): 756–768.

170 El Abbassi, F.E., Assarar, M., Ayad, R. et al. (2019). Effect of recycling cycles on the mechanical and damping properties of short alfa fibre reinforced polypropylene composite. *Journal of Renewable Materials* 7 (3): 253–267.

171 Bourmaud, A., Åkesson, D., Beaugrand, J. et al. (2016). Recycling of L-poly-(lactide)-poly-(butylene-succinate)-flax biocomposite. *Polymer Degradation and Stability* 128: 77–88.

172 Gourier, C., Bourmaud, A., Le Duigou, A., and Baley, C. (2017). Influence of PA11 and PP thermoplastic polymers on recycling stability of unidirectional flax fibre reinforced biocomposites. *Polymer Degradation and Stability* 136: 1–9.

173 Ngaowthong, C., Borůvka, M., Běhálek, L. et al. (2019). Recycling of sisal fiber reinforced polypropylene and polylactic acid composites: thermo-mechanical properties, morphology, and water absorption behavior. *Waste Management* 97: 71–81.

174 Sahmaran, M., Yildirim, G., Hasiloglu Aras, G. et al. (2017). Self-healing of cementitious composites to reduce high CO_2 emissions. *ACI Materials Journal* 114 (1): 93–104.

175 Ashori, A. (2008). Wood-plastic composites as promising green-composites for automotive industries! *Bioresource Technology* 99 (11): 4661–4667.

176 Sharma, V. and Agarwal, V. (2011). Polymer composites sustainability: environmental perspective, future trends and minimization of health risk. In: *2nd International Conference on Environmental Science and Development*, p. 4. IPCBEE.

177 Witik, R.A., Teuscher, R., Michaud, V. et al. (2013). Carbon fibre reinforced composite waste: an environmental assessment of recycling, energy recovery and landfilling. *Composites Part A: Applied Science and Manufacturing* 49: 89–99.

178 Neşer, G. (2017). Polymer based composites in marine use: History and future trends. *Procedia Engineering* 194: 19–24.

179 Collins, R. (2018). Multifunctionality is the future for polymer composites. *Idtechex* accessed by https://www.idtechex.com/en/research-article/multifunctionality-is-the-future-for-polymer-composites/15408.

180 Tyrrell, M. (2020). Carbon fibre reinforced plastics worth $64bn by 2030. *Composites in Manufacturing* accessed by composites.media/carbon-fibre-reinforced-polymers-25092020/.

181 Reidel, H. (2017). Current and potential applications of carbon nanotubes. accessed by https://www.prescouter.com/2017/03/applications-carbon-nanotubes/.

182 Grigore, M.E. (2017). Methods of recycling, properties and applications of recycled thermoplastic polymers. *Recycling* 2 (24): 1–11.

Index

a

accelerating model   14, 15
additive manufacturing (AM)   234, 235, 272, 277
Akron Extruder M-PAK150   239
amorphous polymers   9, 215, 216, 324
  properties of   9
angular potential, of coarse-grained PVA model   285, 287
anisotropic polymer melts   255
  injection moulding   264–266
  preferred anisotropy level evaluation   256–258
  sheared polymer melts   260–264
atactic PP (aPP)   34, 35
Avrami analysis   96
Avrami equation   16, 17, 92, 93
  parameters   93
Avrami index   92, 93, 95, 96

b

benzoate derivatives   44
benzoate salts   44
bimodal polyethylene materials   98
binder jetting (BJ)   234, 235
biobased nucleators   134
biocomposites based on biopolyethylene (BioPE)   238
biodegradable biopolymers   130, 136
bioplastics   124, 131
1,3:2,4-bis(3,4-dimethylbenzylidene) sorbitol (DMBS)   44, 45

blow moulding procedure   5
bone engineering   220–221

c

calcite   43
calcium stearate (CaSt)   47
Ca-pimelate   46
carbon black (CB)   238, 247
  containing ethylene-octene copolymer   329
carbon fibre reinforced plastic (CFRP) waste recycling   344
carbon nanotubes reinforced crystalline composites
  future prospects of   344–345
Carreau–Yasuda parameters   93
Cartesio 3D printer   239
cast film extrusion of iPP homopolymer   56
cellulose fiber reinforced PU polymer matrix   330
cellulose nanocrystal (CNC)
  in PCL crystallinity   204
  reinforced poly(butylene succinate-co-1,2-decylene succinate)-(PBDS)   325
cellulose nanofibers (CNF)   246
  as nucleating agent for PLA   134
CF reinforcement effect   247, 329
chain-folding principle, of polymer crystallization   286, 288
chemical recycling process   340–343

chemical structures, of repeating units of polymer 1
chitosan methyl phosphonate (CMP) 134
Cinquasia red 46
classical nucleation theory 99
classical Ozawa, Flynn and Wall (OFW) method 26
classical thermodynamic nucleation theory 285–286
clay-induced nucleation, on PLA sample 132
CNT nanofiller 111
^{13}C nuclear magnetic resonance (NMR) spectroscopy
   polymer characterization method 39
co-crystallization 35, 49, 103–109
composite filaments 236, 241–243, 245
compression moulding 140, 245
computer-aided design (CAD) files 233
computer software 233
consistency constant 93, 95
Continuous-Cooling-Curve (CCC) diagrams 52, 108
copolymers, crystalline polymers 233
core-shell filaments 239
Craftbot Plus (Craftbot ltd, HU) printer 238
crystalline PA66 based glass fiber filled composite 341
crystalline polymer composites 326
   automotive applications of 331–334
   biomedical applications of 334–335
   civil engineering applications 339
   defense and aerospace applications of 335–339
   electronics 339
   environmental impact and safety issues of 343–344
   filler/reinforcing agent size and type influence on 328
   future trends of 344
   with inorganic based reinforcement materials 329–330
   marine applications 339
   with natural reinforcement materials 330–331
   with organic based reinforcement materials 328–329
   reinforcement materials 323
   technological residence level (TLR) 345
crystalline polymers 2
   applications 8–9
   categories of 323
crystalline regions 215
crystallisation analysis fractionation (CRYSTAF) 103, 104
crystallization enhanced development mechanism 4
crystallization process 13
   measurement methods 13

d

deaccelerating model 14–15
degree of crystallinity 3, 324
   chain crosslinking 324
   cooling rate of melt-phase polymer 325
   density of polymers 324
   molecular weight 324
   reinforcement presence 325
   tacticity 324
density function theory (DFT) 99
dibenzylidene sorbitol
   in polycaprolactone 270
   in polyethylene 268–270
dibenzylidene sorbitol (DBS) 180–181, 266, 268–271, 277
differential scanning calorimetry (DSC) 14, 103, 246
   application 15
   basic principle 14–15
   plots of pure PLLA and PLLA/PDLA blends 170
differential thermal analysis (DTA) 246
diffuse interface theory (DIT) 99
dimethyl 5-sulfoisophthalate sodium salt (SSIPA) 133

dioctyl phthalate (DOP) plasticized PLLA   182
direct digital manufacturing (DDM)   272
disordered molecular segments   89
dispersion behaviors, of layered clay particles   327–328
double notched impact strength, of non-nucleated and β-nucleated iPP homopolymers   46

e

elastic modulus   136–137, 139, 182, 215
electron spun triethoxysilane-terminated PCL (PCL-TES) - bioactive glasses (BG) fiber composites   203
equivalent undeuterated system, shear flow on crystallization behaviour of   263
expandable drug delivery system   224

f

fast scanning chip calorimetry (FSC)   48, 49, 55, 63, 64
feed spacers   242
fermentation process   121, 124
fiber reinforced crystalline polymer composites
   future prospects of   344–345
fiber spinning   59
Filabot EX2 commercial extruder   237
finite-extensible nonlinear elastic (FENE) Lennard-Jones coarse-grained polymer model   285
flow-induced polymer crystallization   283, 301–308
flow-induced polymer nucleation   301–306
foam-over-wire (FOW) embolization device   226
four-armed diblock copolymers of poly(ε-caprolactone)-b-poly (D-lactic acid) (4a-PCL-b-PDLA)   202
Fourier-transform infrared spectroscopy   13
full notched creep test   95
fused deposition modeling (FDM)   224, 234–244, 247–248, 272, 273
fused filament fabrication (FFF)   234
   3D printing process   238
fused granular printing   273

g

gastric retention   222, 224
gastroretentive drug delivery systems (GRDDSs)   222
Gibbs free energy variations   99
Gigabot X   240
glass fiber embedded crystalline Elium thermoplastic composite   341
glycerol monostearate (GMS)   47
GO containing ultra-high molecular weight PE   329
grafted lignin, in PLA/PCL blend   205
grafted polymers on nanofillers   293
   chain length effect   293–295
   grafting density effect   293
graphene nanoplatelets (GNPs)   110, 236, 239, 241, 328, 329, 335, 272, 277
green nanotechnology   345

h

heterogeneous nucleation   13, 20, 49, 124, 126, 133, 134, 144, 208, 288, 289, 291, 295, 305–306
heterophasic copolymers of iPP (HECOs)   46, 47
high density polyethylene (HDPE)   239
   composites with spherical hollow sphere $CaCO_3$ particles   330
   matrix   110
   -modified ICP   59
   polymer matrix, GNP incorporation to   329
highly crystalline polymers   8, 324
   properties of   9

high melting temperature PLLA (hPLLA)
    fibres   181
high melt strength (HMS) PP   42
human hair-reinforced polymer
    composites   330

i

impact resistance   88, 131, 138, 144, 182,
    215, 240
indirect heating methods   218
injection moulding   6, 57, 58, 123, 130,
    131, 140, 183, 234, 244, 264–267
inorganic nucleating agent   43, 125,
    130–133
intramolecular chain-folding nucleation
    286
intermolecular fringed-micelle nucleation
    286, 287, 298
intramolecular nucleation model
    287–288
iPP blends
    with C_2/C_n plastomers   50
    with conventional amorphous
        ethylene-propylene rubber   49
    spherulitic growth   49, 50
iPP crystallization and morphology
    investigation methods for
        home-made instruments   63
        microscopy techniques   60, 61
        spectroscopy-based techniques
            63
        thermal analysis   63
        X-Ray Diffraction   61
iPP homopolymers
    peak-time of crystallization of   49
    β-phase formation   41
    weight average molecular weight on
        flexural modulus   41
iPP impact copolymers (ICPs)   47
    comonomer content variation in EPC
        48
    structural parameters for   47
    systematic crystallization study   48
iPP/sPP blends   35–36
Irgaclear XT 386   46

isoconversional principle   18, 26
isotactic polypropylene (iPP)   241
    alpha (α) monoclinic form   37
    beta (β) phase   37
    C_2 content on melting point   39
    crystallinity and polymorphism, chain
        structure and molecular weight
        effects for   37–42
    crystallization, long-chain branching
        (LCB) effect on   42
    enthalpy (diamonds) of C_2C_3 random
        copolymers   39
    epsilon (ε) form   38
    flow induced crystallisation (FIC)   58,
        59
    gamma (γ) phase   38
    and (HD)PE, epitaxy between   52
    matrix crystallization   49
    morphological scales, characteristic
        hierarchy of   34
    nucleation of   42–47
    origin of   33
    paracrystalline smectic modification
        39
    polymorphism of   34
    processing effects and material
        properties   55, 60
    reinforcement for   54
    tacticity influence on   324
    total chain defect content on melting
        point   39
isothermal crystallization   92, 93, 95
    behaviour of pure PLLA   167
isothermal polymer crystallization   15
    activation energy characterization
        18–19
    analysis   16–17
    crystal geometry   17–18
    performance   16
    polymer composites   19–20
    rate characterization   18

j

jute reinforced polymer composites   330

k

kenaf fiber reinforced starch-based film 330
Kevlar fibers 237, 329
kinetic term model 14
Kissinger–Akahira–Sunrose (KAS) method 26

l

lactic acid 121
   chemical structure of 162
   structure and isomers of 122
lactides, chemical structure of 162
lamellar thickening 125
lamella structure 325
laminated object manufacturing (LOM) 234, 235s
Langevin dynamic simulations 291
Lauritzen and Hoffman theory 164–166
layer-by-layer manufacturing approach 233
LCB-PP 42
Liestritz-Micro-27 237
life cycle assessment (LCA) technique 344
light-depolarizing microscopy technique 96
linear low-density polyethylene (LLDPE) 50, 87–90, 104–106, 239
liquid crystalline polymer nanocomposites
   GNPs reinforcement to 329
low density polyethylene (LDPE) 239
   based MWCNT reinforced composites 342

m

macromolecule crystallization 123, 125
maleic- and acrylic acid-grafted PP 53
Mankati E360 236
MarkerPi 3D printer 221
Markforged X7 system 236
Matrix filaments 237
mechanical recycling process 340, 341
medical stents 218, 221–222, 224
metallocene-catalysed heterophasic iPP copolymers, crystallization kinetics of 49
methylene sequence length (MSL) 104
micro-injection moulding 266
miktoarm star copolymer (μ-PEG-PCL-PLLA) 202
modified Ozawa model 23–25
molecular dynamics (MD) simulations 283, 284
   coarse-grained polymer model 285, 287
   vs. Monte Carlo simulations 284
   united atom chain model 285
molecular simulations 283
   flow-induced polymer crystallization 301–308
   flow-induced polymer nucleation 301–306
   grafted polymers on nanofillers
      interfacial interactions effect 295
   grafted polymers on nanofillers, crystallization of 293
      chain length effect 293–295
      grafting density effect 293
      interfacial interactions effect 295
   nanofiller-induced block copolymer crystallization 291–292
   nanofiller-induced homopolymer crystallization 288–291
   polymer crystallization at quiescent state
      crystal nucleation 285, 287, 288
      intramolecular nucleation model 287–288
   random copolymer nanocomposite crystallization 293
   stereocomplex crystallization of polymer blends 295–300
      chain length effect 297–298
      chain structure effect 300
      chain topology effect 299–300
      nanofillers effect 298–299
      simulation details 296–297

molecular simulations (contd.)
  stretching effect 298
  stretch-induced crystalline structure changes 306–308
molecular weight
  effect on crystallinity
    poly (p-phenylene sulfide) 324
    poly(3-hexylthiophene) 324
  Monte Carlo (MC) simulations 283
    lattice polymer chain model 284
    micro-relaxation step 284
    vs. molecular dynamics simulations 284
Morphology Mapping 256, 275–277
multi jet fusion (MJF) techniques 237
multi-wall carbon nanotubes (MWCNT) 238, 328, 342s

n
NA-21 44–45
nanocellulose (NC) 203–205
nanocomposites, PE 109–112
nanofiller content 109–110
nanofiller-induced block copolymer crystallization, molecular simulations of 291–292
nanofiller-induced homopolymer crystallization, molecular simulations of 288–291
nanohybrid epitaxial brush (NHEB) structure 291
nanohybrid shish-kebab (NHSK) structure 111, 180, 289–291, 293, 306
nanosepiolite 132
natural fibers 134, 139–140, 330, 339, 343
natural flax fibers reinforced PLA-PBS polymer composite, recyclability of 343
neutron scattering techniques 256
N, N′-bis (benzoyl) hexanedioic acid dihydrazide (BHAD) 133–134

N, N′-bis (stearic acid)-1,4-dicarboxybenzene dihydrazide (PASH) 134
N, N′-oxalyl bis (piperonylic acid) dihydrazide (PAOD) 133
non-destructive techniques 2
non-isothermal crystallization, PE 98–99
non-isothermal polymer crystallization process 15
  activation energy determination 26–27
  crystal geometry analysis 21
    Jeziorny-modified Avrami equation 21, 22
  Mo model 25–26
  Ozawa model 21, 23–25
  nonlinear crystallization modelling 20
  performance 20
  relative crystallinity analysis 27
nucleating agents, categorization of 43
β-nucleating agent 46–47
nucleating agents, for PLA 127
  efficiency 130
  inorganic
    carbon nanotubes 132
    clay 131, 132
    sepiolite 132
    talc 130, 131
    talc and titanium dioxide combination 131
    talc, calcium carbonate and LAK 131
  organic
    BHAD 133
    cellulose nanofibers 134
    chitosan methyl phosphonate 134
    PAOD 133
    PASH 134
    pectin 135
    SSIPA 133
    wood flour 134
nucleating effect, of PCL-grafted-CNC, and non-grafted-CNC 205

nucleation 99
   of iPP 42–47
   theory 99
nylon(s) 8, 233–234, 236, 237, 246–248
Nylon 6 236–237, 331
Nylon66 236, 246, 326

o

optically active lactide types 124
organically montmorillonite (OMMT)
   incorporation effect, on PLA/PBAT blend 330
organic particulate nucleating agents 44
organophosphate NA-11 44–45
organophosphates 44
Ozawa model 21, 23–26, 28

p

PA-6 and iPP 52, 53
particulate nucleating agents 44–45
PCL-b-PLLA di-block copolymers 202
PCL-Macaiba fibre composites 206
PDMAEMA-b-PLLA-b-PDMAEMA copolymers 177
PE/carbon nanotubes 111
pectin 135
PEG plasticized PLLA 182
PET plastic salad containers (rPET) 240–242
plastic wastes recycling 340
PLLA-b-PCL diblock copolymers 176
PLLA-b-PEG diblock copolymer 177
PLLA/f-CNFs composites 174
PLLA/g-CNFs composites 174
PLLA/GONS composites 174
PLLA/LDH nanocomposites 172
PLLA/multiwalled carbon nanotubes (MWCNTs) composites 173
PLLA/PDLA stereocomplex 135
PLLA/PPZn nanocomposites 172
PLLA/TSOS (trisilanolheptaphenyl POSS) nanocomposites 175
PLLA/twice-functionalized organoclay (TFC) nanocomposites 172
PLLA/Zinc Oxide (ZnO) composite 174
polarized optical microscopy (POM) 13–14, 100, 133, 171, 174, 177, 196–197, 200, 205–207, 326
polyamide nylon (PA12) 234, 237
polyamides 234–238
   composites 237
   foliated graphite crystallization mechanism 24
polybutylene succinate structure, graphene nanosheets incorporation on 329
poly(ε-caprolactone) (PCL)
   based multiphase polymer systems 199–207
   crystallization types 199
   biomedical applications 195, 215
   crystallinity 195–198
   degradation rate 196
   molecular weight 195
   rice husk incorporation into 206
   synthesis methods 196
poly(ε-caprolactone) (PCL) crystallization
   polymer blends 199
      amorphous chlorinated polyethylene 200
      ethylene/octene multiblock copolymer effect 200
      PCL/crosslinked carboxylated polyester resin (CPER) binary blends 200
      styrene-co-maleic anhydride (SMA) copolymer effect 202
      with starch 199
   behaviour of
      bamboo-root flour effect 206
      block copolymers 202, 203
      cellulose nanocrystals on 205
      chitin nanocrystals on 205
      effect of fillers 203–207
      functionalized MWNTs (f-MWNTs) 206
      graphite oxide effect 205
      polyamide 6 (PA6) particles 203
      polymer blends 202
      reduced graphene oxide effect 205

polydispersity 41–42, 56, 58
polyesters 1, 6, 123–124, 126, 138–139, 195–196, 200, 233, 240, 339
polyethylene (PE) 233, 238–240
 blends and co-crystallization 103–109
 crystallization 91
  simulation based on the united atom chain model 285, 286
 crystallization kinetics of
  isothermal crystallization 93–96
  non-isothermal crystallization 96–99
 nanocomposites 109–112
 nucleation theory 99
 reactor 239
 structure and morphology 87–91
 theory of crystallization and kinetics 92–93
poly(ethylene oxide) (PEO)/PLLA solution-cast blended films 169
polyethylene terephthalate (PET) 7, 240, 241
 Ozawa model for 23
polyethylene terephthalate glycol (PETG)
 SEP filaments 241
 sepiolite composites 241
poly (lactic acid) (PLA) 122, 233, 243
 applications 130
 biodegradability 123
 composites, ramie fibers addition 330
 crystallization kinetics, improvement of 126–130
  annealing 126, 130
  nucleating agent 126, 127
  plasticizers 126, 129
 drawbacks 140
 films 123
 as food packaging material 123
 with malonate oligomers, plasticization of 182
 matrix and blends, organic and inorganic nanoparticles in 143
 matrix, titanium dioxide-coated flax fibers into 330
 molecular weight 121

natural fiber composites 139–140, 141
nucleating agents 130–133
 inorganic 130–133
 organic 133–136
nucleation 130–136
optical isomers 161
organoclay nanocomposites 131
plasticisation 182
production pathways, from lactic acid 121–122
resin quality 123
poly (L-lactic acid) (PLLA)
 chemical structure 162
 concentration gradient 7–8
 crystal structure 162
 glass transition temperature 162–163
 and graphene nanosheet (GNS) composites 174
 hydrolytic degradation 161
 melting temperature 163
poly (L-lactic acid) crystallization
 in block copolymer 175–178
 calorimetry method 166–168
 classical Lauritzen and Hoffman theory 164–166
 fullerene (C_{60}) loading 172
 in nanocomposites 172–175
 after nucleating agent addition 178–182
  BTA-cyclohexyl (BTA-cHe) 180
  1, 3:2, 4-dibenzylidene-D-sorbitol 180
  N,N'-Bis(benzoyl) suberic acid dihydrazide 180
  N,N,N'-Tris(benzoyl) trimesic acid hydrazide 181
  N,N,N'-Tris(1H-benzotriazole) trimesinic acid acethydrazide 181
  organic salt (bistrifluoromethysulfonyl) imide lithium salt 181
  PDLA-b-PM-b-PDLA triblock copolymer 180
  stereocomplex crystallites 179
 PLLA/PCL blend 170

PLLA/PDLA blend   168, 172
PLLA/PEG blend   169, 171
PLLA/POM blend   171
by solvent evaporation   8
through spherulite growth   163–164
poly (L-lactide)-*block*-methoxy poly(ethylene glycol) copolymer (PLLA-*b*-PEG)   176
(PLLA-*b*-MePEG) diblock copolymers   176
poly (L-lactide)-*block*-poly (vinylidene fluoride)-*block*-poly (L-lactide) (PLLA-*b*-PVDF-*b*-PLLA)   176
poly (L-lactide)-*b*-poly(ethylene oxide) (PLLA-*b*-PEO) block copolymers   176
poly (L-lactide-*b*-2-dimethylaminoethyl methacrylate) (PLLA-*b*-PDMAEMA) copolymers   176
polymer(s)   215
  blending technique   199
  chains   215
  crystallinity   215
  molecular chains, arrangement of   3
polymer crystallization
  characterization   15
  crystal growth
    and amorphous phases   103
    chain immobility   100
    characteristic values of   101
    iPP spherulite growth   100
  kinetics   93
  from solution   7–8
  strain-induced   5–7
polymer simulation systems
  molecular dynamics simulation   283
  Monte Carlo simulation   283
polymeric stimuli responsive materials   215
polypropylene (PP)   233, 241, 242
  composites   247
  submicronic-talc composites   330
polystyrene-*b*-poly (L-lactide) (PS-*b*-PLLA) diblock copolymers   175

poly(vinyl alcohol) (PVA)
  coarse-grained polymer model   285, 287
  MWCNT polymer composite   328
polyvinylidene fluoride (PVDF) matrix   26
  GNFs addition to   329
pseudo-monoclinic polyethylene   90

q
γ-quinacridone   46

r
RAMAN spectrum   13
random copolymer nanocomposite crystallization, molecular simulations of   293
random-type morphology   98
raw materials   234, 237, 240–242
recycled high density polyethylene (rHDPE)   239
recycled polypropylene (r-PP)   242
recycling process
  carbon fiber reinforced crystalline polymer composites   341–342
  carbon nanotubes reinforced crystalline polymer composites   342
  chemical   340
  flax fiber reinforced PA11-PP composites   343
  glass fiber reinforced crystalline polymer composites   340–341
  mechanical   340
  of natural fiber-reinforced crystalline polymer composites   343
  of sisal fiber incorporated PP-PLA composites   343
  thermal   340
relative crystallinity of PEG   27
relative crystallinity of PW   27
Rheomex 252p   239
Rice husk reinforced rPP filaments   243
ring-opening lactide polymerization   121

S

secondary crystallization process 103
selective laser melting (SLM) 234–235, 272
selective laser sintering (SLS) 234–235, 237, 242, 245
  printing technology 234
self-nucleation 124, 179
self-nucleation annealing (SSA) 103–105
self-seeding 124
semi-crystalline polymers 1, 8, 92, 100, 215, 216, 233, 323–325
  crystallinity values of 327
  properties of 9
shape memory alloy (SMA) 202, 236, 238
shape memory effect 215–227
shape memory materials (SMMs) 240
shape memory polymers (SMPs) 215, 227
  biomedical applications of
    bone engineering 220–221
    drug delivery application 222
    medical stents 221–223
    tissue engineering 218, 220
  cryogel scaffolds 221
  mechanism of 217
  self-healing materials 222, 224–226
  shape memory cycle 216, 217
  types of 218, 220
  vascular embolization 226–227
shape memory stent (SMDES) 222
sheared polymer melts 260, 264
  with nanoparticles 271–272
  with nucleating agents 266–271
short carbon fiber (SCF) 236, 242
short glass fibres 43, 331
sigmodal model 14, 15
single walled carbon nanotube (SWCNT) reinforced PP composite 328
slow crack growth (SCG) 93, 95, 105
small-angle neutron scattering
  of deuterium labelled polyethylene mixtures 260
  sample characteristics used for 261–262
small-angle X-ray scattering (SAXS) 103, 259
  for 3 Cl-DBS/PCL system 271
  for isotactic polypropylene after injection moulding 266–267
  PCL extruded filament 275
  for PCL/graphene sample 272
  3D printing of LDPE 275
sodium benzoate (NaBz) 47
soluble nucleating agents 43, 45, 58
solvent evaporation 7, 8, 172
spherulite structure 203, 325
spherulites collide 90
spherulites morphology, of pure PLLA and PLLA/CNTs-g-PLLA nanocomposites 174
step crystallisation (SC) 103, 104, 296–300
stereo complex crystallite (sc), in PLLA/PDLA blends 169
stereocomplex crystallization, of PLLA/PDLA blends 296
  chain length effect 297–298
  chain structure effect 300
  chain topology effect 299–300
  nanofillers effect 298–299
  simulation details 297
  stretching effect 298
stereolithography (SLA) 233–235, 272
stereorandomness 215
strain-induced crystallization in natural rubber 258, 259, 277
  X-ray diffraction study 259
strain-induced polymer crystallization 5–7
stretch-induced crystalline structure changes 306–308
surface free energy, of PE folded-chain nucleus 286
syndiotactic PP (sPP) 34
  form I 35
  form II 35
  form III and IV 35
  limitations 35

t

tacticity 166, 324
talc 43, 130, 131
   calcium carbonate and LAK 131
   -nucleated iPP compositions 44
   and titanium dioxide combination 131
temperature rising elution fractionation (TREF) 103, 104
temperature term model 14
template crystallisation 259
tensile testing 244
thermal recycling process 340
thermodynamics
   on crystallization of polymers characteristics 4–5
thermogravimetric analysis (TGA) 246, 247
thermomechanical pulp (TMP) fibres 238
thermoplastics 93, 130, 144, 218, 248, 273, 331, 340
thermosets 218, 340
3D printed crystalline polymers
   material and process
      nylon and polyamides 234–238
      polyethylene 238–240
      polyethylene terephthalate 240–241
      polylactic acid 243
      polypropylene 241–243
   mechanical properties/characteristics 244–246
   thermal properties/characteristics 246–247
   tribological properties/characteristics 247–248
3D printer 221, 233, 236–243
3D printing
   using extrusion 272–275
   in-situ studies of polymer crystallisation 273–275
   using pellet fed extruder system 273
3D spherulitic structure 96

tissue engineering 130, 137, 195, 218–220, 227, 334
transcrystalline layer (TCL) formation 54, 108
1,2,3-tridesoxy-4,6:5,7-bis-O-[(4-propylphenyl) methylene] nonitol 45

u

Ultimaker 2 236
ultrahigh molecular weight polyethylenes (UHMWPE) 88
   composite 238
uniaxial deformation experiment, with WAXS patterns 258–259
united atom chain model 284, 285, 308

v

vascular embolization 218, 226

w

WANHAO Duplicator 4/4x 236
water vapor transmission rate (WVTR) 123
wide angle X-ray scattering (WAXS) patterns 103
   for BPE and HPE samples 263
   of polyethylene specimens sheared in melt 269
wood-plastic composites 343

x

X-ray diffraction (XRD) technique 2, 13, 60, 63, 169, 200, 202, 206, 258
X-ray scattering technique 96, 256, 265

y

yield stress 215

z

zero shear viscosity 93, 95
Ziegler–Natta catalyst (ZNC)
   based PP random copolymers, increasing C_2 content effect in 40